Famine Early Warning Systems and Remote Sensing Data

Molly E. Brown

Famine Early Warning Systems and Remote Sensing Data

Springer

Molly E. Brown Ph.D.
2933 Melanie Lane
Oakton VA 22124
USA
molly.brown@nasa.gov

ISBN: 978-3-540-75367-4 e-ISBN: 978-3-540-75369-8

Library of Congress Control Number: 2008926124

Cover design: deblik, Berlin

Printed on acid-free paper

9 8 7 6 5 4 3 2 1

springer.com

This book rests upon the work of the dedicated FEWS NET staff, its representatives, affiliated scientists and network partners. It draws heavily on the reports, analysis and knowledge generated through its work. The book is dedicated to the FEWS NET staff and contributors who are focused on their mission of providing accurate, informative and life saving information to humanitarian aid donors for their decision making processes. In particular, I would like to thank my colleagues James Verdin and Gabriel Senay at USGS EROS, Christopher Funk, Gregory Husak and Diego Pedreros at the University of California Santa Barbara, and Gary Eilerts, Richard Choularton, Laura Glaeser and Patricia Bonnard at Chemonics International for their insights, arguments and responsiveness to my many questions throughout this process. They have been generous with their time and patient with my ignorance, providing constant redirection and material for this project. They are my close collaborators and have been at the center of the generation of many of the ideas in this book. I would also like to thank my colleague Dr. Brent McCusker whose valuable and timely comments have made this a better book.

Finally, the book is also dedicated to my supportive husband Devin Paden and my children, Ben and Phoebe, who have given me the time, love and encouragement in which to complete this project. Without them I would not be nearly as productive as I am, nor have as interesting a life.

Preface

This book describes the interdisciplinary work of USAID's Famine Early Warning System Network (FEWS NET) and its influence on how food security crises are identified, documented and the kind of responses that result. The book describes FEWS NET's systems and methods for using satellite remote sensing to identify and describe how biophysical hazards impact the lives and livelihoods of the population where they occur. It presents several illustrative case studies that will demonstrate the integration of both physical and social science disciplines in its work. FEWS NET's operational needs have driven science in biophysical remote sensing applications through its collaboration with the US Geological Survey, the National Aeronautics and Space Administration, National Oceanographic and Atmospheric Administration, and US Department of Agriculture, as well as methodologies in the social science domain through its support of the US Agency for International Development, UN World Food Program and numerous international non-governmental organizations such as Save the Children, Oxfam and others.

Because FEWS NET is an organization that must provide a global picture of food insecurity to decision makers, the information it relies on are by necessity observable and able to be documented. Thus many aspects of traditional livelihood analysis, for example, cannot be used by FEWS NET as they rely upon relationships, and ways of expressing power and knowledge at the local scale that cannot be easily scaled up to express variations in access to food at a community level. The book focuses on the ways that remote sensing information is transformed into an understanding of the actions that must be taken in order to ensure that lives and livelihoods are protected, including describing the remote sensing observations and models needed to identify hazards and the information gathering requirements and analytical frameworks needed to understand their impact. Its focus is primarily analysis conducted in Africa, but also touches upon FEWS NET's work in Central America, Haiti and Afghanistan. As an organization that seeks to integrate social and physical science methodologies and strategies into its work on a daily basis, it is a fascinating and rich example of interdisciplinary knowledge generation and innovation.

Contents

Acronyms

ADDS	Africa Data Dissemination Service (data server at USGS)
ADEOS	Advanced Earth Observation Satellite
AET	Actual Evapotranspiration
AFWA	Air Force Weather Agency
AGRMET	Agricultural Meteorology Modeling System
AIRS	Atmospheric Infrared Sounder
AMSR-E	Advanced Microwave Scanning Radiometer for EOS (NASA's Earth Observing System)
AMSU-A	Advanced Microwave Sounding Unit-A
AVHRR	Advanced Very High Resolution Radiometer
BRDF	Bidirectional Reflectance Distribution Function
BERM	Basin Excess Rainfall Maps
CDC	US Centers for Disease and Control
CERES	Clouds and the Earth's Radiant Energy System
CHARM	Collaborative Historical African Rainfall Model
CHG	Climate Hazard Group (at UCSB)
CILSS	Comité Permanent Inter-Etats de Lutte contre la Sécheresse dans le Sahel (Permanent Interstate Committee for Drought Control in the Sahel)
CIRES	Cooperative Instituted for Research in Environmental Sciences (University of Colorado at Boulder/NOAA)
CMORPH	NOAA Climate Prediction Center 'Morphing' Technique Rainfall
COLA AGCM	Center for Ocean-Land-Atmosphere Studies Atmospheric General Circulation Model
CPC	Climate Prediction Center (at NOAA)
CSIR	South Africa's Council for Scientific and Industrial Research
DCHA	Democracy, Conflict and Humanitarian Assistance Bureau (USAID)
DFID	United Kingdom's Department for International Development
DLIS	Desert Locust Information System

EC	European Commission
ECOWAS	Economic Community of West African States
EMD	Empirical Mode Decomposition
ENSO	El Niño-Southern Oscillation
EOS	End of season (in the context of crop phenology)
EROS	Center for Earth Resources Observation and Science (USGS)
ESRI	Environmental Systems Research Institute
ET	Evapotranspiration
ETM	Enhanced Thematic Mapper Plus Landsat 7 data
EU	European Union
FAO	United Nations Food and Agriculture Organization
FAS	Foreign Agricultural Service (USDA)
FEWS NET	Famine Early Warning Systems Network
FOODNET	Foodborne Diseases Active Surveillance Network
FFP	Office of Food for Peace (USAID)
FPAR	Fraction of Photosynthetically Active Radiation
GeoWRSI	Geospatial Water Requirement Satisfaction Index model
GDAS	Global Data Assimilation System
GFS	Global Forecast System
GIMMS	Global Inventory Mapping and Modeling Systems (at NASA)
GIS	Geographic Information Systems
GLAM	Global Agricultural Monitoring
GMT	Greenwich Mean Time
GPCP	Global Precipitation Climatology Project
GPM	Global Precipitation Mission
GSFC	Goddard Space Flight Center (NASA)
GSOD	Global Surface Summary of Day
GTS	Global Telecommunication System
HYDRO1K	Elevation Derivative Hydrological Database at 1km resolution
ICPAC	IGAD's Climate Prediction and Applications Centre
IGAD	Intergovernmental Authority on Development
IIED	International Institute for Environment and Development
IKONOS	a commercial earth observation satellite at 1 and 4 meter resolution
IMF	International Monetary Fund
IPA	International Production Assessment
IPC	International Phase Classification
ITCZ	Intertropical Convergence Zone
LAI	Leaf Area Index
LEWS	Livestock Early Warning System
LIDAR	Light Detection and Ranging
LIS	Lighting Imaging Sensor
LP DAAC	Land Processes Distributed Active Archive Center
LST/E	Land Surface Temperature and Emissivity
LTDR	Long Term Data Record

MI	Moisture Index
MJO	Madden Julian Oscillation
MODIS	Moderate resolution Imaging Spectrometer
MODIS CMG	MODIS Climate Modeling Grid (0.05 degree resolution)
NASA	National Aeronautics and Space Administration
NDSI	normalized difference snow index
NDVI	normalized difference vegetation index
NEPAD	The New Partnership for Africa's Development
NESDIS	National Environmental Satellite, Data, and Information Service (NOAA)
NGO	Non-Governmental Organization
NIR	near-infrared reflectance
NPOESS	National Polar-orbiting Operational Environmental Satellite System
NOAA	National Oceanic and Atmospheric Administration
NPP	NPOESS Preparatory Project
NMCP	National Malaria Control Program
NVAP	NASA Water Vapor Project
OFDA	Office of Foreign Disaster Assistance
OLR	outgoing long wave radiation
PET	Penman-Monteith potential evapotranspiration
PPT	Precipitation
PR	TRMM Precipitation Radar
Radar	RAdio Detection And Ranging
RATES	Regional Agriculture Trade Expansion Support
RATIN	Regional Agricultural Trade Intelligence Network in East Africa
REDSO	USAID Regional Economic Development Services Office
RFE	Rainfall Estimate
SADC	Southern African Development Community
SLC-off	Scan Line Corrector off mode for Landsat 7 ETM+ data
SOS	Start of Season (in the context of crop phenology)
SPI	Standardized Precipitation Index
SPOT-VGT	French coarse resolution mapping spectrometer (1 km)
SSM/I	Special Sensor Microwave Imager
SSTA	Sea Surface Temperature Anomaly
SWI	Soil Water Index, modeled crop parameter
TRMM	Tropical Rainfall Monitoring Mission
TMI	TRMM Microwave Imager
UCSB	University of California Santa Barbara
USAID	United States Agency for International Development
USDA	United States Department of Agriculture
USGS	United States Geological Survey
VAC	Vulnerability Assessment Committee
VAM	Vulnerability Assessment Mission

VI	Vegetation Index
VIIRS	Visible Infrared Imaging Radiometer Suite
VIRS	Visible Infrared Scanner
WFP	World Food Program (United Nations)
WHO	World Health Organization (United Nations)
WIHA	Weather Impacts Hazard Assessment
WMO	World Meteorological Organization (United Nations)
WR	water requirement for crop yield estimation
WRSI	water requirement satisfaction index

Section I
Background

Chapter 1
Introduction

Remote sensing is the use of environmental sensors placed in orbit around the earth to observe climate, vegetation and rainfall dynamics, enabling a daily global assessment of ecosystem health. These observations have formed the foundation of famine early warning systems by providing quantitative assessments of food production across large areas. Although this information is a critical piece of famine early warning, food security is rarely ensured by adequate food production alone. The political and economic situation in which the production occurs is usually far more important in determining the food security. The focus of this book is on remote sensing measurements and how they are used to identify, quantify and motivate a response to variations in climate in an integrated system where the social and political context in which these variations occur, which are not observable from space, are a central concern.

The severe drought and resulting widespread crop failure in West Africa in 1984 and 1985 precipitated an intense, highly publicized, and extremely devastating famine. Hundreds of thousands of people died directly or indirectly due to the combination of the effects of the drought and the lack of government intervention. Globally, the lack of information which would allow for effective response to extreme and widespread drought became obvious. Local and regional shortages of food caused enormous increases in prices, with simultaneous decreases in purchasing power by a large portion of the population in the affected region. Response to the famine was slow and assistance arrived late, much too late to save local lives and livelihoods.

The US Agency for International Development (USAID)'s designed the Famine Early Warning System in 1986 to provide information on food security of communities in semi-arid regions so that such a widespread catastrophe does not occur again. Early and effective intervention could break the link between climate extremes and famine, where the role of government is to provide economic and political safety nets that can defuse a crisis (Wisner et al., 2004). The objective of early warning systems are to monitor both the production of food as well as a wide range of other indicators meant to measure demand, the ability of a population to purchase food, and other political and economic pressures that may affect the food security of a region (Davies et al., 1991). The development of remote sensing systems to monitor environmental conditions has provided, for the first time, a way to monitor current climate variations over entire continents for very little expense.

With the development in the early 1980s of an index that could measure vegetation health using meteorological satellites which could see the entire globe every day, remote sensing became a cornerstone of the early warning activity (Tucker, 1979). These observations enabled the monitoring of remote areas of the world that have never had consistent and real-time meteorological observations. Operational famine early warning systems, staffed by social scientists but collaborating with physical scientists, followed shortly thereafter.

This book examines how remote sensing has been used in USAID's Famine Early Warning Systems Network (FEWS NET). Remote sensing is used by humanitarian organizations to identify a hazard, guide decision making and to build consensus around whether or not a region has a food security problem. The primary objective of early warning systems is to elicit an appropriate response to an identified problem. Often, this requires that everyone, from the various donors of food aid, to regional organizations made up of numerous national governments, to national and local governance structures in the country in question agree that there is a problem and understand and concur on its severity. This consensus building is based on datasets which it collects and analyzes in the field, including information on market prices, agricultural production statistics, population and nutrition of the most vulnerable groups. Using experts at a local, regional and international level, these datasets are combined with quantitative remote sensing data to estimate the extent and severity of a particular food security crisis.

In this process, it is often only the quantitative remote sensing imagery that provides irrefutable evidence of a significant reduction in food production. Other sources of information can be highly contested, the result of political processes or machinations of dysfunctional governments that typify regions with food security problems. Although everyone agrees that political and economic factors are usually far more important in determining food access and that these factors are obvious to those living and working in the region, it is often the biophysical evidence derived from remote sensing that all parties can agree upon as being 'real', valid, and conclusive. This puts remote sensing at center stage in famine early warning systems.

Most famines are no longer primarily caused by environmental conditions: other factors, such as conflict, political crisis, or economic collapse are far more important in the development of such crises than simply crop failure (Dilley and Boudreau, 2001). These problems are obviously not detectable by environmental sensors orbiting the earth. Even in systems where there is perfect information, early warning of an impending food security crisis does not guarantee success. A coordinated effective government response is the critical ingredient required to respond to crisis. Governments are loathe to expend resources on potential crises a priori, thus having as much evidence as possible on the situation is a critical part of decision making, one which FEWS NET does for its partners in the countries it works in as well as for the United States Government, its humanitarian agencies and representatives.

In the most recent edition of their book 'At Risk', Wisner et al. (2004) focuses on the fact that biophysical hazards are usually the least important factor in disasters that can claim hundreds of thousands of lives, particularly in slowly evolving crises such as famine. Usually it is the economic and political situation surrounding the biophysical event that is the most important determinant for how a severe drought or flood affects a community (Wisner et al., 2004). Despite two decades of development and implementation of carefully crafted indicators of nutritional status, food access, and metrics to measure the severity level of a food security crisis, remote sensing is still at the forefront in decision making. The reason for this has more to do with the decision makers in the US and Europe who early warning systems are targeting than with any reality on the ground.

As Roger Pielke Jr. states in a recent review in *Nature*, 'quantitative predictions fulfill important political and social roles, regardless of their quality, accuracy or appropriateness, and will continue to be demanded by decision-makers and produced by scientists.' Remote sensing provides evidence that is believable, understandable and incontrovertible for decision makers at a variety of levels (Pielke Jr, 2007). Remote sensing also provides evidence of an impending problem significantly earlier than any other indicator that FEWS NET has at its disposal because it measures crop production factors during the growing season instead of measuring the impact of a poor season after the harvest. Regardless of the appropriateness and relevance of remote sensing measurements, they will continue to be the focus of decision making in the context of food security. This is attributable to the nature of the humanitarian aid system that FEWS NET serves and to the compelling nature of the evidence of a real hazard that remote sensing maps provide. Because USAID's response to a food security crisis is overwhelmingly food aid, which can ameliorate primarily food availability problems, remote sensing data remains at the center of FEWS NET's work despite the evolution of the nature of food security crises to be more centered around food access and food utilization than food availability.

Scientists involved with FEWS NET work to provide biophysical, quantitative metrics that are as close to measuring the actual problem as possible. For example, instead of providing rainfall anomalies, the data is ingested into a model that shows the impact of these anomalies on maize, millet and other crop yields at each stage of their productive cycle. Instead of describing food prices in one market, FEWS NET provides analysis that shows the impact of food prices on the livelihood of pastoralists in a region. Maps and images displaying areas that deserve increased attention are incredibly powerful tools for communicating a problem, much more than tables and long, wordy descriptions. FEWS NET has continued to increase the use of maps to communicate food security problems, even those with complex and diverse origins.

The intense collaboration between social and physical scientists in the context of early warning has influenced the development of knowledge in both domains. By working with a real-time system that examines each remote sensing image and comparing it to what is seen on the ground, early warning systems have been a source of enormous change and development for scientists who work on producing the remote sensing datasets. The influence of this ongoing and continuous quality checking against ground truth cannot be overstated in the development and maturation of remote sensing datasets. For social scientists, the ability to measure and quantify variations in food production from year to year has enabled the development of robust systems that assess whether or not a community is food insecure in the face of a hazard.

In this book, a description of both the social and physical science basis for FEWS NET and its widely reaching systems and processes that help inform and respond to food security are provided. The book is divided into four sections. The first section will introduce remote sensing and describe FEWS NET and its approach. This is followed with a summary of remote sensing data and products that FEWS NET uses in its day-to-day operations. The third section focuses on the networks developed to gather information on the specific social and political context in which food production occurs in order to understand the overall food security outcome of a particular growing season. In the last section, case studies of how remote sensing data was used in particular crises are presented. In the concluding chapters, an analysis of the politics surrounding the use of biophysical data in a fundamentally social process is presented, as well as a summary of what the future holds for FEWS NET.

1.1 Famine Early Warning in a Modern Age

Famine early warning systems are implemented by organizations that use social and political information about the ways people gain access to food, combined with spatially extensive biophysical information to determine the onset of severe food insecurity. Famine is a slow-fuse disaster, a social catastrophe that takes many months or years to develop, the consequence of multiple failures on many levels before famine takes hold (von Braun et al., 1998). Early warning of this process should, therefore, be straightforward, but because there is little agreement on exactly how to measure changing food systems, and the because famines can occur not only when there is no food but when food is plentiful but access to it has been disrupted, it is not.

If it takes such a long time to occur, then why are early warning systems needed? Early warning of such a slow, multi-year process involves two aspects: first an adequate capability to detect and document a crisis, and advance preparation by international, national and local organizations for an effective response to an identified crisis. The role of early warning systems is to identify and allow governments the time and information needed to deter these crises from occurring, preventing the destruction of the lives and livelihoods of countless people as well as the social and economic systems on which they depend. Thus, effective early warning revolves around prior agreement as to what constitutes a crisis, and what responses will occur when such crises occur. These responses tend to be very expensive, both economically and politically, and they will not occur if there is not consensus on what needs to happen. Alternatively, if no response occurs, that can also be extremely expensive in the long run. Agreement on what is a proper response and how quickly a response should occur is difficult to achieve, especially given the diversity of local, national, and international actors involved. Famine early warning systems provide the forum both for agreement on the signs of an impending crisis, and the platform for mobilizing the preparation for response on multiple levels.

When the U.S. government responds to disaster internationally, the primary institution for managing humanitarian assistance is the U.S. Agency for International Development (USAID). Provision of this assistance is a core activity of USAID and is recognized as a strategic goal (USAID, 2007). For this goal, one of the highest priorities is to 'prevent and mitigate disasters'. For prevention and mitigation of disasters, USAID specifically cites the Famine Early Warning System Network (FEWS NET) as the prime example of how it is achieving this strategic priority. USAID clearly values the role this early warning system plays in reducing risk of famine, hunger and food insecurity, and, ultimately, in reducing the human and financial toll of famine.

FEWS NET is only a small part of the overall larger geo-political system that has grown up around food aid, humanitarian programs and overseas development aid. Many who are familiar with USAID's programs believe that food aid is used too often and in too many places where it cannot ameliorate the underlying long term problems. FEWS NET works to ensure that decisions regarding food aid are made with the most accurate information possible about the impact of both action and inaction is available to the decision maker. That said, there is much that can be improved in the larger food aid system and with development assistance in general. This book will explore efforts to improve the advice that FEWS NET gives, and ways that FEWS NET can be made better, but does not seek to critique the larger humanitarian aid system. In order to be effective, FEWS NET must work within that system to retain the trust of those in decision making positions.

FEWS NET's personnel are predominantly social scientists, trained in the humanitarian response field, anthropology, economics and other social sciences. They are deeply committed to improving the response to international food security crises. They are not, however, experts in remote sensing. Using satellite-derived remote sensing information to inform social science discourses requires an intense interaction between the physical scientists who develop and present the data and the social scientists who use it in their work.

FEWS NET defines its goal as working to 'strengthen the abilities of countries and regional organizations to manage risk of food insecurity through the provision of timely and analytical early warning and vulnerability information' (FEWS, 2005). FEWS NET collaborates with international, national, and regional partners to provide timely and rigorous early warning and vulnerability information on emerging or evolving food security issues. FEWS NET professionals in the US, Africa and other regions monitor various data and information – including remotely sensed data and ground-based meteorological conditions, crop and rangeland health – as early indications of potential threats to food security. It then analyzes that information in the context of the existing livelihood strategies of the people who live in the region.

FEWS NET uses a modified food entitlements approach derived from Sen (1981) and Dreze and Sen (1989)'s work. This theory describes how famines develop by focusing on variations in people's entitlements to food instead of overall food production. Entitlements are how people gain access to food, through a combination of growing it themselves, purchasing it on the market with funds derived from activities such as wage labor in trade-based or production-based activities, through transfers from other households or from the state, or creating it through access to common land, such as the production of wild foods or gathering of fire wood for sale (Sen, 1981). Famines are caused by a series of events that can effect various aspects of these entitlements, and thus it considers not only supply-side causes (such as drought), but also demand-side causes (high prices), policy causes (such as restrictive import or excessive export policies), and any other cause of a change in entitlements (Dreze and Sen, 1989). FEWS NET terms examination of entitlements 'livelihood analysis'.

Livelihood analysis provides information about how, and why, people survive or fail to survive difficult times (Boudreau, 1998). The product of a livelihood analysis may be as simple as a national livelihood zone map, or may include in-depth baseline profiling of livelihood patterns among different wealth groups in selected livelihood zones. The objective of the livelihood information is to provide an understanding of what 'normal' is, so that when a disruption occurs due to drought or flood, or to unusual or adverse political or economic events, then FEWS NET can interpret the impact of these events on the ability of the people in the region to continue living their lives. If a drought is severe enough to disrupt not only the most vulnerable who may need help every year, but also the average citizen and the most well-off, the impact of the crisis will be much more severe and prolonged and the response to it needs to be correspondingly larger.

FEWS NET's primary responsibility is to provide information – credible, understandable, relevant information – for policy makers and networks through which policy makers can gather information about events and their impact on food security. The ultimate objective is to illicit an effective and appropriate response. FEWS NET seeks to assess emerging or evolving food security problems and define their severity and spatial extent across regions and countries. The information products that FEWS NET produces focus on the information necessary for decision-makers to intervene effectively. These decision-makers are an incredibly diverse group – from farmers and cooperative members in remote localities, to the ministers and executives in the host government in the affected region (Table 1.1).

FEWS NET information is important for decision makers in the United States and in the United Nations and other donor nations who will fund and mobilize a response, and in private voluntary organizations such as Oxfam and Save the Children. The information must be credible, as many of the required decisions, such as whether or not intervention is needed and in what form, are subject to intense scrutiny and are surrounded by politics, media and carry great risks to those who make them. Without sound, overwhelming, and relevant information, decisions made may either be wrong or may not be made at all. Thus FEWS NET acts as an expert advisor, whose information can be trusted by multiple parties, and whose advice can be used to provide impetus for consensus that will lead to intervention.

Table 1.1 Use of remote sensing-based data by people in different communities, at different scales (from R. Choularton, FEWS NET web site)

	Local	National	Regional	International
Community	Individuals with access			
Civil Society	Local non-governmental organizations (NGOs)	National NGOs	Regional NGOs	International NGOS (Save the Children, Oxfam)
Government or intra-governmental	Municipal or departmental government	Ministry of Agriculture, Health Parliament	ECOWAS SADC African Union	United Nations General Assembly
Private Sector	Local shop owner or trader	National Companies	Regional Companies	Transnational Corporations
International organizations			CILSS	FAO WFP
Donors		National Government Private sector	African Development Bank	USAID DFID EU

1.1.1 Lives and Livelihoods

In order to create policy-relevant information, FEWS NET must know how events will affect food security. Perturbations in the climate or rainfall are only one of many factors that are important. Figure 1.1 shows a summary of studies in household food security in southern Africa, where climate/environment was just one of 33 drivers of food insecurity mentioned as important by householders. The impact of

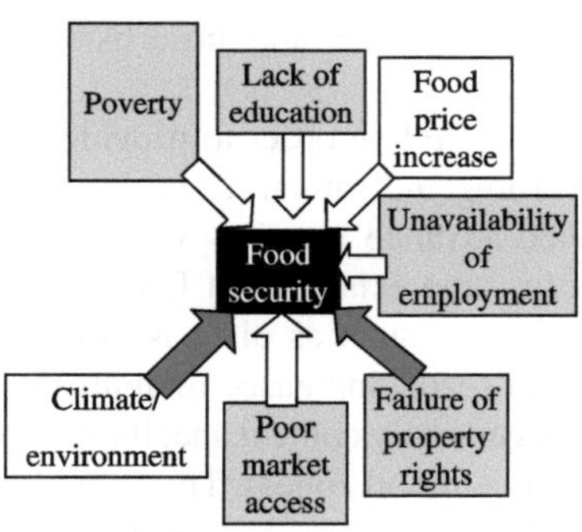

Fig. 1.1 The seven most frequently cited drivers in 49 studies of household-level food insecurity in southern Africa, derived from 555 citations of 33 possible drivers. The drivers shaded in grey were noted as being chronic, while those in white indicate drivers that were experienced as ‘shocks’. The shaded arrows indicate drivers that acted primarily via reductions in food production, while the white arrows indicate those which acted by restricting access to food. Derived from (Cooper et al., 2004; Gregory et al., 2005)

sudden drought, for example, is felt on top of ongoing long-term stresses and the inability to cope with such shocks and to mitigate long-term stresses means that the coping strategies, such as short-term employment, may not be available, and thus the impact for one household may be far greater than another. FEWS NET needs to know and understand about the entire complex picture as well as all the potential shocks to the system in order to provide accurate and useful information about how to intervene.

The conceptual framework in which FEWS NET combines information derived from remote sensing with information from livelihood analysis. Livelihoods are the sum of ways in which people make a living. In most communities in low-income countries, poor families balance a set of food and income-earning activities. Acute food insecurity results when the failure of one or more of these strategies or entitlements cannot be compensated for by other strategies. Livelihoods provides the context about how, and why, people survive (or fail to survive) difficult times.

People survive primarily through their own labor (growing crops; hunting and gathering; working for a meal or sack of grain; trading and bartering; income generating activities; etc.). However, there are also other small, critical ways the poor get by (a debt incurred or forgiven; a child sent to relatives during lean months). In addition to analyzing food production and acquisition, FEWS NET considers the means by which people secure other basic necessities such as health care, water, clothes or agricultural inputs. Livelihood analysis takes it all into account, adding up each food and cash source until the analyst can show how different wealth groups in a particular area survive in a given year.

Different options for survival are available to different people depending on where they live (the agro-ecological zone) and what resources they have (cash, savings, loans, labor, and so on). The range of options available to most households are rather limited. People produce food; they exchange things for food; or they earn cash to buy food. Patterns become evident. Most people's livelihoods can be characterized by a predominant activity, which is then supplemented by several other, less dominant, activities (Boudreau, 1998). Describing the dominant activities is the focus of livelihood zoning. Major rural livelihood types include those listed in Table 1.2.

The livelihoods framework was adopted by FEWS NET in order to provide essential baseline material for interpreting early warning indicators. There are two types of early warning indicator, those that provide advance warning of a famine (indicators of imminent crisis) and those that confirm the existence of famine (indicators of famine). The latter group includes indicators such as distress sales of productive assets (e.g. plough oxen), consumption of seeds, increased malnutrition and increased mortality. Indicators of famine are not generally context specific (i.e. a single list could be prepared that would apply to all livelihood zones). They are also of little use in predicting or preventing severe food shortage or famine, and therefore are not used in the livelihood profiles (Cuny and Hill, 1999). By categorizing people into regions with similar livelihood activities, then analysis over much larger areas becomes possible and the use of broad remote sensing indicators in conjunction with them also makes sense.

Table 1.2 Description of a few main rural livelihoods (from adapted from FEWS NET livelihood profiles)

Rural Livelihoods	Description
Agriculture	Commercial agriculture Subsistence grain farming (highland, midland, lowland) Mixed cash- crop/grain farming Agri-horticulture
Livestock Production	Commercial ranching (dairy/meat farming) Pastoralism
Agro-Pastoralism	Mixed agriculture-livestock production
Fishing	Off-shore (ocean-based) In-shore (lake/river based) Mixed agro-fishing
Labor-Based	Plantation/ranch/commercial farm worker Migratory labor Mining labor
Hunter-Gatherer	Forest-based subsistence economy

1.2 What is Remote Sensing?

In 1985, Compton Tucker published remotely sensed, continental-scale, ten day composite images of the vegetation in Africa on the cover of Science magazine (Tucker et al., 1985). This event represented the maturation of a previously obscure branch of biology and signal processing, announcing that remote sensing science had advanced to the point where science quality, continental-scale, biophysical measurements were possible (Tucker and Choudhury, 1987). For the first time, there was the possibility of using these measurements for simultaneous, continuous and current images for monitoring food production across vast portions of the globe. Development of famine early warning systems, staffed by social scientists collaborating with physical scientists followed shortly there-after. This collaboration with data users in the humanitarian community has been a key source of both funding and feedback for the vegetation remote sensing community.

FEWS NET was formed as an inter-agency partnership between National Aeronautics and Space Administration (NASA), Oceanic and Atmospheric Administration (NOAA), and the US Geological Survey (USGS). NOAA provided rainfall expertise, but these data have been until recently less reliable and more problematic than visible spectrum-based remote sensing approaches due to the lack of sufficient direct observations of rainfall across the continent, and the difficulty of determining the rainfall intensity from the observable cloud-tops. For much of its existence, FEWS NET has used vegetation data derived from the National NOAA series of satellites carrying the Advanced Very High Resolution Radiometer (AVHRR) to estimate interannual variations in food production. Although the data that comes from this sensor are too coarse to determine how a particular crop or community's fields were doing, it can provide an overview of how the growing season is progressing over a region. Vegetation estimates do not allow specific estimation of crop yields, as the information from both agriculture and fallow vegetation and trees are combined together into a single observation. However, by comparing a given period of the current year with those from previous years when conditions were known, or with the mean of all previous years, a reasonably reliable estimate of the quality of the growing season and ultimate yield can be developed. Thus, using satellite remote sensing FEWS NET can determine if the cropping season in an area will be better or worse than last year or from the average (Hutchinson, 1998).

In addition to AVHRR data, vegetation data from the MODerate Resolution Imaging Spectroradiometer (MODIS) and SPOT-Vegetation sensors have moved center-stage since 2000, with their higher resolution and improved calibration and processing. NOAA provides Rainfall Estimation (RFE) imagery based on multiple satellite inputs, which has become as reliable as the vegetation data and more widely used. The RFE is an automated (computer-generated) product which uses Meteosat infrared data, rain gauge reports from the global telecommunications system, and microwave satellite observations within an algorithm to provide rainfall estimates in millimeters at an approximate spatial resolution of 10 km. The main use of these data is to provide input for hydrological and agrometeorological models which enables the user to evaluate much more directly food crop health than simply vegetation index or rainfall. This is done by integrating the RFE measurements into crop models which provide crop-specific growth requirements driven by information about soil type, evapo-transpiration and temperature information.

Agricultural monitoring does not occur just in the context of early warning. Many nations monitor and record statistics on the progress of their agricultural crops, and FEWS NET both learns from these efforts and contributes to them. Monitoring the diversity of agricultural systems can be difficult, as field sizes and the confounding aspects of native vegetation are a continuing problem. Table 1.3 shows three distinct types of enhanced agricultural information needs as a function of development status. FEWS NET operates primarily in nations with agricultural systems of Type 1, but shares information and methods with scientists who work with data derived for Types 2 and 3.

Buchanan and Davies' 1994 book focused on the difficulty that early warning systems have had in motivating a proper response. In the past decade, FEWS NET has worked to improve response to the information it provides by increasing its visibility within USAID, providing comparative reports that seek to identify the most severe food security problems where an effective and early response will be most beneficial, and most importantly, focus on evidence-based reporting with concurring consensus of all parties in the region (local, national, regional, international at a variety of levels) so that response to an appeal for financial and food donations is parties in the region (local, national, regional, international at a variety of levels) so that response to an appeal for financial and food donations is encouraged. Unfortunately, when there are many emergencies occurring simultaneously creating pressure on resources, response can be slower and less than desired than in other years with lower demand.

Table 1.3 Characteristics of agricultural regions worldwide, from (Justice and Josserand, 2006)

	Type 1	Type 2	Type 3
Reason for enhanced observation	Agricultural commodity supply/demand assessment; practice verification; damage assessment; precision farming	Regional food security assessment; insurance/micro-finance programs; irrigation monitoring	Food insecurity assessment; famine early warning
Development context	Developed countries & high income developing countries	Middle income developing countries	Least developed countries
Geographic Region	North America, Europe, Australia, northern Asia, Near East	South America, East Asia, North Africa, Pacific, Central Asia	Sub-Saharan Africa, South Asia, Central America
Dominant climate type	Temperate, continental, Mediterranean, Arid	Tropical, sub-tropical, arid, semi-arid	Tropical, sub-tropical, arid, semiarid
Farming Systems	Highly specialized, mechanized, irrigated, high input & productivity, few holdings	Mixed technology, large acreage under irrigation	Low technology, rain-fed, largely subsistence, low productivity, many farm holdings
Institutional Capacity	High capacity, high technology, large data sets (depth & breadth), reliability	Mixed capacity, large potential, rapid growth in data reliability & quantity	Very low capacity, low technology, low levels of reliability

1.3 FEWS NET as an Multi-Disciplinary Project

Today, mitigating food insecurity and preventing famines remains a real challenge for national governments and the international community, despite the significant progress made in the last 25 years. FEWS NET's early warning analysts face a number of defining challenges which have shaped the innovations that have been made in recent years. These challenges include increased complexity and uncertainty in the local conditions, the difficulty of striking a balance between the need for rigor and the need for immediate action, the limited capacity of our information users to absorb the data available to them, and the new demand for comparable analysis across very different and data poor contexts. The humanitarian community needs to respond to climate-caused crises in a diverse and nuanced way, not just with one solution. Remote sensing tends to provide information only on issues relating to food availability, but food security crises are increasingly caused by problems of food access as well as food utilization due to disease.

FEWS NET has focused on four areas in an effort to improve their ability to prompt effective and appropriate action. First, they have worked to align analysis systems with decision maker needs in a systematic way. Second, they have worked to penetrate decision making processes, proactively feeding analysis into the organizational processes where decisions are made and resources allocated at USAID. This was done by moving FEWS NET to the Food for Peace office where resource allocation decisions regarding humanitarian assistance are made. Third, they have pushed food security analyses towards explaining causes and implications that can form the basis for defining the right action that should be taken in a particular situation. By increasing the nuance and specificity of policy recommendations, FEWS NET can contribute to a better and more effective humanitarian response beyond food assistance. Finally, they have standardized a minimum set of reporting outputs to provide a consistent picture of food security across all the countries they cover, reengineering its information management systems in the process.

Generating early action to respond to an emerging short term crisis is a challenge. Underlying trends in vulnerability, such as population growth, urbanization and trends in rainfall and temperature due to global environmental climate change have a strong influence on events. FEWS NET is recognizing that today's crises are rooted as much in these trends as it is in environmental extremes. Short term climate stress will continue to be as much part of the foundation of the tomorrow's problems as climate change itself. Remote sensing will remain at the center of FEWS NET's efforts to identify and respond to these short-term crises. More and more, however, remote sensing data will also be used to identify and understand long term trends in climate. These data, coupled with analyses on how climate interacts with larger social and political processes, will remain at the core of early warning of food security.

1.4 Summary of Chapters

The book is organized into five sections. The first section, which includes this chapter, provides background information on FEWS NET, its organization, systems and processes, and the conceptual frameworks that it uses to incorporate remote sensing data into its analysis. Section II describes in detail what remote sensing is, summarizes how the measured biophysical estimates are created and used, and how model derivatives of these measurements are estimated using state of the art science. The third section describes how the remote sensing data are put to use in FEWS NET through analytical applications in the domains of food security monitoring and contingency planning processes. Section IV will provide case studies describing how remote sensing was used in the early warning and response to a specific food security crisis. Section five concludes with an analysis of how remote sensing is used today and how FEWS NET is changing to meet new challenges.

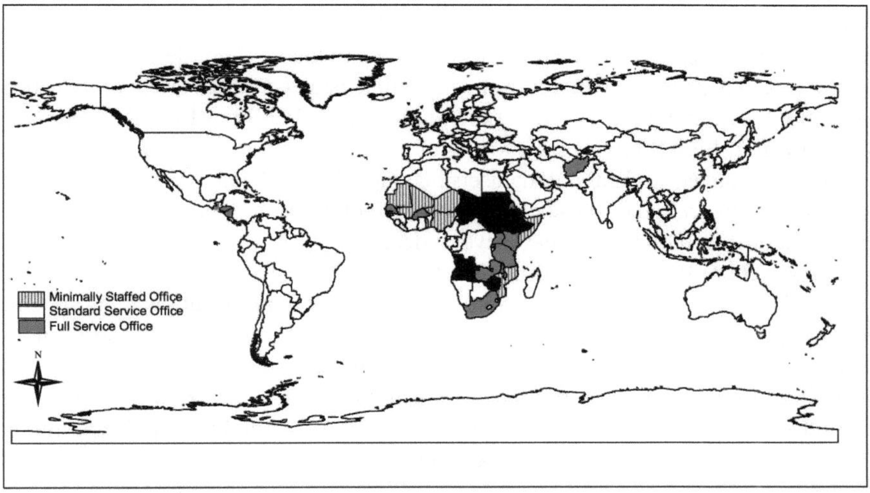

Fig. 1.2 FEWS NET country locations and levels of services, as of 2007

Chapter 2 in this first section focuses on describing key concepts of food security, utilization, access and availability, and how these concepts are used in FEWS NET. The framework in which FEWS NET works will be examined in the light of the larger process of early warning and humanitarian assistance, and how biophysical hazards are identified, assessed and translated into estimates of need. These processes are conducted with an understanding of vulnerability and the diversity of economies, cultures and biophysical environments in which it occurs. A description of the limitations of FEWS NET systems and processes is also presented in Chap. 2, including its geographic extent (Figure 1.2). Chapter 3 focuses on the structures that have been put into place through interagency agreements that gives FEWS NET access to the expertise and data from remote sensing scientists. FEWS NET has partner organizations that provide key remote sensing datasets, analysis tools and expertise that is then used by the food security analyst in the field. The chapter also provides a broad overview of all the social and biophysical datasets that the organization uses to provide actionable food security information in its myriad of products.

1.4.1 Remote Sensing for Early Warning

The second section of the book focuses on the biophysical data and models that have been developed for use in FEW NET. It starts with a presentation of the rainfall datasets that FEWS NET uses to monitor growing conditions in FEWS NET regions. The Rainfall Estimate (RFE) is a merged gauge-satellite product created by NOAA's Climate Prediction Center for FEWS NET's use in Africa. The problems and advantages of using the daily rain gauge data product that is delivered by the

World Meteorological Organization, the Global Telecommunications Systems data, is used to train the satellite remote sensing inputs to the RFE are described. Certain countries and regions are well represented by this data and others are not on the network at all. These differences affect the quality and reliability of the RFE product, thus impeding FEWS NET's monitoring of food production in regions that are not on the grid. Longer-term rainfall datasets that include data back to the 1950s are also presented in this chapter. These datasets enable comparison of current conditions to those that occurred during the past several decades.

Chapter 5 presents integrated models that ingest merged satellite-gauge rainfall data model outputs and evapotranspiration estimates with information about specific crop water requirements and soil type. The water requirement satisfaction index (WRSI), moisture index and other modeled parameters allow the analyst to estimate the effect of rainfall on crop production much more easily than rainfall alone. The strengths and weaknesses of these products are discussed and their utility in monitoring agricultural production described in the context of smallholder agriculture in Africa.

Biophysically measured vegetation indices are the focus of Chap. 6, detailing what they are, how they are measured and the various uses of the information in the context of FEWS NET. The biophysical basis of vegetation indices is reviewed and the scalability of the measurement discussed. Low and moderate resolution imagery from AVHRR and MODIS sensors are presented, along with examples and discussion of how images from multiple periods are constructed to create a time series, enabling anomaly images to be made for analysis. FEWS NET uses anomalies of vegetation images in conjunction with rainfall and crop models to monitor the progress of the growing season during operational hazard assessment conducted weekly by FEWS NET partners.

Climate models are important tools for FEWS NET for its focus on providing guidance about future changes in the food security situation to USAID and its other partners. Chapter 7 summarizes the various models and methods that FEWS NET uses to obtain forecasts of climate conditions. Numerical weather forecasts for predicting rainfall conditions a week in advance are summarized. To extend these short-range forecasts with those that go out several months, FEWS NET partners have developed models that use canonical correlation analysis to relate variations in sea surface temperature with rainfall in Africa. Global forecasts from the International Research Institute at Columbia University are also used routinely to understand future variations in climate. To help analysts understand these forecasts, a tool called the Forecast Interpretation Tool (FIT) was developed that translates forecasts expressed in probabilities into the actual rainfall amounts that they may represent. To complement information from climate models, FEWS NET is developing statistical projections of its two most commonly used indices: rainfall and vegetation data. These projections are based on models that use observed conditions to estimate future variations due to persistence in the atmosphere. These forecasting products are integrated into FEWS NET's systems to provide a quantitative basis for the Food Security Outlook six months into the future.

1.4.2 Food Security Analysis

The third section of the book begins with Chap. 8, which privides a description of the primary analytical areas that FEWS NET uses to identify and analyze the proximate causes of food insecurity. This knowledge base includes livelihood analysis, monitoring of markets and trade in commodities, climate, crop production, livestock, conflict/security, health and nutrition, policy, water and sanitation, natural resources and other causes of food insecurity. When appropriate, each of these specific areas is mentioned in reporting from each region or country in order to identify possible sources of food insecurity. Different levels of geographical focus, from municipal and community specific to multi-nation region are examined using this knowledge base. Subsequent chapters in this section focus on specific data and analyses that permit the further understanding of the impact of variations in production as measured by remote sensing.

Chapter 9 focuses on the reporting mechanisms that FEWS NET uses to provide decision support to decision makers at a variety of levels. Information flows from FEWS NET in country representatives through reports to Washington DC, to tailored products to a myriad of decision makers at the local, national, regional and international levels. The timing of information and the requirements of decision makers was presented, along with a summary of reports that attempt to meet these requirements. A summary of assessments, nutritional surveys and other tools that the humanitarian community determines the severity of the crisis is presented. FEWS NET's suite of reports and reporting mechanisms was described along with a summary of their content.

Chapter 10 describes how remote sensing information is used to identify and explain food production deficits in the local FEWS NET offices, and how networks of organizations and individuals work together to determine the impact of these deficits on the food security of particular populations. A key part of FEWS NET's activities is to strengthen these networks. A wide variety of meetings, reports, information products and web site information are used to improve the flow of information into and out of localities. Remote sensing data and products can be greatly improved by incorporating the information and understanding of users who know their environments very well. Decision makers profit from using remote sensing information by having fact-based evidence to illustrate their situation graphically, how it relates to their neighbors, and to previous years. By providing the actual data itself, food security specialists can request and get access to detailed biophysical analysis that can improve their ability to communicate and get an appropriate response to impending food security crises.

The next few chapters in this section focus on the use of particular tools and datasets to better identify and report populations at risk for food insecurity. Chapter 11 summarizes the use of population data in early warning and humanitarian communities in their efforts to estimate the numbers of people at risk during emergencies. FEWS NET typically uses official census data from the local communities in which they work, but would like to use gridded population data, such as that available from LandScan (Bright, 1998). This data allows for much more accurate identification of populations at risk and their locations. FEWS NET will integrate the LandScan dataset with both official statistics and scientifically produced population models that apportion population across the land surface in order to have the most accurate and yet politically acceptable dataset for use in food security work.

In Chap. 12, the focus is the critical role of markets in providing food and the role of prices in food security. Food prices are a key component to food security analysis, because access to food for rural residents is often through small, informal markets where grain is bought and sold. Food prices are influenced in these markets by local availability of cereals and by the price and availability of food produced elsewhere. Cross-border trade and transportation costs are other factors that affect food prices in local markets. Until recently, food price analysis has been at the individual market level and conducted in the context of the monthly reports. Time series of local prices were plotted and the implications of significant increases or decreases on food access noted in the context of an analysis of food security. In the past year, FEWS NET has begun focusing on providing food security outlooks for the next six months. This future orientation creates a need for price projections that can both fill in information that is missing from real-time market reporting, and provide information regarding future price directions. This chapter presents methods and approaches for estimating prices and coupling them with NDVI observations for improved modeling.

Chapter 13 describes the contingency planning and scenario development strategy that FEWS NET uses to plan for problems in the countries they work in. The utility of remote sensing data in contingency plans is described and examples are given for different stages of the contingency planning process. Planning is typically conducted by individuals most involved in conducting response that typically do not have experience with remote sensing data. It is to this audience that FEWS NET targes its products. Recent developments in remote sensing and climatology analysis can support these developments.

1.4.3 Case Studies

The fourth section of the book focuses on specific events and cases that illustrate the use of remote sensing data in FEWS NET. Most of what FEWS NET does is context-specific and thus examining specific events during the past decade can illustrate how remote sensing can be operationally integrated into a primarily political and economic analysis. The underlying vulnerability of the population dictates much about the impact of a particular biophysical hazard. Of course, coping with a large deficit in production due to climate is difficult for any farming community, regardless of its location. The difference between a community who can cope and one that requires significant outside intervention is the role of government and the underlying resiliency of the affected community.

The extraordinary response to the 2002/3 crisis in Ethiopia is chronicled in Chap. 14, which provides a good case study for how FEWS NET uses data to identify and motivate response to an impending crisis. The indicators that were used in November and December 2002 to convince the donor community of the crisis that was emerging in Ethiopia are presented, demonstrating the use of remote sensing data coupled with socio-economic indicators. FEWS NET reporting is summarized, including agroclimatic analyses, food production projections, access to food, threats to agriculture from pests, and the context in which these events were occurring. An analysis produced by FEWS NET-affiliated scientists that seeks to understand and estimate the likelihood that the drying trends observed in Ethiopia will continue. Although FEWS NET employs remote sensing scientists in their four primary regional offices (East, Southern and West Africa and in Central America), these scientists do not have time to do much long term climate change analysis. Thus this kind of assessment is often conducted in collaboration with US scientists. Finally, the chapter concludes with a summary of the Anderson and Choularton retrospective analysis on how to move forward in Ethiopia after an enormous effort by the relief community to respond to the crisis.

Chapter 15 describes the Meso-american food security early warning system or MFEWS, which grew out of decades of work done in Africa. The adjustment to a new agroclimatological system has been accomplished with carefully tuned products that have been validated with the help of partners in the region. MFEWS has also worked to build an understanding of the local and regional factors that affect food security. Remote sensing plays a key role in identifying regions that are influenced by weather-related variations in food production, markets, trade infrastructure and other critical economic assets. These variations may result in food insecurity in the affected zones. The food security outlook shows clearly the impact of variations in the growing season on food security for the poorest people in the region. FEWS NET thus monitors very closely the start of the season in the country, variations in labor demand, maize crop health and the progress of the rainy season every year. Remote sensing products, including the WRSI, NDVI anomaly, soil water index, and rainfall estimates from several sources are used weekly to make these estimates.

In the final case study, Chap. 16 outlines the food security situation of Zimbabwe in 2007, which resulted from a combination of political, economic and biophysical factors. The extremely poor economic condition of the country will adversely affect its ability to import necessary commodities to make up for poor production in the 2006–7 cropping year. Land reforms that transferred highly productive commercial farms to multiple, poor farmers who have few resources and little knowledge about farming has resulted in a huge reduction in the overall ability of Zimbabwe to be self-sufficient. Thus, it is forced to import grain from outside of the region at a time of record international maize prices. The chapter focuses on an analysis conducted in April of 2007 which used MODIS NDVI data to estimate maize production. The Vegetation-Sum method was shown to be able to capture up to 80% of the production variation during the past six years that the data has been available. These NDVI-based estimates were a key component to determining the food aid requirement for Zimbabwe, and thus FEWS NET was able to ensure early consensus due to the analysis provided at a critical moment.

1.4.4 Looking to the Future

The last section includes the final two chapters, which focus on the larger context in which FEWS NET works and where it is going in the future. Chapter 17 is entitled 'Power, Politics and Remote Sensing' and focuses on the politics of food and the role that remote sensing plays in the power struggles surrounding humanitarian food aid at multiple scales. A description of the politics surrounding food at the community scale is provided as an example of the complexity of translating local ways of knowing into international information that is comparable and understandable at the global scale. Remote sensing information itself can be used as a political tool, and in this context, three brief case studies are presented where remote sensing plays a central role in the way decisions are made.

The final chapter of the book focuses on the future of FEWS NET and the challenges it faces in improving the way remote sensing is used. Efforts are being made to improve the integration of remote sensing into larger analytical processes that transform information about threats to food security into understandable knowledge about what kinds of interventions are most appropriate, if any.

References

Boudreau, T.E., 1998. The Food Economy Approach: A Framework for Understanding Rural Livelihoods. RRN Network Paper 26, Relief and Rehabilitation Network/Overseas Development Institute, London.

Bright, E.A., 1998. LandScan Global Population 1998 Database. Oak Ridge National Laboratory, Oak Ridge, TN.

Cooper, J.J., Scholes, N. and Biggs, R., 2004. Ecosystem services in southern Africa: a regional assessment. Council for Scientific and Industrial Research, Pretoria.

Cuny, F.C. and Hill, R.B., 1999. Famine, Conflict and Response: A Basic Guide. Kumarian Press.

Davies, S.M., Buchanan-Smith, M. and Lambert, R., 1991. Early Warning in the Sahel and Horn of Africa: The State of the Art Review of the Literature, Report No.20. Institute of Development Studies, University of Sussex, Brighton, UK, 148pp.

Dilley, M. and Boudreau, T.E., 2001. Coming to terms with vulnerability: a critique of the food security definition. Food Policy, 26: 229–247.

Dreze, J. and Sen, A., 1989. Hunger and Public Action. Clarendon Press, Oxford.

FEWS, 2005. Famine Early Warning System Network Home Page. USAID FEWS NET.

Gregory, P.J., Ingram, J.S.I. and Brklacich, M., 2005. Climate change and food security. Philosophical Transactions Royal Society, 360: 2139–2148.

Hutchinson, C.F., 1998. Social science and remote sensing in famine early warning. In: D. Liverman, E.F. Moran, R.R. Rindfuss and P.C. Stern (Editors), People and Pixels: Linking Remote Sensing and Social Science. National Academy Press, Washington DC, pp. 189–196.

Justice, C. and Josserand, H., 2006. Agricultural monitoring meeting convened for the integrated global observations for land (IGOL) theme, Rome, Italy 8–11 March 2006, Integrated Global Observing Strategy, Rome, Italy.

Pielke Jr, R., 2007. When the numbers don't add up. Nature, 447: 35–36.

Sen, A.K., 1981. Poverty and Famines: An Essay on Entitlements and Deprivation. Clarendon Press, Oxford, 270pp.

Tucker, C.J., 1979. Red and photographic infrared linear combinations for monitoring vegetation. Remote Sensing of Environment, 8: 127–150.

Tucker, C.J. and Choudhury, B.J., 1987. Satellite remote sensing of drought conditions. Remote Sensing of Environment, 23: 243–251.

Tucker, C.J., Vanpraet, C.L., Sharman, M.J. and van Ittersum, G., 1985. Satellite remote sensing of total herbaceous biomass production in the senegalese sahel: 1980–1984. Remote Sensing of Environment, 17: 233–249.

USAID, U. S. D. O. S. A. (2007) Strategic Plan, Fiscal Years 2007–2012: Transformation Diplomacy. Washington DC, US State Department.

von Braun, J., Teklu, T. and Webb, P., 1998. Famine in Africa: Causes, Responses, and Prevention. The Johns Hopkins University Press, Baltimore.

Wisner, B., Blaikie, P., Cannon, T. and Davis, I., 2004. At Risk: Second Edition. Taylor and Francis Books Ltd, Wiltshire.

Chapter 2
Conceptual Frameworks

Early warning organizations use a particular set of conceptual frameworks to transform all kinds of information, from biophysical information derived from remote sensing to market prices, into actionable food security policy recommendations. It is not enough to simply monitor agricultural conditions and to report that, for example, Senegal has experienced a 30% decline in rainfall during the peak growing period or that Zimbabwe has had a slow start to their maize growing season. These facts, although interesting and relevant, are not close enough to the information actually required to be useful. What government officials and other decision makers need to know is the number of food insecure people in a population that is vulnerable to the consequences of these events. Conceptual frameworks provide a context that enables the transformation of biophysical measurements into an understanding of the consequences for social and economic vulnerability of the measurement.

Food security frameworks are ways of estimating the amount of food available in a region and the degree of access the population has to that food. Popularized by Amartya Sen in his groundbreaking research, food access is the ability to either grow, purchase or borrow enough food to maintain a healthy body weight (Sen, 1981). When FEWS NET was first conceived, food security frameworks revolved around food availability. The cereal balance approach involved calculating the gross amount of food available in a country through a per capita calculation. Using this definition, then, national food deficits were considered the food aid needs of the region. Remote sensing of environmental conditions was at the center of this analysis, enabling continental-scale monitoring, where images of vegetation anomalies were transformed into crude percent cereal production deficit across vast areas of a continent. This type of analysis, however, did not address the differential ability of households to acquire food through the market, or the complexity of stores, stocks and flows of food from one region to another. Not only was excess food aid delivered, but the people who needed the aid did not receive it because the tools needed to identify and target the most needy were lacking.

In the past ten years, FEWS NET's conceptual framework has evolved to incorporate both food access and food availability. The goal of vulnerability analysis is to identify populations who are likely to face disruptions in their ability to produce or acquire food and who are likely to lack the resources to absorb such shocks. Moving from cereal balance analyses at the country level to vulnerability analyses at the community level has also improved the ability of early warning organizations to prioritize and compare relative needs across different regions, enabling a more efficient distribution of assistance.

Today, FEWS NET uses remote sensing information in the context of a great deal of additional social and economic data on the ways people make a living in a region to determine food aid needs in as small a region as possible. Although there has been an improvement in how well vulnerability to food insecurity is measured, there is still a large gap between the assessed needs and the actual provision of food aid, even when there is excellent information and evidence for that need. Humanitarian assistance, in the form of cash and cereal shipments, is determined by social and political processes and is part of a complex and ongoing negotiation based on imperatives that have little to do with the actual need of the local people. This is particularly true during emergencies when humanitarian aid meets only a portion – however vital – of total needs (Barrett and Maxwell, 2005; Choularton, 2007).

The global humanitarian aid system is extensive and complex, encompassing organizations with multiple, and sometimes conflicting, agendas. The main actors of the system are governmental organizations that act as international donors, international implementing agencies (such as United Nations bodies, the International Committee of the Red Cross (ICRC) and international non-governmental organizations (NGOs)), national and local NGOs, as well as the country governments (TransparencyInternational, 2006). Humanitarian aid essentially consists of a one-sided transfer of resources, usually from industrialized countries into poorer ones. Aid providers largely act voluntarily, while aid recipients are often dependent on external assistance. This power imbalance provides difficult conditions for accountability: aid recipients have very few powers of sanction in relation to aid providers, while donors can largely choose for themselves the level at which their work is subjected to scrutiny. Even within international agencies, few controls are put on the types and distribution of aid, thus action or inaction occurs through general consensus of all parties, with no one party or organization in control of what occurs, and thus no one is responsible or accountable. More on this topic can be found in the last section of this book.

For FEWS NET, this means that although its stated goal is to prevent famine and destitution, all it really can do is provide information and foster consensus; it has no direct ability to make organizations act, even in the face of an obvious and impending emergency. Early warning systems have been criticized by authors such as Buchanan-Smith and Davies in their 1995 book as being unable to get the international relief system to move quickly enough to prevent problems from becoming emergencies.

In this chapter, we will review concepts and ideas that guide FEWS NET's vulnerability analysis and describe how remote sensing data is incorporated into the information stream. These frameworks will be examined in the light of the larger process of early warning and humanitarian assistance, and how biophysical hazards are identified, assessed and translated into estimates of need. These processes are conducted with an understanding of vulnerability and the diversity of economies, cultures and biophysical environments in which it occurs.

2.1 The Process of Early Warning

Early warning is defined as the provision of timely and effective information, through identified institutions, that allows individuals exposed to a hazard to take action to avoid or reduce their risk and prepare for effective response. The early warning process, as described by Twigg, 2003, has three inter-related stages:

1. Observation, forecasts and quantitative description of the consequences of the forecasts based on scientific expertise and advanced technologies. Remote sensing and local observations are combined into credible, scientific predictions of food production characteristics for a region.
2. Warnings and dissemination of the predictions of the consequences of the observations, transmitted to the appropriate personnel and agencies as recommendations for action. Investment in this stage has resulted in modern information methods and procedures, including bulletins, web sites, media broadcasts, briefings, and networks into civil society at a variety of levels. Early warning has institutional and political characteristics that enable broad dissemination offindings.
3. Response, where the warnings are turned into actions. The actors in this process are more numerous and diverse than the agencies and personnel who receive the warnings in Step 2. They include officials at international, national and local levels, non-governmental organizations (NGOs), communities and individuals. This third stage sees the institutional and political aspects of early warning broadening out and the early warning process acquiring a social dimension where risk perception and decision making play a crucial part. This stage has not received as much investment as the previous two stages, and it is where the gap between assessed needs and actual aid provided occurs (Twigg, 2003).

The two key elements of a successful early warning system are that the forecasts of the human consequences of an event are accurate in predicting its location, time and severity, and that these warnings are disseminated in time for populations at risk to do something about it. FEWS NET uses remote sensing observations for step 1, but successful early warning requires that steps 2 and 3 are also taken. In order to transform observations into communications that food security professionals can act on, FEWS NET has undertaken to understand the underlying vulnerability seen in the populations where they are responsible for monitoring.

The USAID Policy Paper entitled 'Food Aid and Food Security' identifies a range of important issues which lead to the food insecurity of households and individuals in the developing world (USAID, 1995). These include, among others:

- chronic poverty
- rapid population growth

- declining per capita food output
- poor infrastructure
- ecological constraints
- limited arable land
- inappropriate policies
- disease
- poor water and sanitation
- inadequate nutritional knowledge
- civil war, and
- ethnic conflicts.

This wide range of social, political and environmental phenomena are all important to food security outcomes. To determine how these various factors are affecting food security, a system of measurement needs to be instituted. Two types of indicators have been developed to measure food security outcomes: those that describe food supply and access, referred to as process indicators, and those that describe food consumption, referred to as outcome indicators (Maxwell and Frankenberger, 1992). Because this book is focused on using remote sensing in the context of early warning, we are interested here in process indicators that can provide information on the current situation, enabling early warning of a decline in food security due to a complex interaction of factors, one of which may be the environmental situation (Chung et al., 1997). Process indicators can be extremely complex and involve a wide variety of permutations, from the 23 that Maxwell and Frankenberger (1992) listed, to the 450 listed in Chung et al. (1997).

Because it would take an entire chapter to do justice to these indicators, they will not be detailed here. Instead, this section will describe how FEWS NET uses information to help answer two sets of decision-maker questions. The first set involves determining which population groups, if any, are likely to be food insecure, why, and for how long? The other set of questions seek understand what the best ways are to mitigate the effects of long-term trends or short-term shocks on the food security of affected populations. What available response options (both food and non-food) would be most appropriate to address the problem? FEWS NET uses a livelihoods approach to answer these questions. Before the workings of this approach are described, the concept of food security and how it is used in early warning will be outlined.

2.2 Concepts of Food Security

USAID defines food security as 'when all people at all times have both physical and economic access to sufficient food to meet their dietary needs for a productive and healthy life' (USAID, 1995).

By this definition, food security is a broad and complex concept which is determined by the interaction of a range of agro-physical, socioeconomic, and biological factors. Like the concepts of health or social welfare, there is no single, direct measure of food security. However, the complexity of the food security problem can be simplified by focusing on three distinct, but inter-related dimensions: food availability, food access, and food utilization. These can be defined as follows:

- *Food availability* is achieved when sufficient quantities of food are consistently available to all individuals within a country. Such food can be supplied through household production, other domestic output, commercial imports, or food assistance.
- *Food access* is ensured when households and all individuals within them have adequate resources to obtain appropriate foods for a nutritious diet. Access depends on income available to the household, on the distribution of income within the household, and on the price of food.
- *Food utilization* is the proper biological use of food, requiring a diet providing sufficient energy and essential nutrients, potable water, and adequate sanitation. Effective food utilization depends in large measure on knowledge within the household of food storage and processing techniques, basic principles of nutrition and proper child care, and illness management (USAID, 1995).

FEWS NET addresses access and availability of food, but cannot monitor utilization. To monitor utilization would require a much more individual-level analysis of millions of households, billions of individuals across the Africa and other regions. This is beyond the scope and financial ability of FEWS NET and nearly all other organizations that provide information on food security. There are few recognized and widely available information sources on food utilization that are comparable and standardized. Those that are available are highly contested, and thus utilization is an area where much progress still needs to be made (see Chap. 8 Sect. 8.7).

Figure 2.1 shows the concepts of food availability, food access and food utilization and how they are related. Based on the large literature on famine and food insecurity, FEWS NET's unit of analysis is the household and not the individual (Corbett, 1988; Friedmann, 1980; Haddad et al., 1994; Maxwell and Frankenberger, 1992). Household resources, categorized at the bottom, are the basis of production of cash crops, food crops and cash income. Some cash is used to purchase food on the market, and some of the food crops are consumed in the household. Food can also be obtained with transfers or loans from friends, relatives, or through government agencies, NGOs or other organizations. Food availability is the physical presence of food in the market or on the farm – without availability no one has effective access, no matter how much cash they have on hand. Finally, food utilization is the ability of each individual to effectively digest their food and turn it into energy. This utilization is mediated by health status, the quality of the food and of the care the person experiences (ie the cooking of food into nutritious meals). FEWS NET strives to determine how households and communities gain access to food and the level of food availability in the household, leaving evaluation of food utilization to other organizations (Riely et al., 1999).

There is both a long-term and a short-term aspect to food security. Chronic food insecurity is defined as a persistent inability of a household to meet the food requirements of its members over a long period of time marked by continuous ups

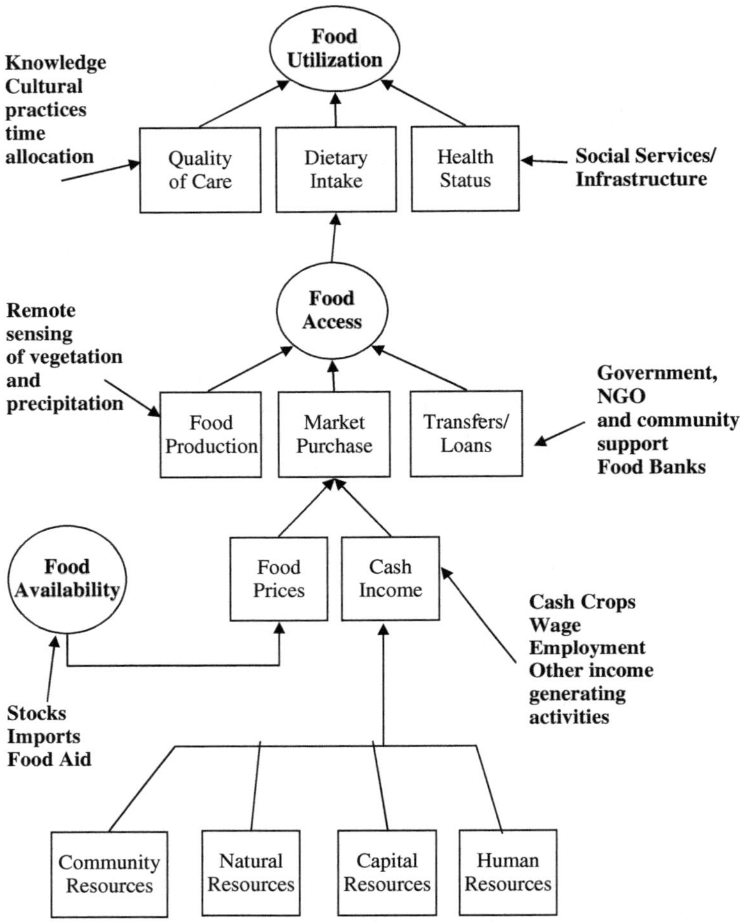

Fig. 2.1 Conceptual diagram showing aspects of food security and monitoring; derived from (USAID, 1995)

and downs. A short-term problem may afflict any household, regardless of whether it has a chronic problem or not. A short term problem, known as transitory food insecurity, may be caused by crop failure, seasonal scarcities, temporary illness of a productive household member, an emergency need for cash, or other causes. A household is said to be food secure only when it is protected from both kinds of insecurity. The average access to food over the long term should be nutritionally adequate, and a household should be able to cope with short-term crises without sacrificing the nutritional needs of any of its members.

FEWS NET operates in regions where 33% of the population is chronically malnourished, and the average number of food emergencies has tripled since the mid-1980s (OXFAM, 2006). This means that although FEWS NET and its partner agencies and governments are providing enough information to capture and respond to these crises, it does not appear to be doing enough to improve the underlying vulnerability. The focus of FEWS NET due to how it reports in FEWS NET is on informing food aid deliveries, but not other ways of supporting livelihoods in the longer term, thus it is limited in the kind of responses it can recommend. The magnitude of shock required to push thousands of households into a food insecure state is getting smaller due to the degradation of the resilience of the underlying household economy. Without a comprehensive approach to development of the local economy, FEWS NET is fighting a losing battle against food insecurity.

2.3 Agroecosystems and Food Security

In Central America, where FEWS NET has recently begun a program called the Meso-American Food Early Warning System (MFEWS), a decline in agricultural production has a direct impact on the ability of many rural people to make a living. Only Nicaragua was able to meet its cereal consumption requirements through domestic production in 2003. Of the other countries in the program, Guatemala was only able to meet 70% and Honduras 67% of its cereal requirements (FAO, 2000). Any decrease in domestic production, through for example excessively wet or dry conditions, will increase the percent of cereal that needs to be imported in order to meet consumption requirements. Thus ensuring food security at the national level depends not only on domestic production, but also on the capacity of countries to import food to make up any shortfalls. Food security strategies may focus on enhancing import capacity rather than trying to achieve food self-sufficiency and enhancing access to food by increasing income for the most vulnerable.

Guatemala, Honduras and Nicaragua are characterized by a mix of large-scale plantation agriculture, extensive cattle ranching (both primarily for export) and small, very poor, subsistence farms. There are hundreds of thousands of poor rural families who farm basic grains on ecologically fragile land on hillsides and at the frontier edges of the large holdings. Unable to make a living on their small plots, these farm families survive by engaging in a myriad of other seasonal, part-time, informal, rural and urban work. Family members of these small farmers are the farmhands, construction workers, maids, custodians, drivers, craftspeople and salespeople in the markets and on the urban streets. At the core of this extensive and thoroughly embedded diversified livelihood system is the farm, be it owned, rented or sharecropped (Holt-Gimenez, 2005).

Although the population in Central America is far more urban than those found in Africa, the economy is still very dependent on primary production from these marginal agroecosystems. After the impact of Hurricane Mitch where a great deal of the fragile tropical soil was washed away, the food security situation of the rural poor became significantly worse. The rural poor in Nicaragua, El Salvador and Guatemala will attempt to compensate for the loss of the key component to their livelihoods by accessing the extensive family and village networks of immigration to the US, as well as through humanitarian assistance such as that provided by FEWS NET.

2.3.1 African Agroecosystems

African farmers pursue a wide range of crop and livestock enterprises that vary both across and within the major agro-ecological zones. There is a large diversity in both the number of different crops being cultivated as well as the varieties available for each crop. Even at the level of the individual farm, many farmers cultivate 10 or more crops in diverse mixtures that vary across soil type, topographical position and distance from the household compound (Alberts and Mehta, 2004). By describing the dominant agricultural systems that FEWS NET is usually monitoring, a better understanding of the effectiveness of remote sensing tools that have been developed can be achieved.

Farming systems in Sub-Saharan Africa include many root crops, especially cassava. The main cereal crops are coarse grains like millet and sorghum, followed by maize. The International Model for Policy Analysis of Agricutural Commodities and Trade (IMPACT) developed by the International Food Policy Research Institute (IFPRI) to project the future demand for these commodities, estimated that the per capita demand for cereal crops will increase in Sub-Saharan Africa by some 4.9% per year between 1997 and 2020, with the main increase in wheat and rice (Rosegrant et al., 2001). Part of the increase will be due to greater demand for animal feed. The demand for root and tuber crops will increase by about 65%, more or less evenly spread over all species. These increases are based primarily on overall population growth in the region. Due to the poor technological level of development of the agriculture sector in most of the region, it is entirely possible that this demand can be met by local farmers, given proper investment in the sector.

Livestock are an integral part of the agricultural systems of Africa and are especially important to the poor, who derive a larger proportion of their incomes from livestock than do the wealthier (Delgado et al., 1999). According to Perry et al. (2002), in the mixed crop-livestock systems of the arid/semi-arid, humid/subhumid and tropical highlands of Eastern, Central and Southern Africa, cattle are judged of greatest importance to the poor, followed by sheep and goats, poultry, horses, donkeys and mules, and finally pigs. In similar systems in West Africa, sheep and goats are the most important, followed by poultry and cattle, then horses, donkeys and mules, with pigs again last. In the pastoral rangeland-based systems in Africa, sheep and goats are generally regarded as of highest importance to the poor, followed by cattle, camels and horses, donkeys and mules (Perry et al., 2002). Integrated livestock and agricultural systems provides a way to transform crop residues and low-rainfall areas into productive regions, enabling farmers to accumulate wealth, but also provides fertilizer that enriches agricultural fields.

Mixed rainfed crop-livestock systems in Africa, found in the arid/semi-arid and the humid/subhumid tropics (ILRI, 2000), support more than 70% of the estimated 280 million poor people in Sub-Saharan Africa (Thornton et al., 2002). Pastoral rangeland-based systems support around 10% of the poor. In North Africa the mixed irrigated arid/semi-arid crop-livestock system comprises 44% of the total poor in the region, while the three mixed rainfed crop-livestock systems represent only 25% (Alberts and Mehta, 2004). All of these systems are vulnerable to drought.

Despite an ever-increasing area under cultivation, countries across Africa have not managed to keep up with very large rates of population increase. Over the past forty years, per capital cereal production has declined, although tuber production has rebounded since the 1980s (Fig. 2.2). Sub-Saharan Africa is the only region of the world with declining per capita food production, forcing a continued investment in monitoring and aid programs. This decline is due to an increasing population with a static amount of natural resources and a declining investment in agriculture. With an average fertility rate exceeding five children per woman during the entire period, the population in Africa has grown the fastest of all regions in the world. There are almost three times as many Africans alive today (767 million) as there were in 1960 (FAO, 2000). This puts a tremendous amount of pressure on food production systems and leaves many of the poorest with the least amount of access to resources chronically food insecure.

Local food security is often equated with agricultural production outcomes. Hence, a chronic or temporary production deficit against local food requirement is immediately translated into chronic or temporary food insecurity. Consequently most early warning and food security monitoring systems draw heavily from two information sources: (i) crop and/or livestock production data; and (ii) market price

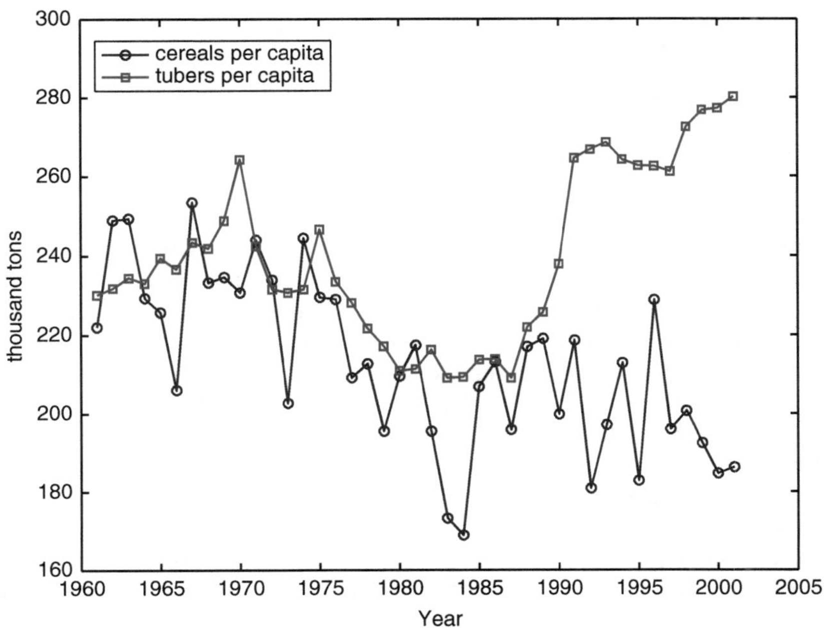

Fig. 2.2 Per capita cereal and tuber production for sub-Saharan Africa, from FAO statistics

information. This is almost never the whole story. A full account of the 'food economy' addresses both food availability – that is, what food people produce – and food access – what cash people earn to purchase food. Data on casual employment or wild foods, or charity from relatives or the sale of handicrafts may be equally important to the livelihood story as data on crop and livestock production, and knowledge of the relative importance of these can guide the design of more appropriate monitoring systems and better rapid emergency assessments.

2.4 Early Warning of Food Insecurity

FEWS NET's goals are to strengthen the ability of national governments and regional organizations to manage risk of food insecurity through the provision of timely and analytical early warning and vulnerability information. FEWS NET works with a wide range of local, national and international governmental and non-governmental organizations to ensure that the information they provide on food access is used in a timely way. This is most often through food aid programs, but can also be through a wide variety of other types of assistance, including food for work programs, cash assistance, among others. In the United States, these other, non-food assistance programs have very limited funding due to the budgetary process that USAID goes through every year. Due to the rapid increase in both demand and the price for corn (maize) in the past few years, increasing farmer's profits, and recent work in the US Congress to reauthorize Public Law 480 which funds food aid, there is hope that this will change. FEWS NET is advocating that the law be changed to be less focused on the grain itself and more on providing appropriate assistance according to what is known about the causes of a particular crisis.

Figure 2.3 shows how food aid deliveries fit into the process of monitoring of the agricultural calendar. The figure shows clearly how food security monitoring works on a multi-year time scale, the implications of which are that if a cropping season fails several years in a row then a crisis can become extremely acute very quickly, especially when the underlying vulnerability of a population is already very high. Food aid delivery occurs at the earliest seven months after the failure of the harvest, and often it is a year or more after the initial shock (Dilley, 2000). Remote sensing plays an important role in quantifying the shock, estimating its spatial extent and severity, and providing supporting evidence for the level of food aid.

In a recent study by the New Partnership for Africa's Development (NEPAD), the effectiveness of the ability of governments to respond to environmental shocks such as drought vary from country to country. These shocks are still critical determinants of severe food crises, even after nearly two decades of early warning system development and improvement of appropriate responses. The report shows that the main constraints to reducing the effect of environmental variability on food security appear to be lack of resources, the lack of an appropriate strategy, insufficient political will, and the lack of reliable and timely food security information. Most national

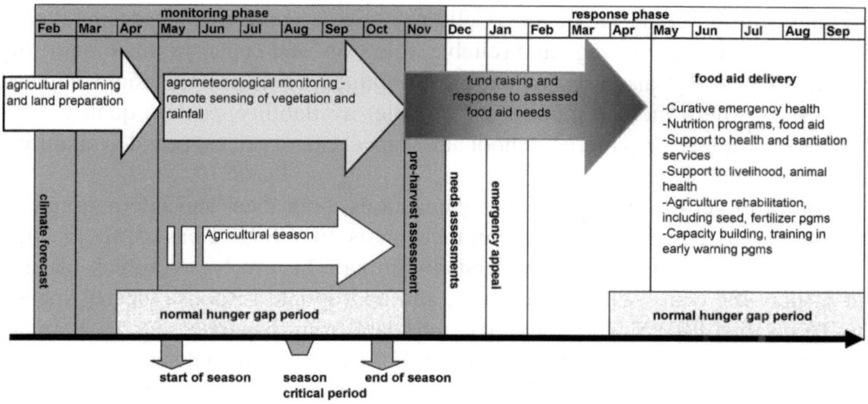

Fig. 2.3 Food security monitoring and humanitarian decision-making framework (from Haile (2005), p. 6). Used with permission

humanitarian response systems are still heavily dependent on donor funding and lack government commitment to allocate adequate resources from its national budget (Haile, 2005). Most countries, however, still monitor food production and food security because it has become a common belief that although drought and food shortages have large budgetary implications, to wait and do nothing may cost even more. Early warning systems are funded so that governments may have the information required to make appropriate decisions, and so that they don't pay a heavy political price for not knowing when a humanitarian disaster is about to occur, on the other side of the globe or right under its nose in its own capital city.

In Fig. 2.4, the data collection cycle is shown as being focused on agricultural production. For households dependent on agricultural production for their incomes, the shortfall in production caused by drought, flood or infestation is a

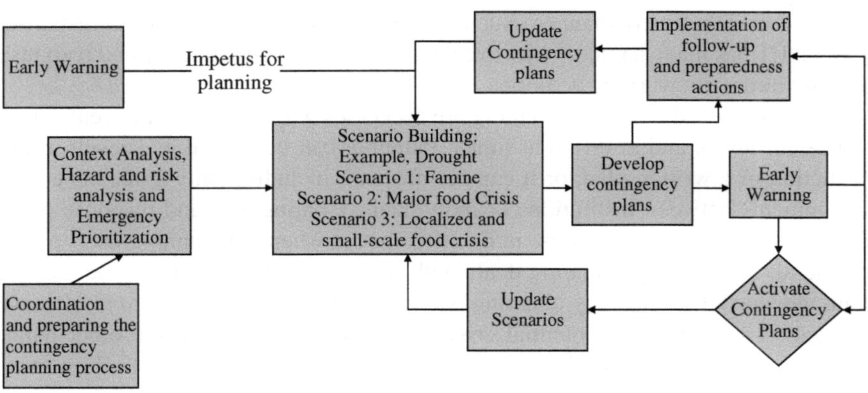

Fig. 2.4 Contingency Planning Process, from Choularton (2007), showing how scenario building and contingency plans are integrated with the early warning process, and are a prerequisite for action

good indicator of the fall in income (Callear, 1997). Because remote sensing data is inexpensive, wide in coverage and reliable, it is front and center in this monitoring scheme. FEWS NET also uses indicators of imminent crisis derived from Cuny and Hill (1999). These include monitoring of water availability, pasture quality, livestock prices, migration patterns, school attendance, cereal prices, cereal availability, and consumption of wild foods.

In order to get governments and organizations, both local and international, to pay attention to the warnings of an imminent crisis, FEWS NET has to provide high quality information in an easy to understand format. This involves graphics, usually, that display the cause, effect, relevance, and appropriate response clearly spelled out. To do this, FEWS NET must integrate data from disparate sources at vastly different resolutions and levels of precision. For example, an analysis of the impact of drought as measured both by vegetation anomalies and rainfall deficits needs to be integrated with information on elevated food prices (point measurements), migration patterns (usually reports from a variety of sources, but without precise geographic detail), and water scarcity (also diffuse reports). To integrate these different types of information takes a great deal of work and experience. The objective is always to get a consistent, coherent response from governments, something that is affected by a wide variety of other factors other than need.

2.5 Humanitarian Response to Early Warning

While gathering information needed to ensure that early warning information is accurate and relevant, FEWS NET also works to ensure that the analysis provided is used to develop appropriate responses to food insecurity. Early warning systems have come under a great deal of criticism during the past decade regarding its continued emphasis on bigger and better tools to detect food insecurity, such as complex remote sensing information systems about which this book is focused, with little emphasis on ensuring that the information is acted upon by governments and organizations in a timely manner (Buchanan-Smith and Davies, 1995). In response, FEWS NET has made a major objective the integration of contingency and response planning into early warning activities.

Contingency and response planning are processes which include different actors at different levels and at different times. Planning can be formal and result in the production of a written plan; or it can be informal, including only agreements and arrangements between institutions or individuals. Contingency and response planning are similar processes, whose main difference is when the planning takes place. In general, contingency planning deals with potential future events or situations – 'contingencies'. Contingency planning is used as a preparatory measure to increase readiness to respond to a potential crisis. Response planning, on the other hand, occurs once a crisis has emerged and planning focuses on dealing with the consequences of the crisis. The main difference is one of certainty of outcome. Response planning tends to be based on assessments of an actual crisis while contingency planning is based on scenarios.

Ideally, these two planning processes are complementary and work together. If a good contingency plan has been prepared then it can easily be transformed into a response plan once the assumptions in the scenario are verified and adjusted according to an assessment. Figure 2.4 illustrates how contingency and response planning are conducted using drought as an example. Early warning and a contingency planning process occur simultaneously. An analysis of the context enables the development of a planning process, including a contingency plan with three scenarios is prepared detailing some of the potential outcomes of a drought. As the drought evolves the contingency plan is updated. Ongoing monitoring through early warning enables the activation of the plan once the drought and its consequences reach a certain magnitude. The activated plan becomes a response plan which is used as a tool for managing the response to the drought. Even though there is an ongoing response, ideally, contingency plans are also maintained to ensure readiness for a change in the situation, such as a further deterioration of drought conditions.

Emergency bulk shipments of cereals are often the direct response to a food security emergency, but there are many alternatives to this direct food donation approach. Providing cash to needy people can ensure that only the most vulnerable receive aid and that the money can be used to purchase local, familiar and appropriate food items. Purchasing grain in bulk in local markets instead of importing internationally can be a powerful tool because not only is far more food obtained for those in need due to the reduction in transportation costs (usually 50% more), but local farmers can be supported contributing to the region's development in the long-term. Local purchases require local currency and systems set up far in advance to facilitate the purchase, transportation and storage of the grain from disparate markets to bring the grain from where it is to where it is needed. Ensuring that imported grain is used to improve food security is only marginally less complicated then local purchase, however, as economies where food crises occur are rarely well functioning.

Even when there is no food security emergency, many regions where FEWS NET works have a great deal of chronic food insecurity. Food aid can then be an important development resource, supporting programs with a wide range of development objectives (Table 2.1, from (Riely et al., 1999)). There are many strategic alternatives to food aid in early emergency intervention, effectively using these strategies requires precise information on local economic and social structures (Jayne et al., 1995). FEWS NET, however, works with USAID's Food for Peace office, which has most of its response capability in food grain, thus its systems are focused on providing this kind of assistance.

The complexity of response to food aid crises is matched only by the complexity of the situations themselves. FEWS NET's information gathering must provide the data needed to both provide early warning of an impending crisis, but also to advise local, national and international governments and organizations on programs to reduce the likelihood that a crisis may occur at all.

Table 2.1 Title II food aid program Types (from (Riely et al., 1999))

Name	Description
Humanitarian Feeding (HUM)	In these programs food (or cash in isolated cases) is distributed directly to disadvantaged groups, or those severely affected by emergency conditions.
Food-for-Work (FFW)	Food-for-Work programs use food aid as payment for laborers in public works programs designed to build and maintain local infrastructure (e.g., roads, dams, wells, latrines, schools). Cash from monetization proceeds may also be used to purchase inputs or as cash wages in cash-for-work (CFW) programs.
Maternal and Child Health (MCH)	In MCH programs, targeted food aid provides supplementary rations in programs seeking to improve the health and nutritional status of, typically, pregnant and lactating mothers and children under the age of five. Most MCH programs combine food aid with other elements such as nutrition and health education, growth monitoring and counseling, and immunization, which may, in part, be funded through monetization proceeds.
School Feeding Programs	School feeding programs provide students with snacks, lunches, and/or breakfasts at schools as incentives to increase enrollment, maintain attendance, and improve the performance of students.
Other Child Feeding (OCF)	OCF programs provide meals to particularly vulnerable groups of children outside the school setting.
Monitization Programs	The sale of food aid through monetization programs provides financial resources for use in a variety of activities, including education and training, health and nutrition, agriculture, rural credit, microenterprise, cash-for-work, and other development programs.

2.6 Challenges and Opportunities for FEWS NET

Although FEWS NET provides detailed information on the cause of a particular crisis and its severity to its prime sponsor in FEWS NET, it does not do a very good job disseminating the information in other areas of USAID and in other areas of the humanitarian community. The information FEWS NET provides focuses on the number of people who are food insecure and the amount of bulk grain shipments required to feed them. This focus on grain misses an opportunity to inform the larger overseas development assistance community that has the ability to use the entire range of both Title II programs listed in Table 2.1 as well as other types of livelihood support. Improving the climate and vulnerability analysis of development programs who seek to fundamentally alter livelihood strategy through agricultural development, small business training and manufacturing could be an extremely valuable service that FEWS NET could provide.

FEWS NET's main sponsor in Food for Peace, however, keeps its focus on providing information which informs a response that is primarily bulk grain shipments. This keeps FEWS NET's integration of information about the ways people make a living and the diverse strategies local people use to cope with a shock limited to analysis. Livelihoods analysis could also provide policy guidance that could inform

Table 2.2 Food Aid Needs and Beneficiaries from the Executive Overview of Food Security in Sub-Saharan Africa of September 26, 2007. PNSP is the Productive Safety Net Programme (PSNP), IDP are internally displaced persons, WFP the World Food Program and C-SAFE is the Consortium for Southern African Food Security

Food Aid Needs and Beneficiaries		
Country	Population at Risk	Food Aid Beneficiaries
Chad	747,850	599,600
Djibouti	150,000	121,250
Ethiopia	8 million (estimated)	7.5 million (PSNP + emerg.)
Kenya	2.4 million	650,000
Mozambique	660,000	192,000 (emerg.)
Somalia	1.5 million (incl. IDPs)	200,000
South Sudan	1.7 million	1 million
Uganda	1,757,000 (incl. IDPs)	1,757,000
Zimbabwe	4.1 million	614,000 (WFP and C-SAFE)

a wide range of responses from multiple agencies and organizations. FEWS NET's focus on grain can be seen clearly in their executive briefing document which is intended for USAID Food for Peace decision makers. To provide useful decision support for USAID, it delivers information such as is in Table 2.2 which provides the number of food insecure people, the basis of food aid calculations, instead of focusing on the diversity of causes and optimal potential interventions which are the outcome of their livelihood-based food security analysis. Because USAID provides over 50% of all emergency aid, this focus on one kind of response has a significant impact on the functioning of the entire system (Barrett and Maxwell, 2005). FEWS NET's challenge moving forward is to try to expand the number of decision makers that it can provide information to so that the type of response that it can engender is also diversified. Integrating its information into longer-term development is becoming more critical, as global climate change and its adverse effects are beginning to be felt.

2.7 Summary

In this chapter we have presented the concepts of early warning and of food security, and a summary of how FEWS NET uses data and information to provide policy guidance that can be acted upon by governments. We defined food security as being the ability of all individuals to access enough food for an active, healthy life. Food security is often analyzed at the household level, although there are some aspects to food security which must be measured for the individual. Three aspects of food security were identified: food utilization, food access and food availability. Food utilization is composed of an individual's ability to digest and transform food into energy for life. Food utilization is affected by quality of care, dietary intake and health status. Food access is composed of the various ways a household, instead of an individual, acquires food for consumption. The elements of food access are production (farming and livestock rearing), market purchase (through cash income) and transfers and loans (including food aid). Food availability is the physical presence of food in the market or at the farm, which is a prerequisite for food access. In the last section, challenges for FEWS NET in their efforts to motivate a timely and appropriate response in the institutional environment where they work are presented.

References

Alberts, B. and Mehta, G., 2004. Realizing the Promise and Potential of African Agriculture. Interacademy Council, Amsterdam, The Netherlands.

Barrett, C.B. and Maxwell, D.G., 2005. Food Aid After Fifty Years: Recasting its Role. Routledge, New York.

Buchanan-Smith, M. and Davies, S.M., 1995. Famine Early Warning and Response. IT Press, London, England, 240pp.

Callear, D., 1997. Can we get policymakers to take notice of information on drought and imminent food shortages? Internet Journal of African Studies, 2.

Choularton, R., 2007. Contingency Planning and Humanitarian Action: A Review of Practice. Humanitarian Practice Network, Overseas Development Institute, London, England.

Chung, K., Haddad, L., Ramakrishna, J. and Riely, F., 1997. Identifying the Food Insecure: The Application of Mixed-Method Approaches in India. International Food Policy Research Institute, Washington DC.

Corbett, J., 1988. Famine and household coping strategies. World Development, 16(9): 1099–1112.

Cuny, F.C. and Hill, R.B., 1999. Famine, Conflict and Response: A Basic Guide. Kumarian Press, West Hartford, CT.

Delgado, C., Rosegrant, M.W., Steinfeld, H., Ehui, S. and Courbois, C., 1999. Livestock to 2020: The next food evolution. Food, agriculture and environment discussion paper 28, International Food Policy Research Institute, Food and Agriculture Organization, and International Livestock Research Institute.

Dilley, M., 2000. Warning and intervention: What kind of information does the response community need from the early warning community, USAID, Office of US Foreign Disaster Assistance, Washington DC.

FAO, 2000. FAO Statistical Database. Food and Agricultural Organization of the United Nations (FAO).

Friedmann, H., 1980. Household production and the national economy: concepts for the analysis of agrarian formations. Journal of Peasant Studies, 7: 158–184.

Haddad, L., Kennedy, E. and Sullivan, J., 1994. Choice of indicators for food security and nutrition monitoring. Food Policy, 19(3): 329–343.

Haile, M., 2005. Weather patterns, food security and humanitarian response in sub-Saharan Africa. Philosophical Transactions of the Royal Society, Series B, 360: 1746–1760.

Holt-Gimenez, E., 2005. Hurricane Mitch: Crisis or Sustainability. University of California Agroecology Research Group, Santa Cruz, CA.

ILRI, 2000. Livestock Strategy to 2010: Making the Livestock Revolution Work for the Poor. International Livestock Research Institute, Nairobi, Kenya.

Jayne, T.S., Rubey, L., Tschirley, D., Mukumbu, M., Chisvo, M., Santos, A.P., Weber, M.T. and Diskin, P., 1995. Effects of Market Reform on Access to Food by Low-Income Households: Evidence from Four Countries in Eastern and Southern Africa. 5, USAID, Washington DC.

Maxwell, S. and Frankenberger, T.R., 1992. Household Food Security, Concepts, Indicators and Measurements: A Technical Review. United Nations Children's Fund – International Fund for Agricultural Development, New York, 129pp.

OXFAM, 2006. Causing Hunger: An Overview of the Food Crisis in Africa, Oxfam London, England.

Perry, B.D., Randolph, T.F., McDermott, J.J., Sones, K.R. and Thornton, P.K., 2002. Investing in Animal Health Research to Alleviate Poverty. International Livestock Research Institute, Nairobi, Kenya.

Riely, F., Mock, N., Cogill, B., Bailey, L. and Kenefick, E., 1999. Food Security Indicators and Framework for Use in the Monitoring and Evaluation of Food Aid Programs, USAID, Washington DC.

Rosegrant, M.W., Paisner, M.S., Meijer, S. and Witcover, J., 2001. Global Food Projections to 2020: Emerging Trends and Alternative Futures. International Food Policy Research Institute, Washington DC.

Sen, A.K., 1981. Poverty and Famines: An Essay on Entitlements and Deprivation. Clarendon Press, Oxford, 270pp.

Thornton, P.K., Kruska, R.L., Henninger, N., Kristjanson, P.M., Reid, R.S., Atieno, F., Odero, A.N. and Ndegwa, T., 2002. Mapping poverty and Livestock in the Developing World. International Livestock Research Institute, Nairobi, Kenya.

Transparency International, 2006. Corruption in Humanitarian Aid.

Twigg, J., 2003. The human factor in early warnings: risk perception and appropriate communications. In: J. Zschau and A.N. Kuppers (Editors), Early Warning Systems for Natural Disaster Reduction. Springer-Verlag, New York, pp. 19–26.

USAID, 1995. Food Aid and Food Security Policy Paper, USAID, Washington, DC.

Chapter 3
FEWS NET's Structure and Remote Sensing

Remote sensing and agricultural monitoring are a central part of FEWS NET's activities. Climatic hazards are difficult to predict but have widespread impacts on many sectors of these economies. Variability in the amount and distribution of rainfall has a negative impact on agro-ecosystems, and is particularly damaging in arid and semi-arid areas with seasonal regimes. In these areas, variability and risk of the disruption of food production is a central part of the lives of farmers.

Estimating the impact of climatic hazards is more challenging than simply providing the necessary physical evidence of an ongoing drought or the severity of a flood. There is large variation in the amount of climatic stress that vulnerable groups can endure before real and widespread destruction of livelihoods occur (Dilley, 2000). Although the physical characteristics of crop yield reductions due to rainfall deficits can be specified, determining the impact of this reduction in the place and time that it occurs is far more difficult. For example, a 50% reduction in millet production in Mali due to erratic rainfall that occurs after several years of good harvests is far less likely to result in sufficient food insecurity to warrant intervention than the same reduction after several years of below-average production. Just as important as the timing element is its spatial extent. Drought occurring in cropping areas has a different impact than those in pastoral lands, and the size of the area affected also can have a significant impact on food security. These complexities make interpreting climate data and linking it effectively to humanitarian intervention very challenging (Moseley, 2001).

This chapter will describe FEWS NET's structure, the process it uses to identify and monitor food security problems and the social and biophysical data that the organization uses to provide actionable food security information in its myriad of products. It focuses on how FEWS NET goes about identifying a problem using remote sensing and other tools in order to illicit an appropriate response.

3.1 FEWS NET Implementation

The first Famine Early Warning System (FEWS) was created by USAID in 1985 in response to a need for better food security and response information for emergencies in Africa. The activity was managed from the United States Agency for International Development (USAID) Africa Bureau using a field activities Contractor to carry out data collection and assessment activities in 5 countries of the Sahel and Sudan. Since that beginning, FEWS has defined, implemented, learned from, and evolved, implementing a number of approaches and methods for early warning, food security and vulnerability monitoring and assessment that have become standard approaches in the field. Nevertheless, every phase of FEWS/FEWS NET has seen substantial evolution in its methods in response to experience, new technologies, a better understanding of hazards, food security, and vulnerability.

Since its early days, FEWS NET adopted a general orientation of working in the field with partners (or groups of partners – networks), and providing capacity-building to FEWS NET partners. These orientations have proven their functional utility over time, and remain key parts of the FEWS NET approach. In recent years, some new technical themes were integrated into FEWS NET in response to several new USAID institutional needs and objectives. Notable among these was the adding of the word 'NET' to the FEWS name to signal an enhanced focus on working with and through country and regional partners and networks.

In 2002, the responsibility for the entire suite of FEWS NET activity contracts, agreements and management was transferred from USAID's Office of Sustainable Development in the Africa Bureau to the Policy and Technical Division of the Office of Food for Peace (FFP) in the Democracy, Conflict and Humanitarian Assistance Bureau (DCHA) of USAID. The intent of this move to DCHA/FFP was to allow FEWS NET to play a more global role in meeting USAID information and decision-making needs in early warning, food security and vulnerability assessment, and in improved humanitarian response. In 2003, FEWS NET received funding from the Office of Foreign and Disaster Assistance (OFDA), the Asia/Near East Bureau, and from several missions in the Latin America and Caribbean, to open new programs in Afghanistan, Haiti, Guatemala, Honduras, and Nicaragua. In line with the new global scope and placement within the Agency, FEWS NET began delivering regional, continental, and global products to AID/Washington directly.

The US Geological Survey at the Earth Resources Observations and Science (EROS) center, the National Aeronautics and Space Administration (NASA), the National Oceanic and Atmospheric Administration (NOAA), and Chemonics International (Chemonics) to provide the data, information, and analyses needed for the FEWS NET project. NASA and NOAA collect and process satellite data that are used to monitor the vegetation condition (Normalized Difference Vegetation Index, or NDVI) and rainfall (RainFall Estimate, or RFE) across the entire African continent. The NDVI and RFE data are but two of the wide variety of tools used by FEWS NET to monitor agricultural conditions in Africa.

The four inter-agency agreements with the US Government agencies are to provide the data described below.

- The Climate Prediction Center (CPC) at NOAA provides technical support in meteorology and climatology using satellite rainfall estimation products for Africa, Central America and the Caribbean, and Central Asia.
- The International Programs office at the USGS EROS provides assistance in developing operational early warning applications and products that use satellite and remote sensing data. USGS also maintains the FEWS NET archive of tabular, vector, and raster datasets and make them available via the web.
- The GIMMS group at NASA Goddard Space Flight Center provides satellite-derived vegetation data products, particularly the Normalized Difference Vegetation Index imagery (NDVI), for early warning activities, as well as conducting research on ways to improve remote sensing estimates.
- The USDA provides FEWS NET with technically-qualified management personnel, as well as access to USDA expertise on agriculture, markets, early warning, and crop estimation. USDA conducts tours that estimate the accuracy of crop models and agriculture production statistics which FEWS NET often participates in.

3.2 FEWS NET Field Offices

The most visible parts of FEWS NET are its field offices and representatives in roughly 31 countries, and a contractor in Washington D.C. office located near USAID that manages and technically directs them. At the time of this writing, the FEWS NET contractor is Chemonics International Inc, who has had the FEWS NET contract since 2000. The contractor is responsible for integrating FEWS NET's global early warning information, resources and training activities, in the field and in Washington D.C., and delivering it into information-gathering and decision-making processes of USAID (in Washington and the field), as well as to a broad range of international partners. At the time of this writing, these offices are in the following locations:

- Africa: Regional offices in Niger, Kenya, and South Africa. National offices in Senegal, Mauritania, Mali, Burkina Faso, Niger, Chad, Northern Sudan (located in Khartoum), Southern Sudan (located in Nairobi), Eritrea, Ethiopia, Djibouti, Somalia (located in Nairobi), Kenya, Uganda, Angola, Tanzania, Rwanda, Malawi, Mozambique, Zambia, Zimbabwe, and South Africa (for coverage of Lesotho, Swaziland and Botswana).
- Central America: National offices in Guatemala, Honduras, Nicaragua
- Caribbean: Haiti
- Central Asia: Afghanistan

Standard tasks for FEWS NET Regional Representatives includes a mix of regional coverage activities that are similar in nature, but of a regional scope, to the FEWS NET representatives tasks. A FEWS NET national or regional office may vary in its complement of staff and operational resources. All representatives are located in a field office of one of three types: fully-staffed, standard-staffed and minimally staffed. A fully-staffed office, for countries with substantial food security issues, has one FEWS NET National Representative, one Deputy Representative, and an administrative staff that includes one Office Manager, one administrative support personnel (accountant/clerk/ book-keeper), and one Driver. The travel and operational budget provided for the office includes sufficient funds to carry out all fundamental FEWS NET coverage services. It allows the creation of, improvements in, and additions to, the country's livelihoods framework for food security analysis. It allows an initial implementation of food market system and vulnerability risk factor analysis services.

A standard-staffed office, which is usually in countries with medium-severity food security issues, has one National Representative, a deputy, an Office Manager, an administrative support person, and a Driver. The travel and operational budget provided for the standard office includes sufficient funds to carry out all fundamental FEWS NET coverage services. This budget also allows maintenance of the country's livelihoods framework for food security analysis, if it exists already. This office's operational budget does not allow for any of the other services described for the fully-staffed office above.

A minimally-staffed office is recommended only for cases where two or more offices utilize shared space and services (e.g. where a regional and a national office are combined). It is also appropriate in exceptional cases where a larger presence is not advisable for non-technical reasons, or when only a limited scope of coverage is desired. Such an office has one national representative, an administrative assistant, and a driver. Local travel and the office's operational costs budget are reduced to a minimum that will only allow a minimal level of standard coverage services to be provided. Figure 3.1 shows a map of the countries that have the above type of staffing levels.

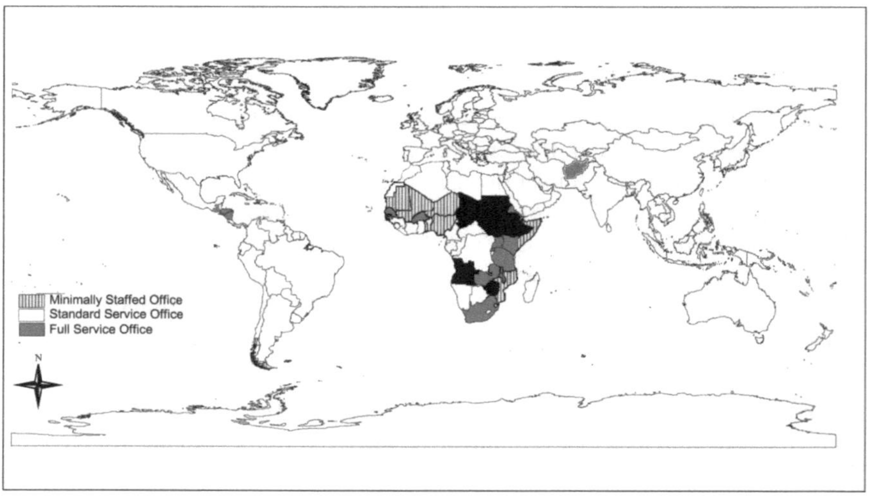

Fig. 3.1 Map of FEWS NET countries with current staffing levels

The principal tasks that FEWS NET manages are one of four general types:

- Managing the FEWS NET field and Washington offices to gather and assess a wide variety of early warning, food security and vulnerability data and information;
- Delivering effective decision-making products to a variety of USAID and US Government agencies, and to country (e.g. Governments, NGOs, and international organizations like the World Food Program who are working in a country where FEWS NET is present), and regional and international organization (e.g. CILSS (Comite Permanent Inter-Etats de Lutte Contre la Secheresse dans le Sahel), the United Nations Food and Agriculture Organization (FAO), Save the Children, etc.), and to other partners;
- Building the human and institutional capacity of country and regional partners/networks through hands-on training, to leave-behind skills and empower effective country and regional-based early warning; and,
- Developing, testing, and implementing new applied tools and methods for more effective early warning, and food security and vulnerability assessment. In the past, FEWS NET field activities were generally oriented towards identifying, monitoring, assessing, and defining hazards, conditions, and outcomes that relate directly to only food-related aspects of household food availability and access. In the time of an imminent crisis or famine, FEWS NET will expand its coverage to include a several non-food-related features of a household's food security, including water availability and specific nutrition and health issues.

FEWS NET provides critical training for both its own staff and for its partners in its field locations. These trainings provide a very important venue for explanation and familiarization with remote sensing data and imagery as well as the methods FEWS NET uses to interpret them. Every professional FEWS NET field office staff member in each region has a planning workshop that allows the continual improvement FEWS NET field personnel's technical skills as well as the development of annual work plans. FEWS NET's field personnel also work to improve USAID's, FEWS NET's, and partners' analytical ability to assess and interpret country and region-specific food security, vulnerability, and famine dynamics/conditions. They do this by collecting country-specific statistical data, archiving and using that data in the context of reporting. Field personnel are be responsible for organizing, documenting, updating, and archiving key statistical, or secondary, datasets (e.g. rainfall, agricultural production, food prices, malnutrition rates, etc). These datasets will be managed and organized within a geographic framework that allows FEWS NET and others to use them for thematic mapping and overlay in geographic information systems (GIS). By having field personnel be responsible for these datasets, they are more likely to include the best local data and be kept up to date. It also means that transferring both the archive and current updates every month from every field office to Washington DC becomes a huge logistical problem, which at the time of this writing is being solved through networking all the computer systems globally.

Another important task for field personnel is to improve the country-specific household livelihoods framework for food security analysis. By developing new or enhancing existing FEWS NET household livelihood-based food security-related zones and profiles in each country, FEWS NET can ensure a continual improvement to its global analysis. These enhancements are made by adding data and information to extend them towards a scenario modeling capability. This capability has been implemented in 2007, and involves estimating the food security situation six months into the future, and is based on potential variations in food production as well as possible changes in the economic or political situation. Contingency planning is described further in Chap. 14.

FEWS NET's field personnel perform regular monitoring and assessment of hazards, of food security conditions, and of changes in vulnerability in each country covered. These include:

- *Hazard monitoring, early warning, and hazard impact assessment:* Field personnel identifies and monitors both natural (e.g. droughts, extreme weather events, frost) and socio-economic hazards (e.g. food price increases, proposed changes in food-related policies, conflict, border closures, infrastructure failures, etc.), with an objective of delivering early warning of an imminent threat to food security. If the hazard actually occurs, field personnel will assess the impacts it has on household food-related livelihood conditions, and market systems.
- *Food security and vulnerability monitoring and assessment:* At least once every month, FEWS NET field personnel will assess and report on food access, food availability, and food utilization conditions, as well as the risks and hazards that affect them, in conjunction with other partners.

FEWS NET's work to improve the capacity of its partners to conduct food security analyses is generally provided in informal settings, and generally revolves around a FEWS NET representative offering hands-on training to a partner through jointly undertaking FEWS NET monitoring and assessment tasks. Field personnel may also offer a mentoring relationship with a partner, and/or a limited amount of participation in FEWS NET conferences where specific training will be provided. FEWS NET field personnel also often initiate, strengthen, and engage network partners to jointly undertake early warning, and food security and vulnerability monitoring and assessment activities, as a form of building the capacity of the country's informal network of early warning practitioners.

All field offices, except for the FEWS NET offices in Central America, produce monthly food security situation reports for each country. Although the level of effort given to these reports usually remains stable, the content and format of these reports may be expected to change in response to:

- an increasing (or decreasing) quality of the baseline data and analytic framework for each country,
- the amount, type, and quality of data available for that country, the conditions of food security that prevail in the country, and the time of year, and
- the expressed information needs of USAID.

The monthly report is focused on providing information on the food security situation, including measures of underlying vulnerability such as child and maternal health reports, requests for assistance in good years, etc. Information on early indicators of a food crisis or on the overall vulnerability of a region includes a seasonal agricultural calendar, information about food availability and food prices across the country and in neighboring regions, school attendance and levels of migration, particularly during seasons when migration is less common. The monthly report is used as the cornerstone of FEWS NET reporting, allowing analysts in Washington DC to compile information. FEWS NET provides color copies of each report per month and per country which are distributed to the local, regional and national government agencies and personnel as needed.

To assist in the integration of geographic information and remote sensing information into standard products and monitoring, FEWS NET has funded four regional representatives through USGS who have expertise in GIS and remote sensing and who can provide assistance in making accurate and effective maps, download and manipulate data and to provide training on new products for FEWS NET technical personnel. There are four USGS Regional Scientists placed in FEWS NET regional offices in the Sahel, Greater Horn of Africa, Southern Africa, and Central America. They provide technical assistance in the use of operational remote sensing products for food security analysis. These Regional Scientists play a very important role in the development of new tools and in understanding the problems and challenges of the FEWS NET representatives in the field in using remote sensing data.

3.3 Food Security Status Alerts

In addition to regular monthly reports, field personnel may also prepare 1–2 page Alert reports when the status of food security in a country or area is determined by USAID to be a problem, using FEWS NET's watch, warning, and emergency criteria. Table 3.1 describes the four alert levels. In the initial stages of a food crisis, FEWS NET issues watches, letting decision makers know that a problem is emerging and that they should start to prepare and develop contingency plans. As a crisis deepens, FEWS NET upgrades its alert to a warning, letting decision makers know they need to respond to the food crisis and prevent it from deteriorating into a more

Table 3.1 FEWS NET Alert levels

Alert level	Description
Emergency	A significant food security crisis is occurring, where portions of the population are now, or will soon become, extremely food insecure and face imminent famine. Decision makers should give the highest priority to responding to the situations highlighted by this Emergency alert.
Warning	A food crisis is developing, where groups are now, or about to become, highly food insecure and take increasingly irreversible actions that undermine their future food security. Decision makers should urgently address the situations highlighted by this Warning.
Watch	There are indications of a possible food security crisis. Decision makers should pay increasing attention to the situations highlighted in this Watch, and update preparedness and contingency planning measures to address the situation.
No Alert	There are no indications of Food Security problems.

severe and wide-spread emergency. The most severe alert level used is the emergency, which is reserved only for situations where immediate action is required.

Alerts are reviewed monthly, and updated as necessary. FEWS NET is vigilant to ensure that countries are downgraded as soon as possible to prevent over-use of the warning and emergency status. At the request of USAID, field personnel occasionally provide briefings, and/or special reports on Alert-related food security issues and conditions.

When Alerts occur, field personnel are often expected to participate in national, regional, and international efforts to improve food aid and food security needs assessment process, and to more effectively move the results of needs assessments into a variety of decision-making and response planning processes in USAID Missions and Bureaus, Food For Peace, and with other partners. The FEWS NET Washington office, and the management and technical personnel in it, carry out a number of assessment and reporting functions for USAID/Washington. This includes working with its partners in NOAA, NASA and USGS to produce remote sensing-based analysis that results in clear, understandable figures and maps for reports. The Washington office provides the strategy for and oversees capacity-building activities that are undertaken by FEWS NET staff, either in the field or in Washington. It is also responsible for managing all network building and strengthening activities that are undertaken by field offices and personnel.

3.4 FEWS NET Livelihood Monitoring and Assessment

In order to determine the impact of particular events on households, FEWS NET uses an integrated approach that identifies homogeneous livelihood zones within each country with explicit knowledge regarding the source of income and food for all demographic groups in the region. By analyzing people's access to food and income as well as their options for coping with adverse events and processes, the livelihoods approach enables FEWS NET to make informed judgments about the resulting likely type and magnitude of the effect on food security of these hazards (Boudreau, 1998; Mathys, 2005). Interventions to reduce food insecurity must be designed in ways appropriate to local circumstances if the planner knows about local livelihoods and whether or not a proposed intervention will build upon or undermine existing strategies. This can only be done if sufficient information is known about these strategies and circumstances across the diversity of agro-ecosystems in each country for which FEWS NET is responsible.

FEWS NET uses four steps to connect biophysical and other hazards with food security outcomes using the livelihood approach. This involves the following:

- Scenario Modeling Baseline – where information about the communities at risk are set out, including livelihood zone mapping, livelihood profiles, demographic profiles and scenario modeling baseline. This provides the local norm from which changes or shocks can be measured, ensuring that these changes can be detected and evaluated.

- Hazard Monitoring – monitor both biophysical hazards (climate) and socio-economic hazards (food prices, production). This is done on a continuous basis with reports and evaluations on a variety of scales.
- Food Security Scenario Modeling – conduct modeling to identify the impact of hazard, the likely food security outcomes for affected populations over a specified time period. Each locality should ideally have its own scenarios and models in place.
- Contingency and Response Planning – work with network partners to identify appropriate response options.

The baseline provides an idea of the 'normal' situation through which anomalies can be interpreted. The baseline consists of livelihood zones and ancillary information called profiles. The zones divide a country into geographic areas with relatively homogeneous ecological and economic characteristics, thus providing a geographic framework and sampling frame for food security analysis (Fig. 3.2). The baseline is developed using both biophysical and socio-economic elements, identifying localized information about how poor, rich and modal demographic groups obtain their cash and food income. These income sources include agriculture, livestock, commerce, services, wage labor and a variety of other activities. Hazard monitoring uses this baseline profile to determine the normal situation from which the impact of both socio-economic and biophysical anomalies can be measured.

3.4.1 Socio-Economic Monitoring

In addition to the extensive data collection required to create the baseline, FEWS NET monitors other aspects of the local economy, including employment demand, population fluxes and school attendance through its partners and connections with local and regional governments. Food prices in markets and measures of the terms of trade (such as the kilograms of corn one head of cattle will purchase) have been recorded in many informal markets across Africa since the mid-1980s. Because rural households in agricultural societies both buy and sell locally produced grain on the market (Jayne and Minot, 1989), the economic impact of food prices is widespread. Markets allow households to obtain cash for household needs from grain production and to purchase grain when stocks run low.

Reliance on agriculture as a primary source of both income and food often leads to vulnerability to seasonal and inter-annual climate variability that affects agricultural yields. Income of food and cash taken together are an important variable for ensuring nutritional sufficiency at the household level because if a household has sufficient income, food can be purchased even if cannot be grown or raised. Local monthly price data on a variety of crops in many markets in the FEWS NET countries of interest is collected and interpreted as a primary input to food security monitoring and decision support system, even if the prices are not integrated into a database as often as may be optimal.

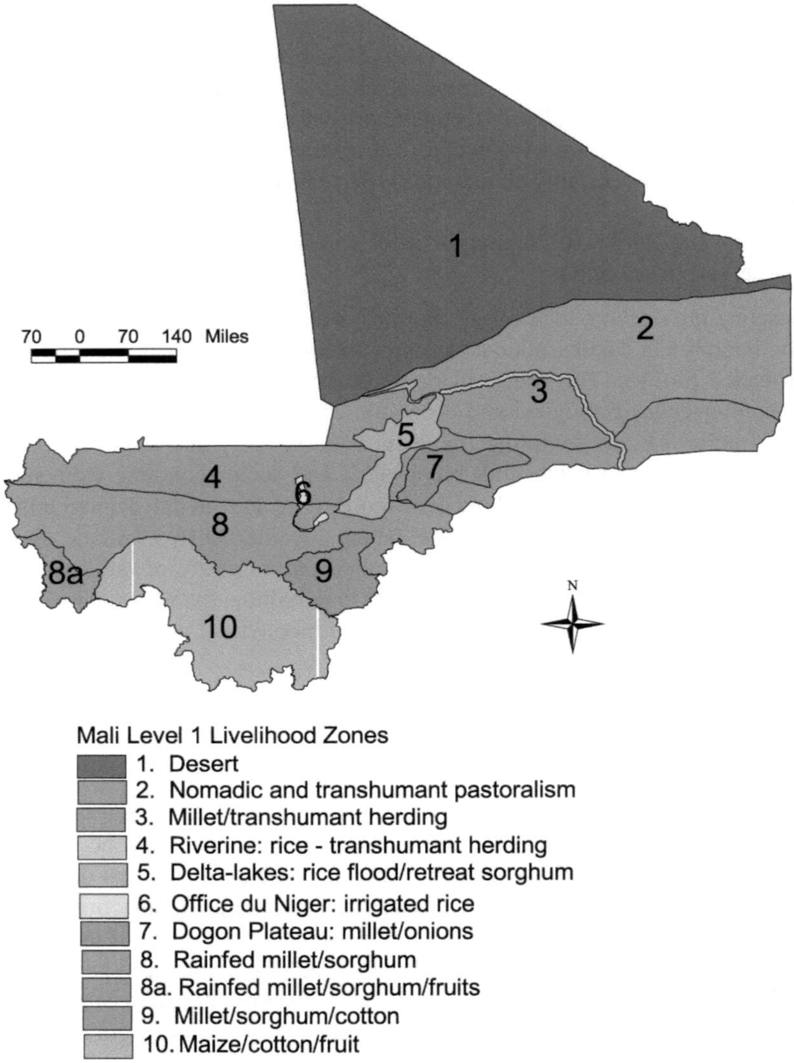

Fig. 3.2 An example of food economy zones from Mali, produced using a livelihood approach in 2005. Each zone corresponds to a generalization about how rural people in the region make a living. From the Mali Livelihood Zoning/Profiles documentation

FEWS NET country or regional representatives write monthly reports that are sent to the FEWS NET Washington DC office before being published. Globally informed analysis and information is added and quality control performed to the products. Analysis of potential impending crises is monitored using these reports and the information transferred to other reports and briefings. Figure 3.3 shows the

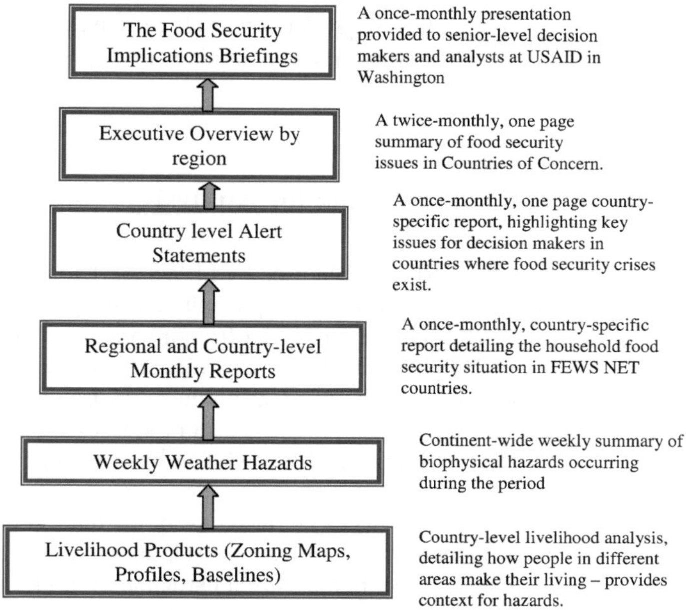

Fig. 3.3 FEWS NET reports from livelihood products to high level policy briefings

general flow of information from the baseline reports, through hazard analysis to monthly reports, alert statements, to the food security implications briefings.

Actual food production figures are a key component to determining overall food deficits for a region, and are the traditional basis for determining if assistance is required. Food production, calculated by multiplying area cropped by yield normally attained in that area, is first estimated throughout the season. Ultimately actual production figures are produced through crop assessments conducted collaboratively with local governments and other large non-governmental organizations. Official figures, produced for national government use and for reporting to statistical bodies such as the United Nations Food and Agriculture Organizations, form the basis for much of FEWS NET's work. Local surveys combined with expert assessment are the also contribute to efforts to quantify the food deficit in a particular region or country. In countries where political and/or economic forces work to motivate these governments to manipulate these figures, FEWS NET has begun to invest in ways to provide more realistic crop production estimates for these regions. Chapter 16 provides an example of efforts to identify ways to improve food production estimates quantitatively through remote sensing in Zimbabwe.

Reductions in food production in a particular locality, however, do not necessarily mean that the region will require international assistance. In regions with long-standing food security issues, the success of coping strategies and internal food movements are monitored by FEWS NET so that the most vulnerable are protected to the greatest extent possible by planning and intervention. Nevertheless, many countries require food aid every year regardless of production levels and have persistent food insecurity which cannot be addressed in the early warning or humanitarian assistance framework.

3.5 Biophysical and Socio-Economic Data for Analysis

Table 3.2 lists the extensive number and type of data used by FEWS NET to summarize the current climatic situation. The data includes precipitation gauges and gridded data from merged satellite models, vegetation data from a variety of sensors, gridded cloudiness products, global climate indicators, precipitation forecasts, modeled soil moisture, gridded fire products, snow extent products, hydrological models for flood forecasting, and seasonal forecasts. These data products were either developed directly by FEWS NET partners for FEWS NET or were adapted to their needs.

The table illustrates how gridded rainfall images produced every ten days have been used to drive a large number of models from a variety of disciplines, including agronomic models specifying the moisture requirements of a particular crop given an underlying soil type (Water Requirement Satisfaction Index or WRSI) and the flooding potential given the soil water holding capacity and the amount of water that has fallen on a given catchment basin (Basin Excess Rainfall Model or BERM), among many others. These modeled products allow social scientists to ask questions regarding the direct effect of a particular rainfall amount on the crop production instead of having to infer from rainfall or vegetation imagery the resulting impact on a particular crop or in a particular flood-prone region.

Vegetation Index data derived from satellites remains an important source of information for the FEWS NET program because it shows results of rainfall on the vegetation. Although rainfall has been used extensively to drive many other models, it is actually far less reliable than directly measured vegetation data as it is prone to errors in approximating the degree of cloudiness, the amount of rain that has fallen from these clouds or the intensity of the rainfall, inadequate capturing of orographic rainfall, sensitivity to the density and accuracy of local rainfall gauge measurements, and other effects which result in significant random error and non-negligible bias (Waymire, 1985; Xie and Arkin, 1997). Vegetation remote sensing measures directly the stable photosynthetic activity resulting from rainfall and is thus can be more precise (Tucker et al., 1991, 2005). Because they measure very different things, both variables continue to be of value to hazard identification.

3.5.1 Weather Hazard Assessments

Members of the many FEWS NET organizations teleconference weekly to discuss and identify potential flood and drought hazards, and then prepare and issue weekly weather hazard reports, which are posted on the FEWS NET site (www.fews.net).

Table 3.2 Current FEWS NET data products and descriptions

Category	Type	Product	Description	Spatial extent	Spatial resolution	Time step	Source (see acronym list for definitions)
Biophysical Monitoring	Precipitation	RFE - Rainfall Estimate	multi-sensor and gauge merged model	Africa, SE Asia, SW Asia	0.1°	daily	NOAA CPC
		TRMM – Tropical Rainfall Monitoring Mission 3b42RT	multi-sensor and gauge merged model	global	0.25°	daily	NASA GSFC
		GTS Station Data – Global Telecom-munication System	station data, daily	global	point	daily	United Nations (WMO)
		CMORPH - NOAA CPC Morphing Technique	multi-sensor and guage merged model (NO gauge data in CMORPH)	global	0.1, 0.25°	daily	NOAA CPC
	Derived Precipitation Products	SPI - Standardized Precipitation Index	18-yr mean standardized anomaly (30-yr mean for Africa SPI, not familiar with the SW Asia product)	Africa, SW Asia	0.1°	10 day, 1, 2, 3, 6 and 12 mon	UCSB-USGS
		SOS – Start of Season	determines beginning of growing season	Regional (Africa, C.America, Haiti)	0.1°	daily, seasonal	USGS
		ITCZ – Inter-Tropical Convergence Zone	estimates onset of rains appro. week before	Regional (Africa)	vector coverage	daily, seasonal	NOAA CPC

Table 3.2 (continued)

Category	Type	Product	Description	Spatial extent	Spatial resolution	Time step	Source (see acronym list for definitions)
		WRSI – water requirement satisfaction index	estimates crop yields by crop type	Africa, SW Asia, C.America, Haiti	0.1°	daily, seasonal	USGS
		Rangeland WRSI	estimates rangeland grass condition	Africa,	0.1°	10-day, seasonal	USGS
	Clouds	OLR – Outgoing long wave radiation	precipitation proxy	global	8 km	hourly	NOAA CIRES Climate Diagnostics Center
		IR – Infrared Temperature	precipitation proxy	global	25 km	hourly	NASA GSFC Global Change Master Directory
		Water Vapor – MODIS	precipitation proxy	global	1 km	daily	NASA GSFC
	Global Climate Indicators	MJO IR – Madden Julian Oscillation/ 200 h/PA velocity potential	upper level convergence, precip predictor	global	25 km	hourly	NOAA CPC
		GFS Vorticity	upper level convergence, precip predictor	global	0.5 and 1.0°	4 × daily	NOAA
		ENSO phase – Sea Surface Temp Anomalies	related to seasonal precipitation in some regions	global	25 km	daily	IRI and NOAA CPC
	Precipitation Forecast	GFS model – Global Forecast System	precipitation forecast – 24-168 hour	global	0.5 and 1.0°	4 × daily	NOAA

Table 3.2 (continued)

Category	Type	Product	Description	Spatial extent	Spatial resolution	Time step	Source (see acronym list for definitions)
		NCEP/Eta model	precipitation forecast – 24- 72 hour	regional models	22 km	3-hourly	NOAA CPC
	Vegetation	AVHRR GIMMS NDVI (normalized difference vegetation index)	vegetation density and health	global	8 km	10 and 15 day composites	NASA GSFC
		AVHRR NOAA Vegetation Health	vegetation plus temperature	global	16 km	Weekly (7 day)	NOAA
		SPOT Vegetation NDVI	vegetation density and health	global	1 km	10 day composites	VITO, FAS-USDA, NASA GSFC
		MODIS NDVI	vegetation density and health	global – limited availability	250 m	16 day composites	NASA GSFC
	Soil Moisture	SSM/I Soil Moisture	soil moisture, vegetation proxy	global	30 km	weekly, monthly	NOAA
		CPC Leaky Bucket model	soil moisture, vegetation proxy	global	25 km	monthly	NOAA CPC
		MI – Moisture Index	estimates available water for crops/vegetation (supply/demand ratio)	Africa, SW Asia	0.1°	daily, 10-day	USGS
		SWI – Soil Water Index	Estimates amount of water available for crops/vegetation	Global, Africa	one deg, 25 km	10-day, monthly	USGS

Table 3.2 (continued)

Category	Type	Product	Description	Spatial extent	Spatial resolution	Time step	Source (see acronym list for definitions)
	Fires	MODIS Rapid Response	fire locations mapped onto true color MODIS imagery	global - limited availability	250 m	daily	NASA GSFC
	Snow	Snow station data	precipitation, snow fall and temperatures	Asia	point	daily	AFWA
		Snow depth grid	Modeled data using SSM/I surface temps + climatology	Asia	48 km	daily	AFWA
		Snow cover	AMSU Microwave from NOAA-satellites 15 and 16	Asia	24 km	daily	NOAA NESDIS
		Snow Water Equivalent	Spatial implementation of the Utah Energy Balance model	Afghanistan	0.1°	daily	USGS
	Hydrology	BERM – Basin excess rainfall model – flooding	basin flood potential driven by NOAA RFE Precipitation	Africa	by basin	daily	USGS
		Reservoir levels	global reservoir and lake elevation from radar	Globe, selected	by water body	monthly	FAS-USDA, NASA

Table 3.2 (continued)

Category	Type	Product	Description	Spatial extent	Spatial resolution	Time step	Source (see acronym list for definitions)
		Cyclone Monitoring	image of cyclone track from Navy	E.Africa	–	daily	JTWC-NOAA CPC
	Seasonal Forecasts	IRI SSTA + COLA AGCM temp and precip predictions	guidance for upcoming agricultural season	global	1–5°	3-month	NOAA-CPC, Columbia IRI
Socio-Economic Monitoring	Agricultural Production	production figures for various commodities	production statistics from selected countries	Africa	–	Seasonal	FEWS/ UN (FAO)
	Market Prices	market prices for various commodities	commodity prices from markets in selected countries	Africa	–	Monthly and/or weekly	FEWS
	Food Economy Zones	Livelihood Zone Maps	map shows division of country into uniform zones	Global	–	static – periodic update	FEWS
		Livelihood Zone Profiles	describes cash income and food production sources	Global	–	static – periodic update	FEWS
		Scenario modeling baselines	describes impact of different shocks	Global	–	static – periodic update	FEWS
	Employment	Monitoring of labor markets	Wage-earning is a critical piece of the local economy in many places	Africa	–	Ongoing	NGOs, local government through FEWS Representatives

Table 3.2 (continued)

Category	Type	Product	Description	Spatial extent	Spatial resolution	Time step	Source (see acronym list for definitions)
	Population	Monitoring of migrant vs permanent population levels	Large movements of populations can signal a food crises	Africa	–	Ongoing	NGOs, local government through FEWS Representatives
	School Attendance	Local Representatives monitor attendance rates	To determine if food crisis is occurring	Africa	–	Ongoing	NGOs, local government through FEWS Representatives
	Infrastructure Maps	Roads, administrative maps, infrastructure maps	enables rapid response in event of emergency	Global	–	static – periodic update	UN WFP/FEWS

The Weather Hazard Impact Assessment regularly identify weather-related threats (what may happen) and hazards (what is happening/not happening) that may significantly alter food security and livelihood conditions. These threats and hazards must be clearly and closely linked to potential losses of food security, or to potential human or animal morbidity and/or mortality, and must be events/conditions that FEWS NET Team members have a first-hand ability to monitor or assess, or an ability to vouch for the accuracy of the data provided by other observers. The impact assessments are oriented in time to the present and to the near-future. Assessment products may continue to report on on-going hazards as long as there is considered to be an anomalous, on-going, and significant impact on food security and livelihoods on the ground.

The impact assessments currently provide coverage of the entire African continent, Central America (between southern Mexico and Panama), Haiti, and a geographic window that extends slightly beyond the borders of Afghanistan. Depending upon needs and resources, other areas may be designated for regular coverage in the future. Even before coverage of other areas is specified by USAID, FEWS NET partners are working to prepare data for a global food security analysis capability. Thus most biophysical monitoring products are either already global in spatial extent or are being extended globally.

The Assessments are delivered to a large local, regional and international audience in Africa, Latin America, Asia, and in the United States. They provide information that indicates where more intense, on-the-ground monitoring should occur. The weekly weather hazards, led by the meteorologists at NOAA's Climate Prediction Center, and is guided by FEWS NET food security experts who orient hazard discussion towards identifying its affect on local livelihoods. By collaborating with scientists from NOAA, NASA and USGS, FEWS NET regional and country representatives, and the USGS FEWS NET Regional representatives, expert FEWS NET personnel work together to determine the impact of these weather hazards on local communities.

3.6 Challenges for FEWS NET

Although remote sensing data is an extremely important resource for FEWS NET, it can also be a challenge to keep the focus on the food security outcome of the hazard that the data identifies, not on the hazard itself. FEWS NET uses a food economy approach and livelihoods analysis that identifies specific causes of a food security crisis for a particular group of people. Because evidence from remote sensing data is so compelling and has been used in some of the regions where FEWS NET works for several decades, it is much easier to focus on the easy to understand hazard and not on the complex and multi-dimensional consequences of the hazard. The underlying assumptions that anomaly maps used in the food security context are both significant and usually unstated, even for FEWS NET analysts. Thus the challenge for FEWS NET is to maintain its focus on the diverse and complex local situation while at the same time providing compelling evidence for action for decision makers.

Another challenge for FEWS NET is the difficulty of finding the resources, time and managerial focus it takes to maintain databases of all the geographic information required to conduct food security analysis. Properly aligned GIS layers of administrative regions, updated livelihood zones, databases of historical livestock and cereal prices, agricultural production data and local rain gauge datasets require management and maintenance. Although USAID does invest in some of this work, much of it is done informally and without explicit funding in the current task structure. Thus FEWS NET needs to reduce the number of steps it takes from data creation to data storage in order to be able to do more with fewer resources. Long term funding remains the primary obstacle, however, to ensure that archiving of currently existing datasets is done in a way to facilitate their integration into modern georeferenced web servers that can distribute the data to all who need it. Expansion of data sources and continual investment in ensuring that livelihood baselines, for example, are current is also required. Adequate funding of the FEWS NET activity would ensure that these tasks are not marginalized in the face of current demands on resources.

3.7 Summary

Chapter 3 has provided a brief overview of FEWS NET, its structure and funding sources, and how it accomplishes its work. FEWS NET has partner organizations, NASA, NOAA, USGS and USDA, as well as a contractor who manages the central Washington DC and the 31 national offices for the funding agency USAID. FEWS NET structure was summarized and a brief description of its activities given. Finally, the social and biophysical data that the organization uses to provide actionable food security information in its myriad of products were listed. Some challenges that this wide variety of activities and datasets pose for FEWS NET are given at the end of the chapter.

In the next section, the remote sensing products will be described in more detail, how they are created and used, and how secondary products are estimated. The third section of the book will describe how the remote sensing data are put to use in FEWS NET through analytical applications in the domains of monitoring, food security monitoring and contingency planning processes. Finally section four will provide three case studies where remote sensing was an integral part of the early warning and response.

References

Boudreau, T.E., 1998. The Food Economy Approach: A Framework For Understanding Rural Livelihoods. RRN Network Paper 26, Relief and Rehabilitation Network/Overseas Development Institute, London.

Dilley, M., 2000. Warning and Intervention: What Kind of Information does the Response Community Need from the Early Warning Community, USAID, Office of US Foreign Disaster Assistance, Washington DC.

Jayne, T.S. and Minot, N., 1989. Food Security Policy and the Competitiveness of Sahelian Agriculture: A Summary of the "Beyond Mindelo" Seminar. Sahel, Club du Sahel.

Mathys, E., 2005. Fews net's Approach to Livelihoods-based food Security Analysis, FEWS NET USAID, Washington DC.

Moseley, W.G., 2001. African evidence on the relation of poverty, time preference and the environment. Ecological Economics, 38: 317–326.

Tucker, C.J., Newcomb, W.W., Los, S.O. and Prince, S.D., 1991. Mean and inter-annual variation of growing-season normalized difference vegetation index for the sahel 1981–1989. International Journal of Remote Sensing, 12: 1133–1135.

Tucker, C.J., Pinzon, J.E., Brown, M.E., Slayback, D., Pak, E.W., Mahoney, R., Vermote, E. and Saleous, N., 2005. An extended AVHRR 8-km NDVI data set compatible with MODIS and SPOT vegetation NDVI data. International Journal of Remote Sensing, 26: 4485–4498.

Waymire, E., 1985. Scaling limits and self-similarity in precipitation fields. Water Resources Research, 21: 1271–1281.

Xie, P. and Arkin, P.A., 1997. Global precipitation: A 17-year monthly analysis based on gauge observations, satellite estimates, and numerical model outputs. Bulletin American Meteorological Society, 78: 2539–2558.

Section II
Remote Sensing for Early Warning

Chapter 4
Rainfall Estimates

Rainfall estimates are one of the most important observational tools for FEWS NET. Although FEWS NET has many ways of determining the progress of the growing season and ultimately the food security of a region, rainfall is the primary determinant of variation in agricultural production from year to year in many of the regions where it works. The distribution, quantity and intensity of rainfall all directly affect the ultimate crop yield. Other factors are also important, such as nutrient availability, crop variety, and weeding techniques, but when there is insufficient water, these other factors decline in importance. In addition, non-rainfall factors do not normally vary substantially from one year to the next. Monitoring interannual variations in production can often be done by monitoring rainfall at continental scales, and thus highly accurate, temporally frequent and spatially extensive data is a FEWS NET priority.

In this chapter, the observed rainfall datasets that FEWS NET uses to monitor agriculture and to derive other products will be described. The primary rainfall dataset that FEWS NET uses is NOAA's Rainfall Estimate product or the RFE. The RFE is currently produced by NOAA's Climate Prediction Center (CPC) at eight kilometer resolution, primarily for the African continent, although it is being extended to Asia for FEWS NET's work in Afghanistan (Xie and Arkin, 1996), and to Central America for use in the Meso-American Food Early Warning System (MFEWS). The second version of the RFE is based on Ping Ping Xie's multiple-sensor integration model (Xie and Arkin, 1996). The RFE is complemented by an orographically enhanced rainfall dataset called the Collaborative Historical African Rainfall Model (CHARM), and the ARC product, an RFE Climatology that begins in 1996. These two products provide contextual information for the RFE2 and enable anomaly products, trend analysis and other analysis for drought detection.

4.1 NOAA's Rainfall Estimate (RFE) Product

FEWS NET currently uses a merged satellite-gauge product for its primary source of information on rainfall in the countries where it works. The product currently being used by FEWS NET is the Rainfall Estimate (RFE) 2.0, which uses several techniques to estimate precipitation while also using traditional cloud top temperature and station rainfall data. The RFE data is particularly useful for FEWS NET because it uses the WMO Global Telecommunication System (GTS) rainfall observation data taken from approximately 1000 stations which are assumed to be the true daily rainfall near each station for each day. Using these observations in the rainfall model produces a dataset which is far closer to the observed rainfall in all locations where observations are taken. This makes working with local Africa meteorological services much easier than if the RFE was based on satellite measurements alone.

The current product's predecessor, the RFE 1.0, used an interpolation method to combine Meteosat and GTS data for daily precipitation estimates, and warm cloud information was included to obtain dekadal estimates. The two new satellite rainfall estimation instruments that are incorporated into RFE 2.0 are the Special Sensor Microwave/Imager (SSM/I) on board Defense Meteorological Satellite Program satellites, and the Advanced Microwave Sounding Unit (AMSU). Both estimates are acquired at 6-hour intervals and have a resolution of 0.25°. RFE 2.0 obtains the final daily rainfall estimation using a two part merging process, then sums daily totals to produce dekadal estimates. All satellite data is first combined using a maximum likelihood estimation method, and then GTS station data is used to remove bias. Warm cloud precipitation estimates are not included in RFE 2.0. In this chapter and in the rest of this book, when the RFE is mentioned it is this second version that is meant as the first version is no longer produced and is no longer used by FEWS NET.

Precipitation has very large spatial and temporal variability, and therefore is very difficult to model. Rainfall measurements need to be very accurate over a wide range of spatial and temporal variability (Zeng, 1999). Figure 4.1 shows the rainfall data from Niamey, Niger compared to the average rainfall accumulation for the previous year. At this station, the rainfall varies from zero to nearly 125 mm of rain per day during the period presented. Capturing this day-to-day and year-to-year variability of rainfall is very difficult in a model. Rain gauge observations are often biased due to the effect of wind and other factors (Sevruk, 1982), but in most areas this bias is relatively small compared with satellite precipitation estimates based on cloud identification or rain rates that either systematically overestimate or underestimate the amount of actual rain falling (Xie and Arkin, 1995). Gauge data are the basis for all methods to estimate precipitation, and therefore the frequency of observation, density of the network and accuracy of each measurement is critical for the quality of the rainfall models, regardless of the other inputs.

Rain gauge data are not available over most oceanic regions and sparsely populated regions. Averaging point values on a sparse, irregular grid into surface means introduces sampling errors that can be significant and non-random. In places with high variability of rainfall and inadequate sampling with gauges of the rainfall that does occur will have systematically incorrect rainfall estimates in the RFE. This is important to keep in mind when using merged gauge-satellite rainfall estimates in real-time for applications such as FEWS NET, because when a country is experiencing a long-term economic crisis that may lead to a food security problem, it is

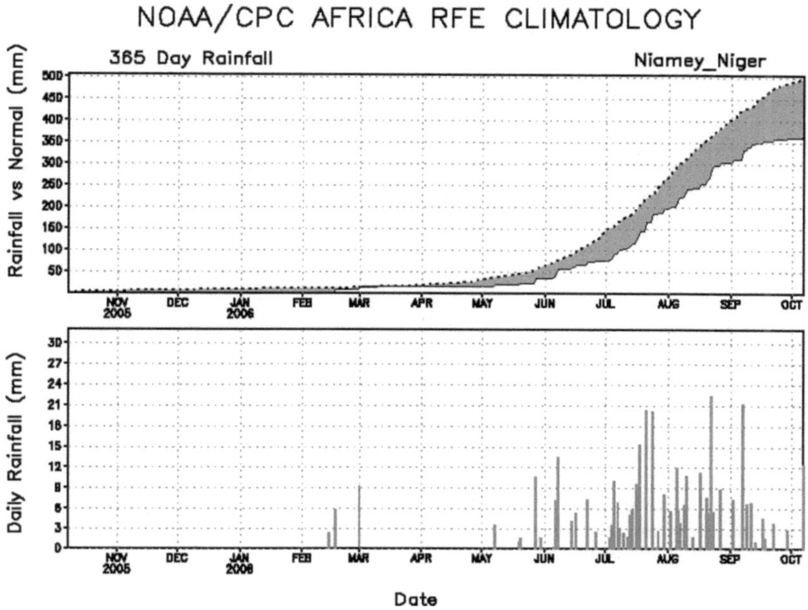

Fig. 4.1 Screen capture of web report of automatically updated rainfall station data from Niamey, Niger from 2005–2006. From the NOAA Climate Prediction Center web site

also likely to not have the resources to adequately monitor and report rainfall for the gauges in its territory. Thus, reliance on the RFE product over that region is likely to increase during the time that its relative reliability declines. Ethiopia, for example, did not report data from its network of rainfall stations for five years, from 2000–2005, reducing the effectiveness of rainfall models over the territory. This was during a time when significant amount of aid flowed into the country due to drought. Thus FEWS NET continues to invest in alternative information sources such as vegetation indices in order to determine the amount of food is being produced in rainfed agriculture.

In addition to gauge measurements, precipitation over the tropics can also be estimated using the cloud-top visible/infrared data from geosynchronous satellites (Adler et al., 1994; Arkin and Meisner, 1987). The sensors detect the very cold tops of clouds that are likely to be producing rainfall. By using data from these visible sensors to determine when and where rainfall is happening, rain can be estimated with every new pass of geostationary infrared data from a variety of satellites, including the NOAA Geostationary Observing Environmental Satellites (GOES), the European Meteosat, and the Japanese Geostationary Meteorological Satellite (GMS) (see Table 4.1). One of these sensors pass over every place on the planet between 60°N and 60°S every three hours.

Using visible and near infra-red data provides a good temporal resolution of precipitation; however, they must be empirically adjusted using observations from a specific region so that the coefficients relating gauge data and the satellite remote sensing estimates might be region and season dependent, as suggested by (Arkin

Table 4.1 Satellite inputs to RFE product. Combined microwave, IR and visible data from these sensors gives a global precipitation estimate at 3-hour intervals

Satellite/Sensor	Product	Channels	Spatial Resolution	Temporal Resolution
AMSU-A/B on NOAA POES 15, 16 & 17	Infrared Rainfall estimates of cloud water	23800.37 MHz (C1) 31400.42 MHz (C2)	50 km at nadir	6-hour intervals
Special Sensor Microwave Imager (SSM/I) on Defense Meteorological Satellite Program (DMSP) 13, 14 & 15	Microwave rainfall estimates	Dual polarizations at 19.4, 22.3, 37.0, and single polarization at 85.5 GHz.	45 km at nadir	6-hour intervals
GOES Precipitation Index (GPI) cold count duration (CCD) estimates from Meteosat 5,6,7	Precipitation estimates using emissivity, or cloud top temperatures	10.2–11.4 μm 11.4–12.55 μm	25 × 25 km (GAC)	30 min

et al., 1994). Satellite microwave (MW) observations provide a physically more direct approach to estimate precipitation through the emission-based retrieval of atmospheric liquid water (over ocean) or scattering-based retrieval of precipitation-sized ice above the freezing level (over land or ocean) (Zeng et al., 1999). Time averages computed only from MW observations suffer from inadequate temporal sampling, since rainfall is extremely intermittent, particularly in regions where much of the rainfall is from thunderstorms and other convective systems. Regardless of the combination of sources from which modeled data originate, the quality of the resulting rainfall estimates strongly depend upon the quality of the individual input data sources (Zeng et al., 1999).

The RFE was developed by Ping Ping Xie and Paul Arkin in 1996 and implemented in real time during NASA's SAFARI 2000 campaign. The product combines data from a variety of sources in order to overcome the shortcomings of each dataset individually. RFE 2.0 uses four types of input data, including three satellite sources, to create the final rainfall estimates. These are: (1) Daily Global Telecommunications System (GTS) automated rain gauge data for up to 1000 stations (2) Advanced Microwave Sounding Unit-B (AMSU-B) microwave satellite precipitation estimates up to 4 times per day (3) Special Sensor Microwave/Imager (SSM /I) satellite rain rate estimates up to 4 times per day; and (4) GOES Precipitation Index (GPI) precipitation estimates on a half-hour basis (Table 4.1). The three satellite estimates are first combined linearly using predetermined weighting coefficients, then are merged with station data to determine the final African rainfall estimate image. Daily binary and graphical rainfall estimate images are produced at approximately 3pm EST with a resolution of 8 kilometer and spatial extent from 40°S–40°N and 20°W–55°E. Additional data sets of 10-day, monthly, and season rainfall totals are created by accumulating daily data.

4.1.1 Global Telecommunications System Gauge Data

Observations of the 24-hour accumulated rainfall for locations throughout Africa are obtained from Global Telecommunications System (GTS) automated rain gauge data system. While approximately 7500 gauges exist globally, the African continent contains roughly 1300 stations, from which between 600–800 report each day. The GTS consists of an integrated network of point-to-point circuits, and multi-point circuits which interconnect meteorological telecommunication centers. The circuits of the GTS are composed of a combination of terrestrial and satellite telecommunication links. They are point-to-point circuits, for data distribution, for data collection, as well as two-way multi-point circuits. Figure 4.2 shows the structure of the GTS.

Meteorological Telecommunication Centers are responsible for receiving data and relaying it selectively on GTS circuits. The GTS has three levels: the main telecommunication network between a few primary countries, the regional network that connects each country to a set of others in its neighborhood and the national networks, which allowed each country to collect data within its own territory before

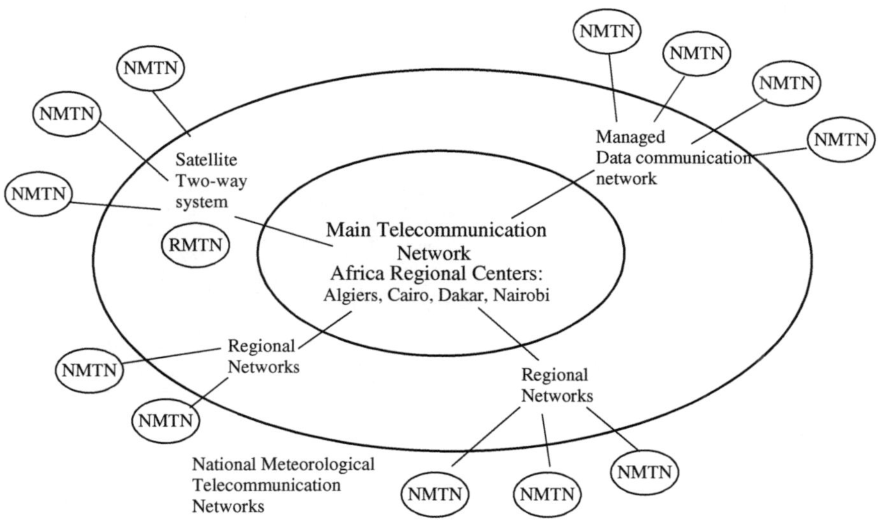

Fig. 4.2 Global Telecommunication System (GTS) Network Structure for delivering real time rainfall measurements (from the GTS web site). NMTN are national metrological telecommunication networks

passing it on. Because the program was built before modern and inexpensive internet network technology, this complicated system of networks was devised to reduce the distance the signal had to travel and improve the likelihood that the data would arrive. This complexity has consequences for the quality of the data as each observation has to pass through multiple points where it is merged with other information and where the data could be lost or corrupted.

The Main Telecommunication Network (MTN) has the function of providing communication service between its centers, in order to ensure the rapid and reliable global and interregional exchange of observational data, processed information and other data. The Regional Meteorological Telecommunication Networks consist of an integrated network of circuits interconnecting meteorological centers, which are complemented by radio broadcasts where necessary. The Regional Meteorological Telecommunication Networks ensure the collection of observational data and the regional selective distribution of meteorological and other related information to Members. There are six Regional Meteorological Telecommunication Networks: Africa, Asia, South America, North America, Central America & the Caribbean, South-West Pacific and Europe. The National Meteorological Telecommunication Networks enable the National Meteorological Services to collect observational data and to receive and distribute meteorological information on a national level. Figure 4.3 shows the global distribution of all GTS stations, with a total of 11393 observation points. The location of the stations and the density of the network are more related to geopolitical and historical reasons than to biophysical requirements.

The ratio of rain gauges which provide easily accessible, daily, near real-time observations to area in Africa is approximately 1:23,300 km^2. This does not account for the Sahara Desert, which occupies about a quarter of the continental land mass,

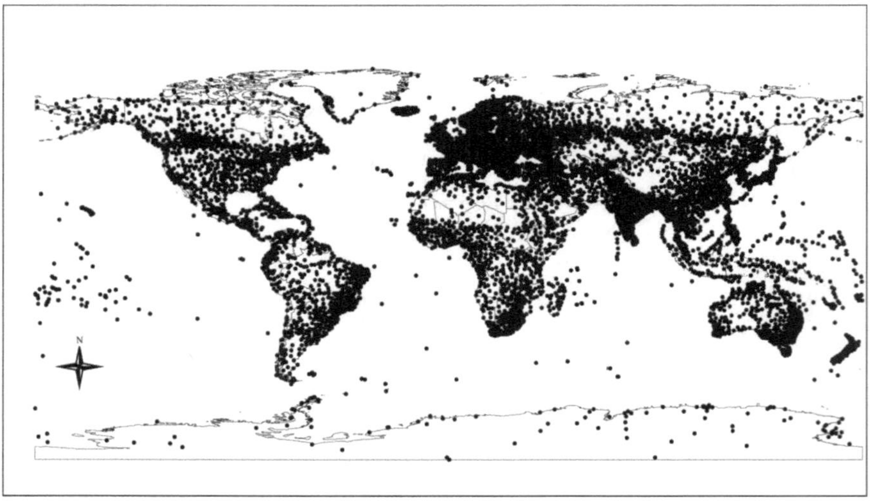

Fig. 4.3 Global Telecommunication System (GTS) station locations (11393 stations total). There are 1162 total listed gauges for Africa

thus the spatial representation of rain gauges throughout the continent is not adequate to allow for detection of local, regional, or even some large scale hydrological and climatic phenomena (Love et al., 2004).

A brief analysis of the daily GTS data from Africa during a five year period from 2001 to 2005 shows that some stations have a higher rate of non-reporting than others. Figure 4.4 shows the percent of days each month that the Africa stations are missing data. Some countries do not report at all during this period, including Ethiopia, Angola, and Botswana. Some countries only report a few stations, leaving most of their territory without observations. These include Nigeria, Somalia and the Democratic Republic of Congo (formerly Zaire). Figure 4.4 shows that the rate of missing observations is non-random. Some regions have very low reporting errors, particularly in North Africa and in the Sahelian countries, while others have a much higher rate (southern Africa and countries along the Guinea Coast). These errors will cause systematic errors in the RFE data in locations with increased missing periods.

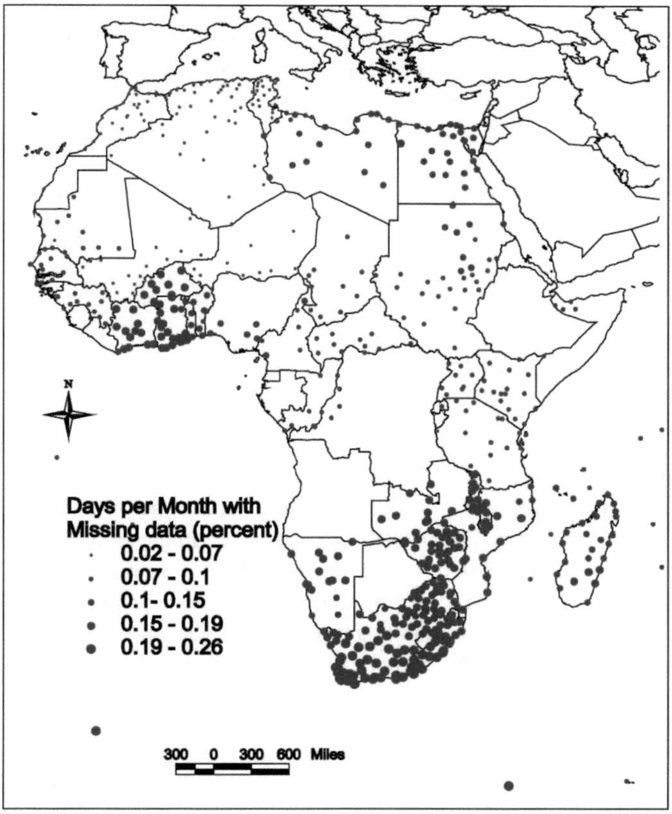

Fig. 4.4 GTS network in Africa, percent of days per month that the station does not report. 651 stations report data 2001–2005 in the Africa region. No point means that there were no reporting stations during the period

Gauge data yield relatively accurate point information through time for a particular location, but when averaged over a surface, the data will give sampling errors that can be significant. In addition, rain gauge interpolation schemes tend to propagate data from semi-arid regions into arid regimes (Funk et al., 2003). Differences in data density also tend to cause non-random errors in the rainfall data models. In Africa, the gauge data network is extremely sparse or nonexistent in some countries that have experienced civil conflict or dysfunctional governments (Liberia, for example), and very dense in others (South Africa). Funk et al. (2003) report the mean absolute error in the precipitation fields have increased recently due to a degradation of the gauge network. Resolving problems with the gauge network is extremely important to maintaining accuracy of rainfall monitoring, since alternate methods of evaluating precipitation still require gauge measurements to produce accurate, unbiased results (Funk et al., 2003). Thus one of FEWS NET's long term goals is to ensure the quality and long-term stability of the meteorological data coming out of each of its countries.

For use in the RFE, NOAA's Climate Prediction Center has a direct line to the GTS system so that it may receive all the GTS data daily. Typically, 400–600 stations report every day within a few hours of observation. The daily accumulation data is ingested, examined for quality, and then gridded to 0.1° resolution. Other station data may also be used as input. In regions critical to FEWS NET operations, such as Ethiopia and Somalia, intensive efforts have gone into obtaining additional datasets directly from the meteorological offices in the countries themselves and using them in the RFE (Funk, 2006). These datasets improve the performance of the RFE considerably. The RFE is not processed if the GTS file from that day is not available.

4.1.2 Satellite Data Input and Algorithm for the RFE

The RFE uses four types of input data, three satellite sources and the GTS station data, to create the final rainfall estimates. The microwave rain rates from the Special Sensor Microwave Imager (SSM/I), humidity profiles from AMSU-A and B, and half-hourly GOES Precipitation Index (GPI) are used to estimate rainfall rates (Fig. 4.5). Three satellite products needed instead of just one because rainfall is extremely variable in both time and space. An optimal rainfall estimate product would need to be based on sub-hourly observations at sub-meter resolution, with vertical profiles of temperature, humidity, and hydrometeors (cloud water and ice, graupel, hail, snow). These observations are not available at the current level of technology, thus current rainfall products are based on three different ways of getting at how much rain fell on a particular day: the temperature of the tops of the clouds (related to how tall the cloud is and, indirectly, how much rain it can potentially produce), the amount of water in the cloud and for tuning purposes, microwave estimates of the actual rate at which the cloud is precipitating. The passive microwave rain rates from SSM/I are far more accurate than the cloud-top and humidity profiles, but the data is too sparse to be used for daily operational rainfall estimation. Thus a combination

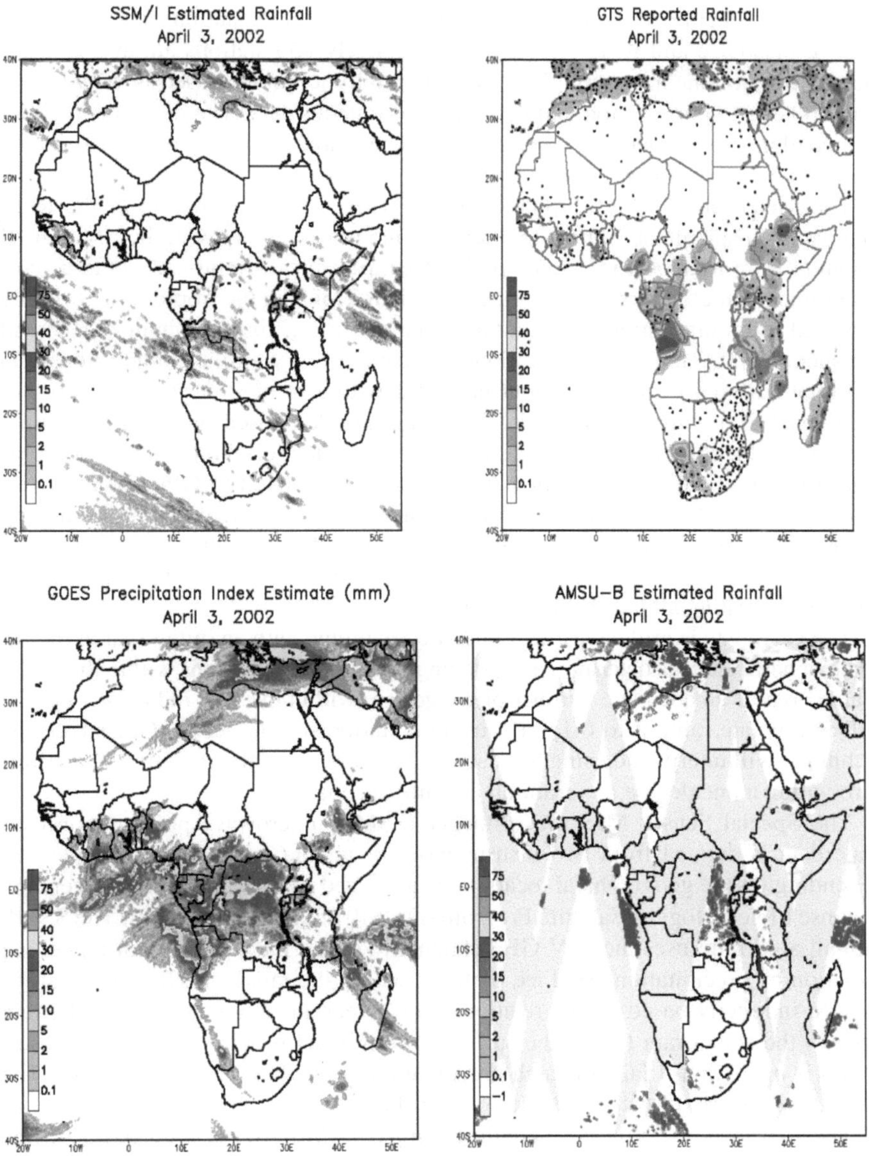

Fig. 4.5 Gridded SSM/I, GTS rainfall, GOES GPI and AMSU-B inputs to RFE for April 3, 2002 (courtesy of Tim Love, NOAA)

of all these products creates a usable, accurate daily rainfall estimate at a suitable resolution for FEWS NET operations.

The RFE obtains the final daily rainfall estimation using a two part merging process, then sums daily totals to produce dekadal estimates. Warm cloud precipitation estimates are not included in the RFE. All satellite data is first combined using a maximum likelihood estimation method, and then GTS station data is used to remove bias (Xie and Arkin, 1996). Approximately 650 stations are available for the African continent on any given day, although the number used is usually less than 500 due to poor station maintenance or erroneous data. The need for satellite-estimated precipitation arises from this non-dependable, poorly spatially distributed rainfall data observation.

The Meteosat GOES Precipitation Index (GPI) is the foundation of the RFE. The data is available at 30 minute intervals, and are combined with hourly rainfall amounts from Meteosat infrared cloud top temperatures. Empirical methods have determined that cloud top temperatures less than 235°K in the tropics are generally expected to produce stratiform rainfall at the rate of 1.5 mm/half hour (Arkin and Meisner, 1987). Thus, all observations are combined by explicit time integration and total daily GPI rainfall is input into RFE with a resolution of 4 km. The files are moved to a computer at NOAA daily and gridded based on satellite position constants. The observation of the temperature of clouds (brightness temperature) is converted into cold count duration (CCD) measurements. The GPI tends to overestimate spatial distribution but underestimates convective precipitationFor quality control, each pixel must have greater than four half hour values, or the pixel is marked as undefined. Greater than 70% of all pixels for that day must be defined after incorporating all half hour data sets.

The GTS data is gridded using a Shepard technique with an initial search radius, where a new radius is determined depending on number of stations within the initial search area. If an adequate number of gauges are within the new radius, the system will interpolate rainfall to 0.1° grid using a station-station vector. Otherwise, the technique will interpolate using least squares regression. If the rainfall in the grid is zero within a one degree box, rainfall at center grid is zero.

The Special Sensor Microwave/Imager (SSM/I) microwave precipitation estimate data is derived from two instruments is available up to four times per day depending on the geographical location, as governed by ascending and descending Defense Meteorological Satellite Program (DMSP) polar satellite tracks. The SSM/I rainfall algorithm uses the 85V GHz channel to detect the scattering of upwelling radiation by precipitation sized ice particles within the rain layer. Rain rate can be derived indirectly based on the relationship between the amount of ice in the rain layer to the actual rain fall on the surface. The product is accurate when an observation is made, but it fails to catch precipitation events between observations. The data is used nearly directly in the RFE algorithm.

The Advanced Microwave Sounding Unit-B (AMSU-B) is a cross-track, line scanned instrument designed to measure scene radiances in 5 channels (Goodrum et al., 2000) and is located on the National Oceanic and Atmospheric Administration (NOAA) polar orbiting satellites. The purpose of the instrument is to receive and measure radiation from a number of different layers of the atmosphere in order to obtain global data on humidity profiles. It works in conjunction with the AMSU-A instrument to provide a 20 channel microwave radiometer. Ninety contiguous scene resolution cells are sampled in a continuous fashion, each scan covering 50° on each side of the sub-satellite path.

These scan patterns and geometric resolution translate to a 16.3 km diameter cell at nadir at a nominal altitude of 850 km. AMSU-B covers channels 16 through 20. The highest channels: 18, 19 and 20, span the strongly opaque water vapor absorption line at 183 GHz and provide data on the atmosphere's humidity level. Channels 16 and 17, at 89 GHz and 150 GHz, respectively, enable deeper penetration through the atmosphere to the Earth's surface. As with SSM/I, the AMSU-B data is available 4 times daily, staggered temporally. This product tends to overestimate most precipitation but does well with highly convective systems.

4.1.3 Rainfall Estimate Algorithm

The RFE combines the three satellite data sets linearly, and the station data are used to remove systematic bias. Combined analysis is a linear combination of each satellite estimate, weighted by the error variance. The variance is computed by first estimating the precipitation for each pixel from mean of all inputs. Then the satellite estimates are compared to the gridded GTS data. Areas without GTS data employ a satellite estimate interpolation. The proportional constant is calculated for every pixel, and bi-linear interpolation used for remaining grids. Greater than 70% of pixels must be defined after combining each input data set.

Output dataset is then ingested into a merging algorithm. The combined output equals the sum of the satellite rainfall estimates divided by the variance of the error. The final image produced by the algorithm is presented in Fig. 4.6. This RFE image from April 1–10, 2002 shows rainfall in the center of the continent, in Ethiopia and Somalia, and along the South African coast. The RFE has become the cornerstone of the hydrometeorological products that FEWS NET uses to monitor the growing season. The RFE is a critical part of monitoring of food security in Africa, as it is used to drive a wide variety of secondary products for agricultural monitoring.

4.2 African Rainfall Climatology (ARC) Data

The African Rainfall Climatology (ARC) data is very similar to the RFE except it uses as input only Meteosat 2 through seven 10.5–12.5 μm wavelength channels, omitting data from SSM/I and AMSU-A and B because they are not available back into the 1980s and 1990s. By using only one source of satellite rainfall information, a continuous, comparable rainfall record can be constructed. Tim Love at NOAA has constructed the Africa Rainfall Climatology or ARC product for use with the RFE, and is available back to 1994. The product can be pushed back to the mid 1980s, but obtaining the required Meteosat data is difficult and time consuming, as the data must be retrieved from an archive that does not have automated or computerized systems. The shorter-term mean from 1994 to 2005 is used operationally by FEWS NET while the longer-term ARC dataset is being constructed.

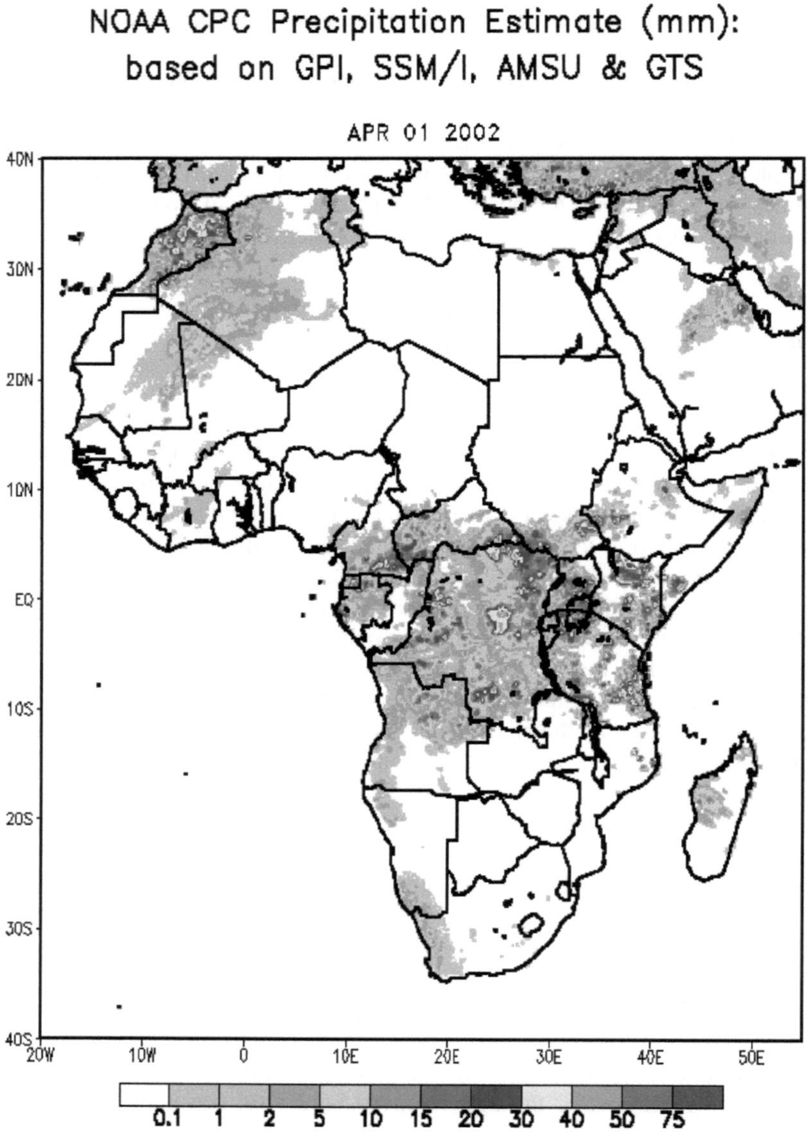

Fig. 4.6 Final RFE image from April 1–10, 2002, showing total rainfall in millimeters

The methodology used to create the ARC product is similar to the RFE2 product (Xie et al., 2002). To promote accuracy in regions exhibiting problematic merged analyses due to poor quality GTS gauge data, a daily threshold of 300 mm rainfall is used. Upon exceeding this constant, the erroneous precipitation grids are replaced by satellite-only estimates (Love et al., 2004). Table 4.2 shows the error between the ARC product and the RFE2 operational algorithm. The ARC algorithm is less able

Table 4.2 December 1999 monthly accumulated rainfall as estimated from the RFE2 algorithm, compared to the ARC algorithm with various inputs (from (Love et al., 2004) page 2)

Data	Bias (mm/day) compared to RFE2 data	Correlation with RFE2
GPI Only	2.26	0.345
SSM/I Only	−0.24	0.321
AMSU-A only	−0.15	0.095
GPI+SSM/I+AMSUA+GTS (RFE Inputs)	−0.15	0.501
GPI + GTS (ARC Inputs)	−0.04	0.467

to remove bias in the output, even when all four RFE2 inputs are used. However, the low bias and relatively high correlation of the ARC inputs to the RFE2 shows that a two-input product is a valuable addition to the FEWS NET suite of tools. The primary use of the ARC product is to produce anomaly images with a longer time series than would be available from the RFE.

4.3 The CHARM Dataset

The Collaborative Historical African Rainfall Model (CHARM) was developed by Christopher Funk at the University of California Santa Barbara. This final rainfall dataset provides a long-term mean from 1960 for analysis of trends in rainfall across the continent. The lack of consistent and sufficiently dense daily station data and the complicating influences of topography make the effort to provide an accurate dataset at the appropriate spatial and temporal resolutions difficult.

The data is described in detail in Funk et al. (2003). The CHARM dataset uses the following inputs:

- Daily 1.9° NCEP/NCAR Reanalysis Precipitation fields
- Monthly 0.5° interpolated rain gauge data
- Daily 0.1° orographic precipitation model

The Climatologically Aided Interpolation (CAI) daily station data was obtained from the University of Delaware, and thus is a unique source of precipitation not used elsewhere in FEWS NET. It provides synoptic scale monthly bias correction with measured long-term climatological rainfall amounts. The NCEP/NCAR Reanalysis precipitation fields provide daily variability across all years and physically-based rainfall gradients that can correct the bias seen in gridded gauge data. Finally, the Orographic Model, developed by Dr. Funk which builds on Sinclair's (1994) diagnostic model, provides internal gravity wave-based approximations of mesoscale orographic precipitation enhancements (Funk and Michaelsen, 2004). The orographic model is important for correcting for bias in the gauge network that tends to follow human populations, usually in valleys and in wetter areas.

The CHARM dataset shows that it is possible to use numerical weather prediction precipitation fields to represent daily rainfall variability at regional scales when the fields are constrained to match monthly interpolated gauge data. This is good news for many food security and hazard preparedness efforts, since monthly interpolated gauge and weather prediction fields are often the only data available in many regions. Although the CHARM is not widely used by FEWS NET at the moment, new research is ongoing to improve the accessibility and reliability of long term datasets in the context of operational monitoring. Particularly in the context of global climate change, these long data record projects will become more and more important for identifying appropriate response to rainfall anomalies.

4.4 Rainfall Datasets in Regions Outside of Africa

Description of datasets up to this point has focused on Africa, as the region where FEWS NET has the most countries to report on, is the most data-poor, and has the longest heritage. However, FEWS NET uses a similar strategy elsewhere. It has implemented the RFE algorithm in Central America and in Afghanistan for daily monitoring, but also uses other products that have advantages in these regions. Because it has strong rainfall patterns caused by both orographic and convective rainfall, the RFE does not always capture the rainfall in all regions accurately. Thus FEWS NET also uses the NASA Tropical Rainfall Measuring Mission (TRMM) Rainfall Estimates to augment the RFE data.

The TRMM data is unique as it uses a Precipitation Radar which was designed to provide three-dimensional maps of storm structure. These measurements yield invaluable information on the intensity and distribution of the rain, on the rain type, on the storm depth and on the height at which the snow melts into rain. In regions like Central America, that have large amounts of rain and very high rain rates, the radar measurements greatly improve the rainfall product. The TRMM rainfall estimates used by FEWS NET are the 3B42 merged TRMM and IR data products, produced 3-hourly, $0.25^\circ \times 0.25^\circ$ degree products that are available from 50°S–50°N.

In addition to the TRMM combined product, NOAA also uses the NOAA CMORPH product to augment analysis with TRMM. The CMORPH is a product that does not use gauge data, is at 0.07277° lat/lon ($\sim$8 km at the equator), has a 30 minute temporal resolution, and extends from 60N to 60S and is thus a global product (Joyce et al., 2004). The technique uses precipitation estimates that have been derived from low orbiter satellite microwave observations exclusively, and whose features are transported via spatial propagation information that is obtained entirely from geostationary satellite infrared data. At present the CPC incorporates precipitation estimates derived from the passive microwaves aboard the DMSP 13, 14 & 15 (SSM/I), the NOAA-15, 16 & 17 (AMSU-B) and the TRMM (TMI) spacecraft. These estimates are generated by algorithms of Ferraro (1997) for SSM/I, Ferraro et al. (2000) for AMSU-B and Kummerow et al. (2001) for TMI. Work is being done to include estimates from the AMSR and AMSR-E sensors aboard Aqua and ADEOS-II, respectively. The rainfall estimate has been produced since December

2002, and has a very different error structure than TRMM and gauge-based datasets, thus is used by FEWS NET as a comparative product to determine how accurate the RFE and other datasets may be. The CMORPH is not used alone for analysis.

4.5 Challenges in Rainfall Monitoring

FEWS NET's RFE is one of the best known merged gauge-satellite rainfall data products in Africa. It is the most commonly used dataset by scientists, practitioners and students in Africa. This high profile means that the RFE is frequently criticized by data users from a wide variety of disciplines and organizations. The RFE can be shown to be less accurate, at a lower resolution and with data over a shorter time period then a number of other widely used scientific products (Dinku et al., 2007), such as those produced by large research groups such as the Global Precipitation Climatology Project from the Laboratory for Atmospheres NASA at Goddard Space Flight Center, by Mike Hulme at Climatic Research Unit at the University of East Anglia, and others (Artan et al., 2007; Hulme, 2001; McCollum et al., 2000; Nicholson, 2005). There are a number of reasons that FEWS NET continues to support a rainfall data product using its own resources instead of migrating to one of these better-known scientific products.

FEWS NET's work with partners in the region means that it must continue to involve the local meteorological organizations in each country in which it works. These meteorologists spend a lot of effort collecting observations of rainfall using a network of rainfall gauges. The RFE algorithm uses the gauge data as the foundation of the rainfall product, statistically altering the satellite estimates to match rainfall observations. This approach means that the RFE tends to be much more similar to the observations made in a particular location then the other global products that are currently available. This is an enormous asset when working and building consensus with local partners because although the RFE may have some statistical biases over the entire series, the rainfall that is observed in a particular locality is reflected in the product.

FEWS NET also continues to support the RFE because it maintains control over both the data quality and the operational production of the data. As the inputs continue to change and scientific advancements continue to be made, the need for data continuity and consistent, prompt production tends to outweigh scientific developments. A scientific product may put considerations other than operational production first, reducing the product's availability. By funding the RFE production directly, FEWS NET avoids many of these problems.

4.6 Summary

In this chapter, the rainfall datasets that FEWS NET uses to monitor growing conditions in FEWS NET regions are described. The Rainfall Estimate (RFE) is a merged gauge-satellite product created by NOAA's Climate Prediction Center for FEWS

NET's use in Africa. The daily gauge data product that is delivered daily by the World Meteorological Organization, the Global Telecommunications Systems data is also described as it is a primary input to the RFE. Certain countries and regions are well represented by daily gauge data and others are not on the network at all. These differences may affect the quality and reliability of the RFE product, thus impeding FEWS NET's monitoring of food production. The ARC and the CHARM rainfall datasets were also presented, which use longer-term datasets to enable comparison of current conditions to those that occurred during the past several decades. These climatology datasets will be important in the next chapter, which will present derived products that bring rainfall analysis closer to the actual requirements of analysts who are ultimately interested in the rainfall's effect on various aspects of food production.

References

Adler, R.F., Huffman, G.J. and Keehn, P.R., 1994. Global rain estimates from microwave-adjusted geosynchronous IR data. Remote Sensing Reviews, 11: 125–152.

Arkin, P.A., Joyce, R. and Janowiak, J.E., 1994. IR techniques: GOES precipitation index. Remote Sensing Reviews, 11: 107–124.

Arkin, P.A. and Meisner, B.N., 1987. The relationship between large-scale convective rainfall and cold cloud over the Western Hemisphere during 1982–1984. Monthly Weather Review, 115: 51–74.

Artan, G., Gadain, H., Smith, J.L., Asante, K., Bandaragoda, C.L. and Verdin, J.P., 2007. Adequacy of satellite derived rainfall data for stream flow modeling. Natural Hazards, 43: 167–185.

Dinku, T., Ceccato, P., Grover-Kopec, E., Lemma, M., Connor, S. and Ropelewski, C.F., 2007. Validation of satellite rainfall products over East Africa's complex topography. International Journal of Remote Sensing, 28: 1503–1526.

Funk, C., 2006. Overview of FEWS NET Rainfall Validation/Enhancement Activities, University of California, Santa Barbara, California.

Funk, C. and Michaelsen, J., 2004. A simplified diagnostic model of orographic rainfall for enhancing satellite-based rainfall estimates in data-poor regions. Journal of Applied Meteorology, 43(10): 1366–1378.

Funk, C., Michaelsen, J., Verdin, J., Artan, G., Husak, G., Senay, G., Gadain, H. and Magadazire, T., 2003. The collaborative historical African rainfall model: Description and evaluation. International Journal of Climatology, 23(1): 47–66.

Ferraro, R.R., 1997. SSM/I derived global rainfall estimates for climatological applications. Journal of Geophysical Research-Atmospheres, 102: 16715–16735.

Ferraro, R.R., Weng, F., Grody, N.C. and Zhao, L., 2000. Precipition characteristics over land from the NOAA-15 AMSU sensor. Geophysical Research Leffers, 27: 2669–2672.

Goodrum, G., K.B.K. and Winston, W., 2000. NOAA KLM User's Guide. September 2000 ed. Suitland, MD US Department of Commerce, National Oceanic and Atmospheric Administration National Environmental Satellite, Data, and Information Service.

Hulme, M., 2001. Climatic perspectives on Sahelian desiccation: 1973–1998. Global Environmental Change, 11: 19–29.

Joyce, R.J., Janowiak, J.E., Arkin, P.A. and Xie, P., 2004. CMORPH: A method that produces global precipitation estimates from passive microwave and infrared data at high spatial and temporal resolution. Journal of Hydrometeorology, 5: 487–503.

Kummerow, C., Shin, D.-B., Hong, Y., Olson, W.S., Yang, S., Adler, R.F., McCollum, J., Ferraro, R., Petty, G. and Wilheif, T.T., 2001. The Evolution of the Goddard Profiding

Algorithm (GPROF) for Rainfall Estimation from Passive Microwave Sensors. Journal of Applied Meteorology, 40: 1801–1820.

Love, T.B., Kumar, V., Xie, P. and Thiaw, W.M., 2004. 20-Year Daily Africa Precipitation Climatology using Satellite and Gauge Data, 14th Conference on Applied Climatology, American Meteorological Society Seattle, WA, pp. 5.4–5.7.

McCollum, J.R., Gruber, A. and Ba, M.B., 2000. Discrepancy between gauges and satellite estimates of rainfall in Equatorial Africa. Journal of Applied Meteorology, 39(5): 666–679.

Nicholson, S., 2005. On the question of the "recovery" of the rains in the West African Sahel. Journal of Arid Environments, 63(3): 615–641.

Sevruk, B., 1982. Methods of correction for systematic error in point precipitation measurement for operational use, Operational Hydrology Report vol. 21, World Meteorological Organization, Geneva, Switzerland, WMO-No. 589, 91pp.

Xie, P. and Arkin, P.A., 1995. An intercomparison of gauge observations and satellite estimates of monthly precipitation. Journal of Applied Meteorology, 34: 1143–1160.

Xie, P. and Arkin, P.A., 1996. Analysis of global monthly precipitation using gauge observations, satellite estimates, and numerical model prediction. Journal of climate, 9: 840–858.

Xie, P., Yarosh, Y., Love, T., Janowiak, J. and Arkin, P.A., 2002. A Real-Time Daily Precipitation Analysis over South Asia. 16th Conference on Hydrology. American Meteorological Society, Orlando, FL.

Zeng, N., Neelin, J.D. and Lau, W.K.-M., 1999. Enhancement of interdecadal climate variability in the Sahel by vegetation interaction. Science, 286: 1537–1540.

Zeng, X., 1999. The relationship among precipitation, cloud-top temperature, and precipitable water over the tropics. Journal of Climate, 12(8): 2503–2514.

Chapter 5
Derived Agricultural and Climate Monitoring Products

FEWS NET uses primary rainfall products derived from local observations and satellite remote sensing products to monitor agriculture in developing countries because food production is a critical factor in food security. The United States Geological Survey's EROS data center in Sioux Falls, South Dakota has been FEWS NET's main center of expertise for the implementation of these rainfall-driven models. Although rainfall data is examined carefully every week in the context of identifying hazards, these new products have become extremely important tools to estimate food production variations from year to year.

Agricultural food production is generally estimated by the formula seen below:

$$\text{Production} = \text{Area Cropped} \times \text{Yield} \tag{5.1}$$

FEWS NET attempts to quantify both changes in the area planted as well as crop yield, but does not have the mandate to measure production directly. That mandate is given to the United States Department of Agriculture Foreign Agriculture Service. FEWS NET therefore estimates variations in changes of yield and of the area planted, but does not produce systematic or official estimates of the amount of food produced in any one area. Personnel from FEWS NET have developed a suite of derived and diagnostic products that enable the monitoring of the progress of the growing season in the context of the Weather Hazard Assessment. These products are either calculated directly from the RFE using models, such as the WRSI or the Start of Season, or are standard products that are already available from another agency. The products can be divided into those that are diagnostic of the climate system and those that measure the impact of the climate on cropping systems, with the objective to determine variations in production. Table 5.1 lists the products that will be described in this chapter.

5.1 Climate Diagnostics

The Madden-Julian Oscillation (MJO), the Global Forecast System (GFS) Vorticity and the Intertropical Convergence Zone (ITCZ) are three products that FEWS NET affiliated scientists at NOAA, NASA, USGS, and other organizations use to monitor

Table 5.1 Climate diagnostic and derived precipitation products for analysis of growing season progress

Product	Description	Region	Spatial resolution	Temporal resolution	Responsible organization
MJO IR – Madden Julian Oscillation	upper level convergence, precip predictor	global	25 km	hourly	NOAA
ITCZ – Intertropical Convergence Zone	estimates onset of rains appx week before	Regional (Africa)	vector coverage	daily, seasonal	NOAA CPC
SOS – Start of Season	determines beginning of growing season	Africa, C.America, Haiti	8 km	daily, seasonal	USGS
WRSI – water requirement satisfaction index	estimates crop yields by crop type	Africa, SW Asia, C.America, Haiti	8 km	daily, seasonal	USGS
SPI – Standardized Precipitation Index	18-yr mean standardized anomaly (30-yr mean)	Africa, SW Asia	0.1°	10 day, 1, 2, 3, 6 and 12 mon	UCSB-USGS
SSM/I Soil Moisture	soil moisture, vegetation proxy	global	30 km	weekly, monthly	NOAA
CPC Leaky Bucket model	soil moisture, available water for crops	global	25 km	monthly	NOAA CPC
MI – Moisture Index	supply/demand ratio	Africa, SW Asia	0.1°	daily, 10-day	USGS
SWI – Soil Water Index	water available for crops/vegetation	Global, Africa	0.1° 25 km	10-day, monthly	USGS
Consecutive Days	Seasonal sums of dry and wet days	Africa, C.America, Haiti	8 km	Seasonal	NOAA CPC

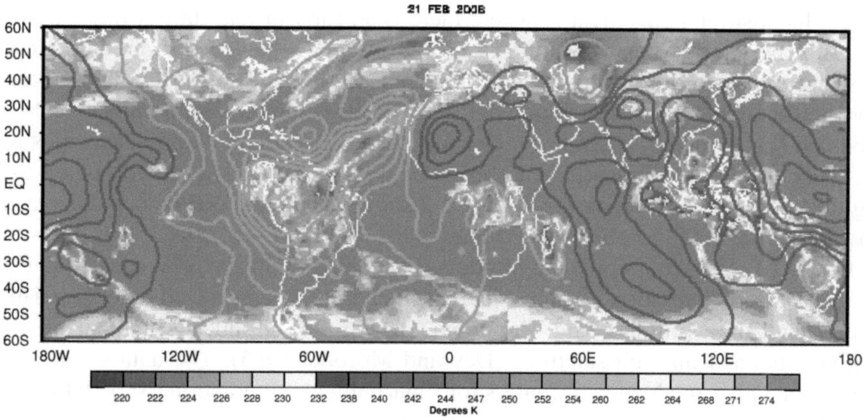

Fig. 5.1 Image of Infrared and 200-hPa velocity potential anomalies (base period 1971–2000) for February 21, 2008. Velocity potential anomalies are proportional to divergence with light contours corresponding to regions in which convection tends to be enhanced, and dark grey contours in regions with suppressed convection

the movement of the monsoon north and south across the African continent and across the Central American peninsula. These products enable monitoring of the movement of moisture into and out of regions of interest.

The MJO, also referred to as the 30–60 day or 40–50 day oscillation, is the main intra-annual fluctuation that explains weather variations in the tropics (Madden and Julian, 1971). The MJO is an equatorial traveling pattern of anomalous rainfall that involves variations in wind, sea surface temperature (SST), cloudiness, and rainfall. Because most tropical rainfall is convective, and convective cloud tops are very cold (emitting relatively little long wave radiation), the MJO is most obvious in the variation of outgoing long wave radiation (OLR), as measured by an infrared sensor on a satellite (Geerts and Wheeler, 1998). FEWS NET is primarily interested in tropical weather, and thus the MJO is an index that is frequently viewed to identify periods of increased convection and rainfall and periods of less rainfall. An example of the MJO field used in FEWS NET for October 14, 2006 can be seen in Fig. 5.1. The areas of suppressed and enhanced convection can be seen in the contours, and these zones move across the globe in waves over a period of 30–50 days.

Another large-scale weather phenomenon that can be use to monitor the progress of the growing season in the tropics is the intertropical convergence zone or ITCZ. As warm, moist air rises from the latitudes above and below the equator due to the action of the Hadley cell, a macroscale atmospheric feature which is part of the Earth's heat and moisture distribution system, it is transported aloft by the convective activity of thunderstorms. The ITCZ is the cause of much of the precipitation in the tropics and is thus monitored closely during the growing season for its position compared to its average position.

The position of the ITCZ is particularly helpful for monitoring the progress of the rainy season in the Sahel, and to a lesser degree in East Africa in the Sahelian zones. Its position helps identify where there is an early or late start or finish to the season. The ten-day mean position of the Africa ITCZ has been monitored since 1988 for the region from longitude 15°W–35°E. Daily analyses of the ITCZ position are found by hand interpolation of the 15° Celsius surface dew point and GDAS modeled surface wind streamlines. Dekadal values are simple ten-day averages of the daily positions. Climatological means of the ITCZ position are computed using the previous years' analyses. Figure 5.2 shows the mean, previous and average position of the ITCZ for the West African Sahel for July 1–10, 2007.

Other NOAA meteorological products, specifically those obtained from the Global Data Assimilation System (GDAS), including Average Daily Relative Humidity, and the Minimum and Maximum Temperature products, augment observations obtained through satellites (Dey and Morone, 1985). The data source for these products is the National Centers for Environmental Prediction (NCEP) GDAS model which contains all assimilated rawinsonde, aircraft, pibal, and other observational information. The data base began in October 1978 when the NCEP model became global. The model products are augmented by other analytical products, including sea level pressure at 1000–500 milibars (mb) height, precipitable water, 925 mb winds, convective available potential energy (CAPE)/lifted index at 800–500 mb, surface relative humidity, 925 mb height winds, vorticity at 300 mb and the latest satellite infrared temperatures. These fields are used for climate diagnostics, and are produced by other branches at NOAA operationally. FEWS NET has links from its pages to the pages of NOAA, removing the necessity to store the data locally. Typically, the diagnostic products are only examined by NOAA meteorologists employed by FEWS NET to produce the weekly weather advisory bulletin, as interpretation and analysis is quite complex and specialized.

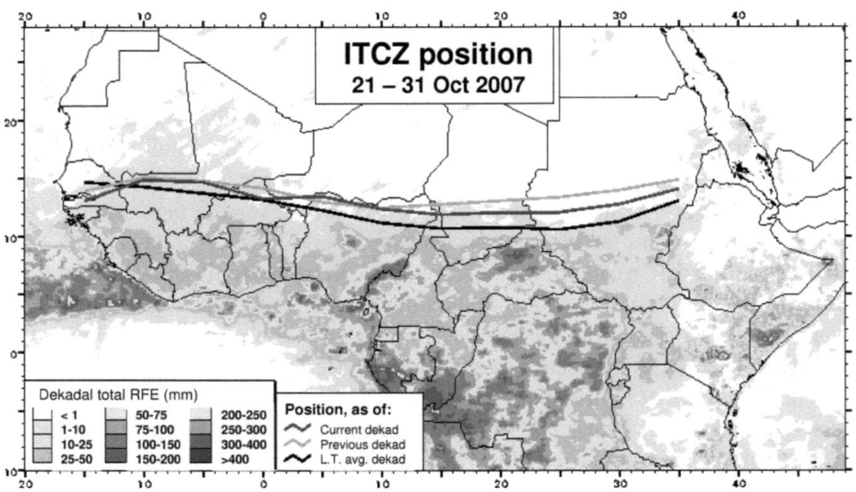

Fig. 5.2 The current, previous and mean position of the ITCZ in West Africa for October 21–31, 2007 from the FEWS NET USGS web site

5.2 Agricultural Season Monitoring Products

Beyond climate level products, FEWS NET monitors rainfall directly with their own rainfall estimate (RFE) as described in the previous chapter. The RFE is transformed into a number of additional products that enhance the utility of the data. In semi-arid agricultural areas, rainfall distribution is as important as quantity. USGS produces images that show the number of days with rainfall and without per month to assist the analyst in identifying locations that have experienced dry spells that can reduce crop yield as well as regions that are having a good year. Both are important pieces of information for food security analysis.

5.2.1 *Start of Season Data*

An important product calculated from the RFE is the Start of Season (SOS) (Fig. 5.3). Derived from the daily RFE imagery, the product shows where the rainy season has begun as compared to the climatological mean. The season is considered to have started when there has been three consecutive ten day periods with more than 20 mm of rainfall. This date is compared to the average date of start during the past ten years during which climatological rainfall data (ARC product) is available. Regions that are experiencing delayed start can then be identified.

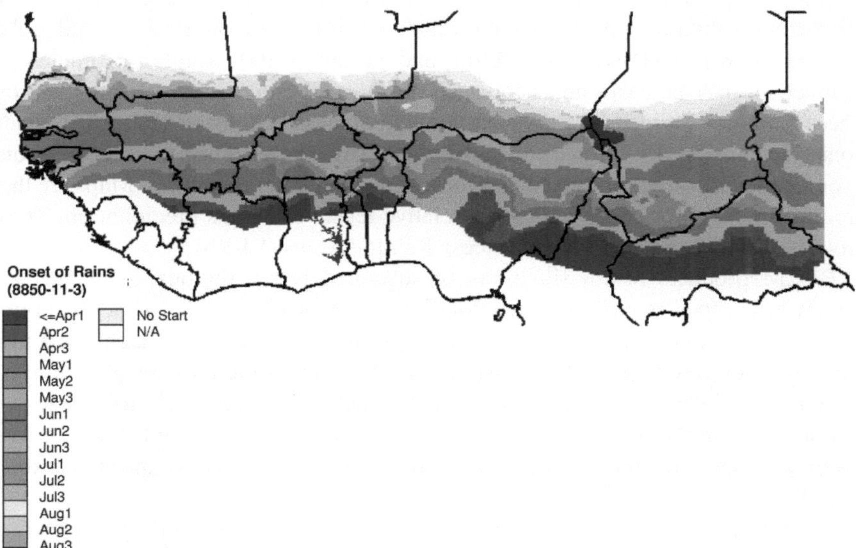

Fig. 5.3 Start of Season image for West Africa, 2007. The horizontal zones show the advancing ITCZ from the more humid regions to the south to the semi-arid desert margin areas in the north. The regions with a later start have a shorter growing season than those with an earlier start

The date that the rainy season starts is critically important for farmers in regions with growing seasons that typically do not exceed 90 days in length. If the season starts late, it is unlikely to end late as well, thus yields will be affected due to lack of sufficient growing days. In addition, farmers may have to reseed their fields if the rains start and then stop again, killing newly germinated seedlings through lack of moisture. Finding sufficient resources to obtain seeds and the energy and time to replant fields is often a significant strain on scarce resources which can impact food security. Start of Season indices also are used in the crop water satisfaction index parameter discussed later in the chapter.

5.2.2 Flooding and Runoff Models Driven by RFE Data

Drought is not always the problem in agricultural areas. Monsoonal systems can be extremely intense, and food insecurity can result from widespread crop destruction due to flooding. A simple method for identifying areas subject to problems of flooding or excess moisture has been developed by the USGS in Sioux Falls through the joint use of satellite rainfall estimates and digital maps of basin boundaries and river networks. Maps are produced which highlight basins experiencing above-average rainfall in the previous ten-day period, and river reaches with potentially higher-than-average stream flow. These maps are then used for monitoring and further investigation.

USGS uses digital maps of local basins and river networks, derived from one kilometer resolution topographic data, which are part of a topologically coded global data set known as HYDRO1K (Verdin and Verdin, 1999). Rainfall estimates are summed over river basin areas for each dekad, and cumulatively for the season. These sums are divided by the corresponding values for long-term average conditions (Hutchinson, 1995), and excess rainfall scores are assigned to basin areas and river reaches accordingly - the higher the ratios, the greater the scores. Maps are then produced with color codes indicating relative levels of excess precipitation. These products have been named Basin Excess Rainfall Maps (BERM).

BERM products reveal situations of sustained heavy regional rains that adversely affect food security through flooding and consequent widespread disruption of agriculture, transportation, and market systems. The basin (or catchment) map highlights sub-basins (out of approximately 3,000 across the African continent) receiving above-average precipitation for the dekad and cumulatively for the season by color coding the relevant polygons. The BERM product is coupled with a river flooding image that shows which rivers are likely to flood due to upstream excess rainfall.

Funding for flooding products have recently been discontinued by USAID, as there are many other organizations that produce similar products that FEWS NET analysts can use. Floods do not cause many food security problems compared to other hazards, and thus in a time of limited resources, developing and maintaining stream flow models has not been possible.

5.2.3 Standardized Precipitation Index

The Standardized Precipitation Index or SPI was developed by McKee, Doesken, and Kleist in 1993. The objective was to normalize wet and dry periods according to a long term mean in order to determine the impact of the drought at different time scales. Because the index is normalized, it is equally good at identifying wet periods as dry periods and measures the severity of the drought according to the historical mean (Hayes, 2006). Different time scales measures water storage that respond to rainfall anomalies in different ways. Soil moisture conditions respond to precipitation anomalies on a relatively short scale whereas groundwater, streamflow, and reservoir storage reflect the longer-term precipitation anomalies (Hayes, 2006). Although McKee et al. (1993) originally calculated the SPI for 3-, 6-, 12-, 24-, and 48-month time scales, FEWS NET is interested in rather shorter time periods, and thus calculates the SPI at periods of 10 days, and 1-, 2-, 3-, 6-, and 12-month periods.

The SPI calculation for any location is based on the long-term precipitation record for a thirty year period. The long-term record is fitted to a probability distribution, which is then transformed into a normal distribution so that the mean SPI for the location and desired period is zero (Edwards and McKee, 1997). Positive SPI values indicate greater than median precipitation, and negative values indicate less than median precipitation. The SPI values range from $+2$ to -2, where near normal values are between $+1$ and -1.

The SPI is a classification system that climate analysts can use to determine how intense a drought is. The SPI is both a criteria to estimate the intensity of a drought event for a particular time scales and a way to estimate how rare an event it may be. A drought event occurs any time the SPI is continuously negative and reaches an intensity of –1.0 or less. The event ends when the SPI becomes positive. Each drought event, therefore, has a duration defined by its beginning and end, and an intensity for each month that the event continues (Edwards and McKee, 1997). The positive sum of the SPI for all the months within a drought event can be termed the drought's magnitude (Hayes, 2006). Because the SPI measures both intensity of the event and the probability of it occurring during the period of record (usually thirty years), it is a very useful measure of anomalous precipitation events.

FEWS NET has implemented the SPI in Africa using its ARC and CHARM datasets described in the previous chapter. The Collaborative Historical African Rainfall Model (CHARM) provides a reasonable historical model for monthly precipitation over the entire continent at a 0.1-degree spatial resolution. The monthly accumulations calculated by CHARM can be temporally decomposed to provide dekadal rainfall estimates which are constrained to match the monthly sums. The extensive historical information provided by CHARM allows for an assessment of 36 years of precipitation data. Because of the difficulty in obtaining accurate and realistic data for the required thirty years in Africa, the SPI is calculated using the mean from the ARC dataset during the 1996–2005 period, and the variance from the CHARM data, from 1961 to 1996. The resulting data provides the necessary continuity while being more accurate than would otherwise be possible. Examples of the SPI from Africa and other locations can be found at the NOAA Africa Data Dissemination Service website.

5.3 Crop Models Driven by Rainfall Data

FEWS NET needs to know how variations in rainfall affect food production, but this may not be straightforward. Two seasons with exactly the same total amount of rain may produce very different crop yields due to differences in rainfall distribution throughout the season. To move observations and merged gauge-satellite rainfall products closer to the information needed for food production estimates, FEWS NET has implemented operational measures of the impact of rainfall on specific crops through the Water Requirement Satisfaction Index (WRSI).

FAO studies have shown that rainfall can be related to crop production using a linear yield reduction function specific to a crop (Doorenbos and Pruitt, 1977; Frere and Popov, 1979; Frere and Popov, 1986). More recently, Verdin and Klaver (2002) and Senay and Verdin (2001) demonstrated a regional implementation of this linear approach in a grid cell based modeling environment using the RFE product. The spatially explicit water requirement satisfaction index (WRSI) is an indicator of crop performance based on the availability of water to the crop during a growing season (Senay and Verdin, 2003; Verdin and Klaver, 2002). Specifically, WRSI measures the reduction in yield per unit area due to water deficiencies at specific stages of crop development. It does not attempt to measure any other kind of yield reduction, of which there are many. FEWS NET has operational programs that calculate the WRSI for maize (corn), millet, sorghum and rangeland grasses in Africa and Central America.

WRSI for a season is based on the water supply and demand a crop experiences during a growing season. It is calculated as the ratio of seasonal actual evapotranspiration (AET) to the seasonal crop water requirement (WR):

$$WRSI = \frac{AET}{WR} \times 100 \tag{5.2}$$

WR is calculated from the Penman-Monteith potential evapotranspiration (PET) using the crop coefficient (Kc) to adjust for the growth stage of the crop. AET represents the actual (as opposed to the potential) amount of water withdrawn from the soil water reservoir. Whenever the soil water content is above the maximum allowable depletion level (based on crop type), the AET will remain the same as the water requirement, i.e., no water stress. But when the soil water level is below the allowable depletion level, the AET will be lower than WR in proportion to the remaining soil water content (Senay and Verdin, 2003). When the maximum allowable depletion level is exceeded, then the plant wilts and it experiences structural damage so that it is less capable of producing grain later, thus reducing yields.

The soil water content is obtained through a simple mass balance equation where the level of soil water is monitored in a bucket defined by the water holding capacity of the soil and the crop root depth, as in Eq. (5.3).

$$SW_t = SW_{t-1} + PPT_t - AET_t \tag{5.3}$$

where SW is soil water content, PPT is precipitation, AET is the seasonal actual evapotranspiration and t is the time step index, typically 10 day periods. The soil water index is reported separately and images generated for the regions that are actively growing crops during the period.

The most important inputs to the WRSI model are precipitation and potential evapotranspiration (PET). FEWS NET at the USGS calculates daily PET values for Africa at 1.0-degree resolution from 6-hourly numerical meteorological model output using the Penman-Monteith equation (Shuttleworth, 1992; Verdin and Klaver, 1998). RFE images for the African continent at 8km spatial resolution are used for the PPT estimates. In addition, the WRSI model uses relevant soil information from the FAO (1998) digital soils map and topographical parameters from Digital Elevation Model (DEM) derived data (HYDRO-1K, (Gesch et al., 1999)).

WRSI calculation requires a start-of-season (SOS) and end-of-season time (EOS) for each modeling grid-cell. Maps of these two variables are particularly useful in defining the spatial variation of the timing of the growing season and, consequently, the crop coefficient function, which defines the crop water use pattern of crops. Corn, for example, has a very different water use pattern and sensitivity than sorghum, so each has their own equations for the WRSI. The model determines the SOS using onset-of-rains based on simple precipitation accounting. The onset-of-rains is determined using a threshold amount and distribution of rainfall received in three consecutive dekads. The start of season is established when there is at least 25 mm of rainfall in one dekad followed by a total of at least 20 mm of rainfall in the next two consecutive decades. The length of growing period (LGP) for each pixel is determined by the persistence, on average, above a threshold value of a climatological ratio between rainfall and potential evapotranspiration. Thus, the end of season period is obtained by adding the length to the start of season dekad for each grid cell. The WRSI model is capable of simulating different crop types whose seasonal water use pattern has been published in the form of a crop coefficient in the literature. Such crops include maize (corn), sorghum, millet, wheat, etc.

At the end of the crop growth cycle, or up to a certain dekad in the cycle, the sum of total actual evopotranspiration (AET) and total water requirement (WR) are used to calculate WRSI in a Geographic Information System (GIS) environment at 0.1° (about 10 km) spatial resolution. A case of 'no deficit' will result in a WRSI value of 100, which corresponds to the absence of yield reduction related to water deficit. Of course many other factors can impact yield, including nutrient stress, pest infestation, physical damage due to hail or flooding, etc. These issues are not addressed by the WRSI, but are also far less common than widespread reductions in yield due to water stress. A seasonal WRSI value less than 50 is regarded as a crop failure condition (Smith, 1992).

Yield reduction estimates based on WRSI contribute to food security preparedness and planning. As a monitoring tool, the crop performance indicator can be assessed at the end of every 10-day period during the growing season. As an early warning tool, end-of-season crop performance can be estimated using long-term average meteorological data. Due to differences in the growing season, WRSI maps are generated and distributed on a region-by-region basis (e.g., the Sahel, Southern Africa, Greater Horn of Africa regions). At the end of every dekad, two image products associated with the WRSI are produced and disseminated for the FEWS NET activity: the current WRSI calculated using the RFE2 and the extended WRSI calculated using the ARC and CHARM products combined in order to determine if the current period is above or below the mean from 1961.

Figure 5.4 shows WRSI values for a particular crop from the start of the growing season until this time period. For example, if the cumulative crop water requirement up to this period was 200 mm and only 180 mm was supplied in the form of rainfall, the crop experienced a deficit of 20 mm during the period and thus the WRSI value will be ($[180/200] \times 100 = 90.0\%$). This approach is slightly different from the traditional single update where the cumulative supply-to-date is compared to the total seasonal crop water requirement, instead of the requirement up to the current period. Note that, unlike the FAO update, the current WRSI can increase in value in the later part of the growing season if the demand (crop water requirement) and supply (rainfall) relationship becomes more favorable. However, both the FAO and this approach are mathematically equivalent at the end of the season (i.e., when the current dekad becomes the end-of-season dekad).

5.3.1 Soil Moisture Indices

The Moisture Index (MI) is an intermediate product of the WRSI. As an agrometeorological indicator defined by a simple supply/demand ratio, the formula for MI is rainfall as measured by the RFE2 divided by soil water then multiplied by 100 to provide a percent value. As a stand-alone product the dekadal (every ten day)

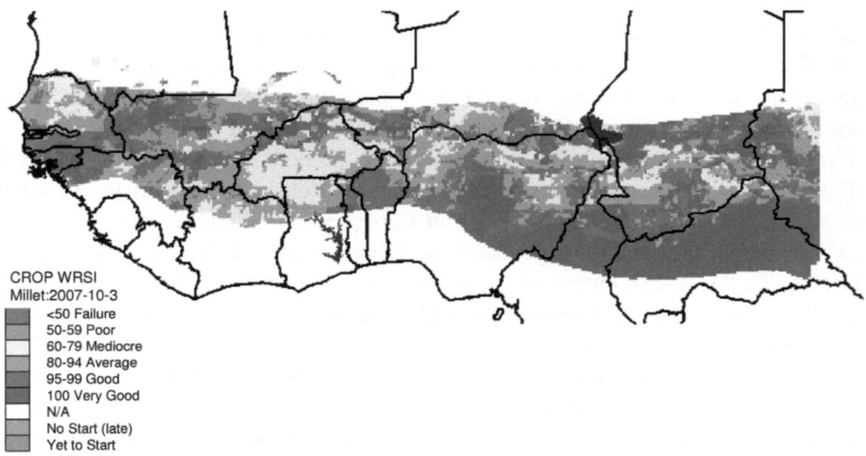

Fig. 5.4 Water requirement satisfaction index for millet for the last ten days of November 2007

Moisture Index images provide additional information concerning moisture available for crops, enhancing the information provided by the RFE images. In order to calculate the dekadal MI on a spatial basis, the RFE images are used directly for precipitation, as are Potential Evapotranspiration grid values computed from the GDAS (Global Data Assimilation System) analysis fields. A simple bucket model (defined by the water holding capacity of the soil and the crop root depth) is used to calculate a new value of SW for each dekad. Soil water in excess of water holding capacity is assumed to be lost as runoff or drainage out of the first 100-cm layer of soil.

Like the WRSI, the Soil Water potential evapotranspiration (PET) inputs are calculated from GDAS analysis fields, generated every 6 hours, using the Penman-Monteith equation (as with WRSI, the formulation can be found in Shuttleworth (1992)). GDAS fields used in the PET calculation include air temperature, atmospheric pressure at the surface, wind, relative humidity, and radiation (long wave, short wave, outgoing, and incoming). PET is computed for each day, and appropriate sums are made to obtain dekadal totals. The spatial variation of soil water holding capacity (WHC) is characterized using the 1997 FAO Digital Soil Map of the World. (FAO, 1997). The scale of the original mapping is 1:5,000,000, and the soil polygons carry attributes that include an estimate of easily available water capacity in the upper 100 cm, based on soil physical characteristics. These values were adopted for calculation of soil water conditions. The FAO soil map has been rasterized at a scale that matches the 0.1-degree RFE grid. The WRSI and Soil water algorithms were developed at USGS EROS where they continue to be used and maintained by FEWS NET-affiliated scientists.

5.4 Limitations of Production Estimates Based on Rainfall

The monitoring products described in this chapter are extremely useful for determining the progress of the growing season in various regions at a continental scale. Through the modeling techniques described above, FEWS NET has been able to improve the relationship between the picture painted by its remote sensing products and ultimate food security outcome through the use of models, but this approach has its limitations. Food production is a function of the amount of area planted and the yield that the crop gives. Rainfall is a primary driver of variations in yield, but there are many other factors as well. Other factors that affect production can be just as important as variations in rainfall, and these interfere with the direct interpretation of WRSI images.

Year to year variations in the amount of area planted is one of the most important aspects to interpreting crop models. Until recently, this portion of the production equation has been neglected by technical experts working for FEWS NET, mostly because in normal circumstances, farmers tend to plant as much land as they can. This is a broad assumption, however, and can frequently be incorrect. The next chapters will describe new approaches that FEWS NET has been developing to estimate directly with remote sensing the area planted in regions experiencing rapid economic and political change. When these estimates are combined with yield estimates, a much more accurate assessment of the amount of food an area has produced can be made.

Another important aspect to crop production that can vary through time is soil fertility. Assessing soil fertility over large areas is very challenging because it is highly variable in both space and time and is affected by management and use, which in Africa can vary by tens of meters due to the very small size of field plots (Stocking, 2003b). Yields decline for many reasons, such as excessive off-take of nutrients in crops without replenishment, pests and diseases, weed infestations, and increasing prevalence of climate change–induced drought (Stocking, 2003a). Variations in soil quality are simply not captured by the models used by FEWS NET. Although differing underlying soil substrates are represented in the models, management techniques, wind and water erosion, and other stressors are not. Thus there are distinct limits to the utility of these products, as they can only detect variations in crop yields due to the amount and distribution of rainfall and the rate at which the rainwater evaporates due to atmospheric forcing (wind and ambient temperature).

Other issues revolve around the ability of continental scale hydrological models to capture soil water and potential evapotranspiration, key inputs to both the WRSI and the Moisture Index. The rate at which soil can retain moisture depends on its organic material and the mixture of clay to silt to sand. These variations are not captured by the soil maps used in the models, thus the models are generalized across highly variable landscapes. Soil moisture estimates require highly accurate measurements at very fine spatial scales, and is the target of much ongoing research and technology development by organizations such as the US Department of Agriculture and NASA. Thus the soil moisture and evapotranspiration parameters used in FEWS NET's models are not very accurate nor are they at the best available spatial and temporal resolution (Walker and Houser, 2004). WRSI is therefore vulnerable to the errors inherent in the estimates of rainfall as a result of problems with the gauge data accuracy and the satellite inputs (Laws et al., 2003), as well as the errors in the modeled evapotranspiration and soil moisture present locally. These errors diminish with larger spatial scales, but they can be significant sources of uncertainty.

Another problem with the WRSI is that it is based on crop parameters and water requirements determined from agricultural research conducted on farms with uniform mono-cropping and inputs rarely seen in practice in the regions where FEWS NET works. Complex interactions between legumes and field crops, traditional agro-forestry techniques and moisture-capture techniques can fundamentally alter a plant's response to rainfall by changing the local moisture availability in the soil. Thus the crop parameters obtained from the literature and used in the WRSI may have only a vague relation to the actual crop stressors happening on the ground across the large and diverse African and Central American continent.

Despite the issues outlined above, rainfall is the primary determinant of cereal production in rain fed agricultural regions. Other factors that affect yield decline in importance as rainfall sufficiency at each stage of the plant's life cycle fails to meet requirements. Thus the approach that FEWS NET takes is to monitor the effect of observable weather conditions on the underlying variability of crop production. Because weather varies on large scales and tends to affect many farmers in the same area at once, it is a significant source of food insecurity for large numbers of people that can be monitored directly. As food production declines, demand for food in the market rises, with subsequent effects on food prices (Brown et al., 2006). FEWS NET therefore invests significant time and resources in monitoring weather in agricultural areas through integrated hydrologic models. Validation efforts have also found that the WRSI has a strong relationship to year-to-year variations in observed crop yield, which have been verified through annual tours of the primary crop production regions of FEWS NET countries in collaboration with the Foreign Agricultural Service at the USDA.

5.5 Summary

This chapter presented the integrated models that ingest merged satellite-gauge rainfall data model outputs and evapotranspiration estimates with information about specific crop water requirements and soil type. The WRSI, MI and other modeled parameters allow the analyst to estimate the effect of rainfall on crop production much more easily than rainfall alone. The details of these models were presented and examples shown. The strengths and weaknesses of the products were discussed and their utility in monitoring agricultural production described in the context of smallholder agriculture in Africa. The next chapter will outline alternatives to rainfall data and in the subsequent chapters how these vegetation-based datasets are integrated with rainfall data to project future conditions.

References

Brown, M.E., Pinzon, J.E. and Prince, S.D., 2006. The Sensitivity of Millet Prices to Vegetation Dynamics in the Informal Markets of Mali, Burkina Faso and Niger. Climatic Change, 78: 181–202.

Dey, C.H. and Morone, L.L., 1985. Evolution of the National Meteorological Center global data assimilation system: January 1982-December 1983. Monthly Weather Review, 113: 304–318.

Doorenbos, J. and Pruitt, W.O., 1977. Crop water requirements, Food and Agriculture Organization, Rome, Italy.

Edwards, D.C. and McKee, T.B., 1997. Characteristics of 20th century drought in the United States at multiple time scales, Colorado State University, Fort Collins, Colorado.

FAO, 1997. Digital Soil Map of the World (1:5,000,000).

FAO, 1998. World reference base for soil resources. Rome, FAO, ISRIC and ISSS.

Frere, M. and Popov, G.F., 1979. Agrometeorological crop monitoring and forecasting, Food and Agriculture Organization, Rome, Italy.

Frere, M. and Popov, G.F., 1986. Early Agrometeorological crop yield forecasting, Food and Agriculture Organization, Rome, Italy.

Geerts, B. and Wheeler, M., 1998. The Madden-Julian Oscillation, Naval Maritime Forecast Center/Joint Typhoon Warning Center (NMFC/JTWC), Pearl Harbor, HI.

Gesch, D.B., Verdin, K.L. and Greenlee, S.K., 1999. New land surface digital elevation model covers the Earth. EOS, Transactions of the American Geophysical Union, 80(6): 69–70.
Hayes, M.J., 2006. What is Drought?: Drought indices, Lincoln, Nebraska.
Hutchinson, M.F., 1995. Interpolating mean rainfall using thin plate smoothing splines. International Journal of Geographic Information Systems, 9: 385–403.
Laws, K.B., Janowiak, J.E. and Huffman, G.J., 2003. Verification of Rainfal Estimates over Africa using RFE, NASA MPA-RT, and CMORPH, 18th Conference on Hydrology. American Meteorological Society, Seattle, WA, pp. 6.
Madden, R.A. and Julian, P.R., 1971. Detection of a 40–50 day oscillation in the zonal wind in the tropical Pacific. Journal of Atmospheric Science, 28: 702–208.
Mckee, T.B., Doesken, N.J. and Kleist, J., 1993. The relationship of drought frequency and duration to time scales. Eig Conference on Applied Climatology, Anaheim, California.
Senay, G.B. and Verdin, J.P., 2001. Using a GIS-Based Water Balance Model to Assess Regional Crop Performance. Proceedings of the Fifth International Workshop on Application of Remote Sensing in Hydrology. Montpellier, France.
Senay, G.B. and Verdin, J., 2003. Characterization of yield reduction in Ethiopia using a GIS-Based crop water balance model. Canadian Journal of Remote Sensing, 29(6): 687–692.
Shuttleworth, J., 1992. Evaporation. In: D. Maidment (Editor), Handbook of Hydrology. McGraw-Hill, New York, pp. 4.1–4.53.
Smith, M., 1992. Expert consultation on revision of FAO methodologies for crop water requirements, Food and Agriculture Organization (FAO), Rome.
Stocking, M.A., 2003a. Erosion and Crop Yield, Encyclopedia of Soil Science. Dekker, New York.
Stocking, M.A., 2003b. Tropical soils and food security: the next 50 years. Science, 302: 1356–1359.
Verdin, J. and Klaver, R., 1998. Grid Cell Based Crop Water Accounting for the Famine Early Warning System, EROS Data Center, Sioux Falls, SD.
Verdin, J. and Klaver, R., 2002. Grid cell based crop water accounting for the Famine Early Warning System. Hydrological Processes, 16: 1617–1630.
Verdin, K. and Verdin, J., 1999. A topological system for delineation and codification of the Earth's river basins. Journal of Hydrology, 218: 1–12.
Walker, J.P. and Houser, P.R., 2004. Requirements of a global near-surface soil moisture satellite mission: accuracy, repeat time, and spatial resolution. Advances in Water Resources, 27: 785–801.

Chapter 6
Vegetation Index Measurements

Before FEWS NET had reliable rainfall data, satellite remote sensing of vegetation was the primary way of identifying regions experiencing climatic stress. Vegetation vigor is monitored at a continental scale using the Normalized Difference Vegetation Index or NDVI, which is a scalable index derived from spectral indices describing the amount of photosynthetically active radiation absorbed by plants on the ground. Because the index measures photosynthesis directly, it works equally well at the leaf scale as well as landscape scale. Radiometric instruments on satellites take measurements which can related to the amount of biomass and photosynthesis occurring on the ground, from which variations in crop yield in semi-arid regions can be inferred. This chapter will review how vegetation data is derived and how FEWS NET uses NDVI datasets. Indices using other spectral bands have been developed which are similar to NDVI but which measure variations in snow extent and snow depth. These will also be reviewed in this chapter.

A wide variety of sensors are currently available that measure radiation in the photosynthetically active region, from which we can observe the activity of plants. Figure 6.1 shows the various data systems that FEWS NET uses to observe agricultural activity for hazard detection. The coarse resolution datasets that have high temporal resolution such as SPOT Vegetation and the Advanced Very High Resolution Radiometer (AVHRR) are used routinely to monitor the progress of the growing season and to detect locations where the vegetation is less healthy than it should be. FEWS NET uses high resolution data from instruments such as Quickbird and IKONOS are used for detecting agricultural fields in regions with a complex agricultural landscape. FEWS NET uses spectral data from the instruments shown in Fig. 6.1 to monitor agricultural hazards. This chapter will review vegetation indices, the sensors from which they are derived, how they are used in early warning systems, and describe the products that FEWS NET uses to monitor food security.

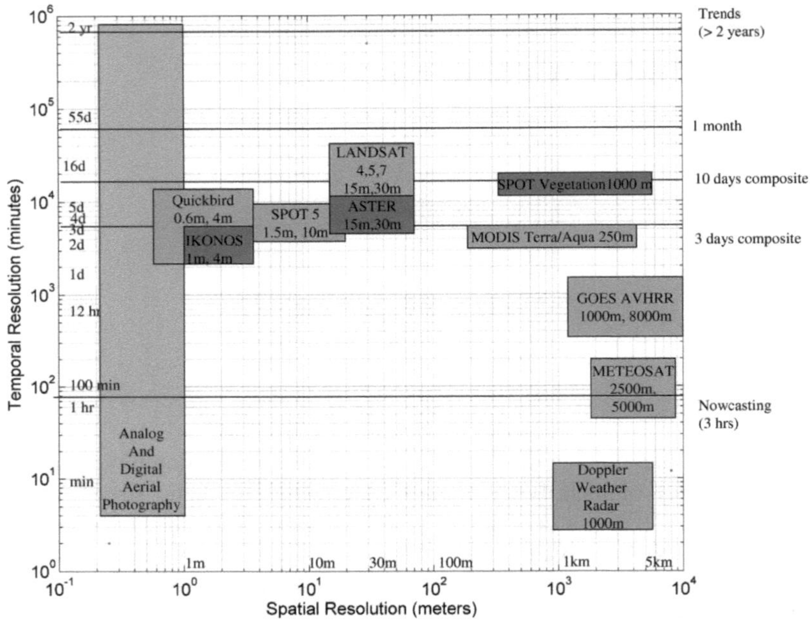

Fig. 6.1 The relationship between the spatial and temporal resolution of various aerial and sub-orbital remote sensing systems

6.1 What is NDVI?

Spectral vegetation indices are usually composed of red and near-infrared radiances or reflectances (Tucker, 1979), and are one of the most widely used remote sensing measurements (Cracknell, 2001). They are highly correlated with the photosynthetically active biomass, chlorophyll abundance, and energy absorption by plants (reviewed in (Myneni et al., 1995)). The index was first developed using hand-held radiometers, and their relationship with aboveground biomass established with extensive destructive measurements in a grassland ecosystem (Tucker, 1977). The use of spectral vegetation indices derived from the Advanced Very High Resolution Radiometer (AVHRR) satellite data followed the launch of the first operational polar orbiting satellite, NOAA-6, in June 1979, followed closely by NOAA-7 in July 1981 (Gray and McCrary, 1981; Townshend and Tucker, 1981). The AVHRR instruments on NOAA-6 and NOAA-7 were the first in the TIROS-N series of satellites to have non-overlapping channel 1 and channel 2 spectral bands. Overlapping red and near infrared spectral bands would make calculating a vegetation index impossible.

The NDVI is calculated as a ratio of the difference of the Near Infrared (NIR) and the Red bands on a sensor: (NIR − Red)/(NIR + Red). Since vegetation has high NIR reflectance but low red reflectance, vegetated areas will have higher NDVI values compared to non-vegetated areas (Fig. 6.2). The longest continuous record of global vegetation observations comes from AVHRR sensor, which has been flown

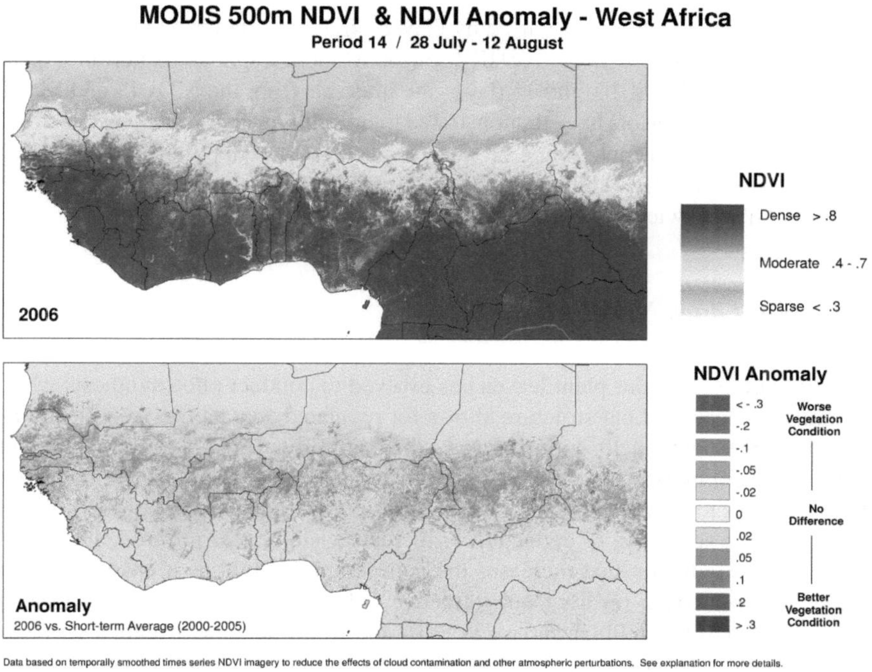

Fig. 6.2 Example of an NDVI composite from MODIS NDVI for West Africa for 28 July – 12 August 2006. The *dark* areas are vegetated while the non-vegetated areas (deserts, clouds) are generally *light*

on operational satellites by NOAA for more than twenty five years. This record is a critical basis for evaluation of current conditions, because of careful calibration of the dataset allows the current image to be subtracted from the long term mean. The Normalized Difference Vegetation Index (NDVI) has become the most used product derived from NOAA AVHRR data (Cracknell, 2001), largely from the use of long records of NDVI images formed via maximum value compositing (Holben, 1986) which minimize clouds and other atmospheric effects. Although the data is quite low resolution, the AVHRR NDVI data have been used extensively since 1981 to study a variety of global land processes (reviewed in (Cracknell, 1997; D'Souza et al., 1996; Townshend, 1994); (DeFries and Belward, 2000), among others). Over 4000 papers have been published using vegetation indices and these papers have been referenced by $>60,000$ other scientific papers (Tucker and Brown, 2008). Compton Tucker from the Global Inventory Monitoring and Mapping Systems (GIMMS) group at the NASA Goddard Space Flight Center has been involved with FEWS NET from its inception and has been producing AVHRR data for their use in monitoring agricultural conditions at continental scales.

New sensors have recently been launched which have been designed specifically for vegetation monitoring, unlike the AVHRR which was originally intended to monitor the weather. MODIS and the European SPOT-4 Vegetation sensors are the two datasets most frequently used by FEWS NET for monitoring at higher resolutions than is possible with the AVHRR sensor. Moderate spatial resolution (250 m to 1 km) and weekly (8, 10, and 16 day) time intervals from the MODIS and SPOT Vegetation (VGT) sensors have demonstrated their utility in characterizing the structure, metabolism, and functioning of ecosystems (Huete et al., 2006). FEWS NET uses primarily AVHRR, MODIS and SPOT Vegetation data because they are global and have daily or twice-a-day coverage.

6.1.1 What does NDVI Measure?

The structure of vascular plant leaves has evolved to conduct photosynthesis while avoiding desiccation. Leaf structure allows for regulated contact between the atmosphere and mesophyll cells within leaves, where atmospheric CO_2 must be in direct contact with chloroplasts, and where chloroplasts are able to shed O_2 resulting from photosynthesis to the atmosphere (Tucker and Brown, 2008). A consequence of leaf structure is the scattering or deflection of incident visible, near infrared, and shortwave infrared radiation that increases the path length of light as it passes through leaves. This scattering results from refractive index differences between hydrated cells ($\sim$1.3) and intercellular air spaces (1.0) and from the many irregular surfaces of the internal leaf labyrinth. Approximately 80–90% of each cell is liquid water. In addition to the scattering or deflection of the incident shortwave solar flux as it passes into leaves, absorption of light occurs depending upon the wavelength, the degree to which the light is scattered or deflected, and the concentration of plant pigments and liquid water content (Tucker, 1980) (Fig. 6.3).These light scattering processes are exacerbated in the plant canopy, where multiple leaves further increase the mean path length of incident solar radiation within the canopy, through multiple leaf interactions. Thus a smaller concentration of spectral absorbers, such as the chlorophylls, the carotenoids, and liquid water absorb a higher proportion of the shortwave incident solar radiation than their concentration and coefficients of absorption would suggest.

The effect of leaf structure, physiological state, leaf density, and plant canopy architecture produces variable spectral reflectance and its compliment, spectral absorptance, from 0.3 to 2.5 μm. The incident shortwave flux that is not absorbed is reflected and it is the light reflected from the canopy that we measure using remote sensing instruments. Many of the processes within leaves have been abstracted into leaf radiation models (Jacquemoud and Baret, 1990; Tucker and Garratt, 1976) that are driven by spectral absorption and scattering or the leaf and canopy mechanisms that increase the mean path length of the radiation through the leaf or plant canopy in question. There is thus a direct and quantitative coupling between the *in situ* spectral reflection of plant canopies and the spectral absorption that occurs; spectral transmission is an intermediate state, ultimately to be reflected or absorbed. Furthermore, the visible light absorbed by plants is the energy that drives photosynthesis (Monteith, 1977) and is referred to as absorbed photosynthetically active radiation.

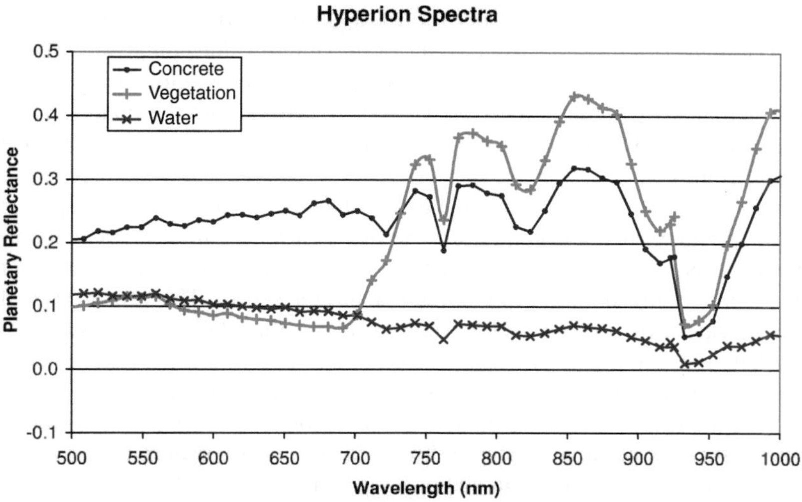

Fig. 6.3 Spectral reflectance from 500–1000 nanometers from concrete, green vegetation, and water derived from Hyperion. Plant pigment absorption in the 500–670 nm range, where concrete is reflecting is responsible for the "green" color of green vegetation. We exploit the difference between the absorbing and reflective portions of this curve to construct the normalized difference vegetation index (from Tucker et al., 2005)

It can be difficult to accurately determine the specific chlorophyll concentration in heterogeneous plant environment, because the various plant species can have different leaf optical properties and/or different canopy architectures. This is not a limitation of remote sensing because the important plant canopy process in question is photosynthesis, and this can be measured or approximated by determining the intercepted or absorbed fraction of photosynthetically active radiation, the energy which drives photosynthesis (Monteith, 1977). The fact that multiple plant species are responsible for the amount of photosynthesis occurring does have implications for using the index to infer crop health, as the absorption of light may be due to non-crop species. The utility of spectral vegetation indices is that they are linearly related to the absorbed photosynthetically active radiation for a wide range of plant species (Calera et al., 2004; Kodani et al., 2002; Trenholm et al., 1999). These relationships are scalable and thus approximation of absorbed radiation globally through time is possible at a variety of scales (Tucker and Brown, 2008). It is the scalability of the NDVI that makes it the foundation for many models and global biophysical research because light interactions at the leaf scale can be calculated and applied at the landscape scale.

6.1.2 NDVI Composites and Time Series Construction

Global NDVI datasets are constructed by taking the maximum value for all days in the period for every pixel. This forms a composite of many days in the same image and minimizes the effect of clouds, dust, and other aerosols and maximizes the signal (Holben, 1986). A long time series of maximum value composite images is available from the AVHRR sensor from July 1981 are formed globally by continent with a ten and fifteen day time step. Data were formed into the images, from the first day of the month through the 15th day and from day 16 to the end of each month for all continents.

Maximum value compositing was used to simultaneously minimize atmospheric and directional reflectance effects (Holben and Fraser, 1984). Atmospheric contamination tends to reduce NDVI values because clouds, aerosols, soot, dust and other constituents reduces the amount of light that reaches the sensor. This affects the red channel to a greater extent than the NIR bands, affecting the NDVI value negatively. When ten daily measurements in a ten day period are used in a composite, it tends to be the day with the least amount of atmospheric interference between the sensor and the ground that has the maximum NDVI value (Los et al., 2000). If it is cloudy for all ten days over a particular location however, either an unnaturally depressed vegetation measurement will be obtained or none at all. Figure 6.2 shows a ten day composite created from AVHRR data. Along the Guinea Coast, where it is raining during October, the NDVI measurements are lower than they should be. Without

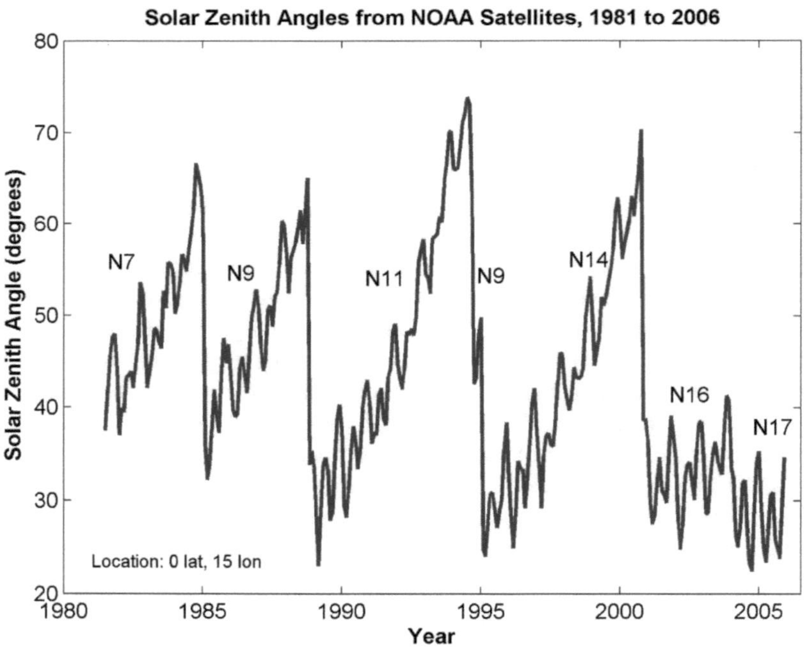

Fig. 6.4 Plot of the changing solar zenith angle, or the angle between the sun and the surface through time from 1981 to 2006 for one pixel at the equator due to orbital drift and change from one sensor to another. The changes in angle reflect changes in time of the image being captured from 1pm to 6pm. If these variations are not removed they result in trends through time in the vegetation data

improved cloud detection and removal algorithms, NDVI data is still not very useful in these highly cloudy locations, even after 25+ years of work to remove these systematic clouds. The lack of consistently accurate atmospheric water vapor fields globally in the earlier part of the AVHRR record prevents the application of explicit atmospheric and directional reflectance corrections.

Another problem with the AVHRR data is the gradual degradation of the NOAA satellites' orbit through time (Cracknell, 1997). This orbital drift significantly affects sun–target–sensor geometry and introduces a trend in the NDVI that varies with latitude, green leaf density, and vegetation structure (Kaufmann et al., 2000) (Fig. 6.4). The trend is due to the change in overpass time of the satellite from early to very late in the afternoon during the life of the sensor. This change in illumination (or, as it is called, solar zenith angle) of the vegetation tends to contaminate the time series with a trend. The GIMMS group performed a solar zenith angle correction to the NDVI data using the adaptive empirical mode decomposition method of Pinzon et al. (2005). The empirical mode decomposition method accounts for local variation in solar zenith angle and embedded nonlinear and non-stationary variations. The GIMMS group post-processed data has been able to account for <90% of the solar zenith angle affect upon the NDVI at all latitudes, remove these effects from the NDVI data, and reconstruct the NDVI data without the solar zenith angle variation. The solar zenith angle corrections were greatest in the tropics for tropical forests, moderate in the tropics for less densely vegetated areas, and lowest at higher northern and lower southern latitudes (Pinzon et al., 2005).

6.2 NDVI Time Series Data for Agriculture Monitoring

The coarse resolution NDVI data from the sensors shown in Fig. 6.1 are useful for agricultural monitoring when augmented with rainfall data and other products. FEWS NET's partner, NASA's GIMMS group in Greenbelt Maryland produces AVHRR data at ten and fifteen day composites the next day after compositing and provides the data and anomaly images for analysis. Anomalies are images that show how the current period is different than the mean of all previous images from the same period. When the current period NDVI values are above the mean then the image is said to have a 'positive anomaly', and when the current period is below it is said to be negative. Vegetation data is very stable and enables comparisons of one growing season to the historical average.

The vegetation anomaly from various sensors is used in conjunction with other products in a 'convergence of evidence' approach, where multiple independent products are used together to estimate the locations, extent and severity of climate hazards on the ground. Because NDVI measures the activity of the vegetation itself and not the moisture component, it is completely independent of the rainfall measures described in previous chapters. This makes the data a valuable addition to the FEWS NET suite of tools for monitoring climatological hazards, even though it is rather harder to interpret than estimates of variations in yield like the WRSI that can be derived from rainfall products.

Development of SPOT and MODIS vegetation data archives for use in FEWS NET has led to significant improvements in spatial resolution. FEWS NET has an archive of 16 day composites at 250 meter resolution from MODIS for Central America and Haiti, and is developing a similar archive for West Africa. Although MODIS data are very useful, the NDVI is not an operational product and therefore each composite is released six days or more after the composite is obtained. This means that for a January 1–16 composite, the image may not arrive for analysis by FEWS NET personnel until January 22 or later. This delay is difficult for an organization such as FEWS NET, who needs much more timely data, and would prefer a shorter, more flexible compositing period. Through collaboration with the US Department of Agriculture's Foreign Agriculture Service (FAS), FEWS NET, the University of Maryland and the NASA GIMMS group will develop a next-day, daily NDVI dataset derived from MODIS, to be put in place by 2009. The Global Agriculture Monitoring (GLAM) Project will produce these new 500 m products, which will greatly enhance the ability for FEWS NET to view changes in vegetation since 2000. The project's objective is to enhance the agricultural monitoring and the crop production estimation capabilities of the USDA using the new generation of NASA satellite observations. Coupled with research that aims to integrate the long-term record from AVHRR with the highly accurate MODIS data, FEWS NET will have much improved data available for monitoring hazards due to variations in climate.

6.3 Comparison Between NDVI Datasets

FEWS NET uses NDVI data from three sensors routinely in its work: global products from AVHRR on the NOAA satellites, SPOT Vegetation (VGT), and MODIS NDVI products. Here, we will compare these datasets to SeaWIFS land product and high resolution NDVI data calculated from Landsat 7's Enhanced Thematic Mapper plus (ETM+). Although FEWS NET does not use the SeaWIFS product, it is an extremely well calibrated sensor with great potential for use in early warning activities. Table 6.1 summarizes the details of these datasets.

For this sensor intercomparison, we examine eight areas where droughts were detected between 2000 and 2004, 13 NASA Earth Observing System validation sites, as well the global distribution in differences between the four sensors. Details on the location of these sites can be found in Brown et al. (2006). For the past decade, newer and more sophisticated sensors are becoming operational providing biophysical measurements that are aimed at addressing various global change related questions. NASA's MODIS sensors on-board Terra and Aqua satellites are providing a series of advanced remote-sensing-based land products (Huete et al., 2002; Kidwell, 2000). However, to achieve any meaningful monitoring of the land surface vegetation, stable, intercalibrated long term vegetation records (a decade or longer) are a key requirement (Vermote et al., 1997). Efforts for using data from MODIS and other sensors with the historic AVHRR vegetation NDVI records are proving to be

Table 6.1 Characteristics of the NDVI datasets presented in this chapter Sensors and the data source, spatial and temporal resolutions, equatorial crossing time, and field of view for each sensor

Sensor	AVHRR	SPOT Vegetation	MODIS	SeaWiFS	LandSAT ETM+
Data Source	GIMMS NDVIg Operational Dataset	FAS-GIMMS, VITO	MODIS-Land and Vermote/ Saleous	SeaWiFS/ GSFC/ GIMMS	Corrected scenes from EOS web
Spatial Resolution	8000 m and 1 degree	1000 m and 1 degree	500 m, 5000 m, 1 degree	4633 m, 1 degree	30 m
Temporal Resolution	15 day and monthly	10 day and monthly	16-day and monthly	monthly	1999–2001
Equatorial Crossing	~9 AM – 6 PM	10.30 AM	10.30 AM	12.05 PM	10:00 AM
Field of View (FOV)	±55.4°	±101°	±55°	±58.3°	±15.4°

challenging. Using the standard products instead of modeled simulations, we were able to intercompare various datasets with the historic AVHRR NDVI record. This analysis revealed that, although relatively large differences existed between the four NDVI datasets, the NDVI anomalies exhibited similar variances.

Composited NDVI images are fairly robust, which can be seen when comparing time series with NDVI from Landsat ETM+ images that have been corrected for atmospheric effects. The AVHRR NDVI dataset provides a bridge between the historical record and the modern satellites allowing an extension of their relatively short records, assisting global change researchers who use vegetation data (IGBP, 1992; Running, 1990). Figure 6.5 shows the time series from AVHRR, SPOT-VGT, and two MODIS datasets for the Bondville, Illinois site (Morisette et al., 2002). This plot shows that all four sensors are able to capture the annual green-up and brown-down of the vegetation to a similar degree.

The similarity of the series is very encouraging and is driving plans to connect the much longer AVHRR record to the other sensors, and eventually to the Visible/Infrared Imager/Radiometer Suite (VIIRS) sensors aboard the National Polar-Orbiting Operational Environmental Satellite System (NPOESS) Preparatory Project (NPP). The VIIRS data will eventually replace the AVHRR sensor as the US national meteorological weather satellite. Data continuity in the vegetation record is central to users' ability to measure the impact of global environmental change on terrestrial ecosystems.

FEWS NET focuses mostly on variations around the mean conditions, or the anomaly of each product. Eight locations where droughts were detected from 2000 to 2004 were examined, and found that the periods of drought were reasonably captured by all four datasets. Figure 6.6 shows the NDVI anomaly from three locations: Louga, Senegal experienced a dry period during the summer of 2002 (FAS, 2002). East Longreach, Australia during the period from 2001–2002 (Beard and Trewin, 2002), and Saskatchewan, Canada during the summer of 2002 (Wittrock, 2004). Although Fig. 6.6 shows three locations that show a diversity of variability,

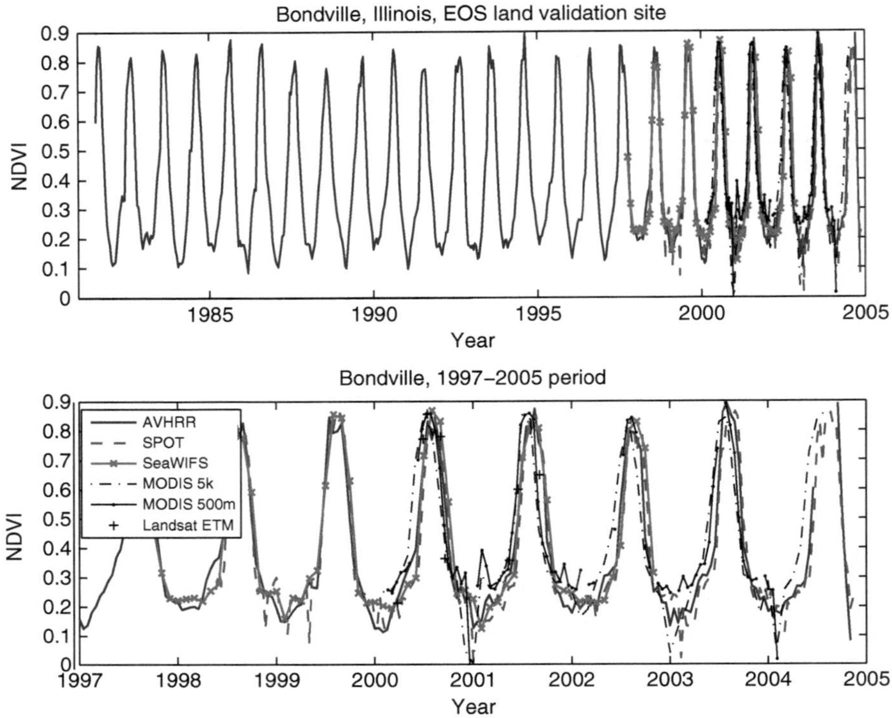

Fig. 6.5 NDVI temporal profiles for the Bondville, Illinois site. 15-day AVHRR, 10-day SPOT VGT, 16-day MODIS 500 m, monthly gridded MODIS 5.6 km, and monthly SeaWiFS data are plotted. The Landsat NDVI averages are also plotted (circles) and were created from atmospherically corrected NDVI subsets around the study sites. All data represent the average NDVI for a 25×25 km window. Top panel highlights the AVHRR 1981 to 2003 record. The bottom panel shows only the overlapping records of all sensors from 1997 to 2004 (Brown et al., 2006). Figure used with permission

the large positive and negative anomalous vegetation periods could be seen in all four sensors.

Figure 6.6 also shows that the 'modified' anomaly (SPOT-AVHRR), which uses the AVHRR mean with the monthly SPOT data, works very well in Louga and Saskatchewan, and shows the overall decline in precipitation during the period in Longreach. The SPOT and SPOT-AVHRR anomalies are both able to detect anomalous vegetation densities in the eight locations examined. The AVHRR anomaly that uses a 23-year mean is higher than the rest due to the inclusion of a much longer historical record in the mean. However, all four series show an overall negative trend in Longreach beginning in January 2001 and continuing through to 2004, reflecting the dry trend reported in the media.

Although the anomaly using SPOT is presented in Fig. 6.6 that was created by subtracting the SPOT data for the years in question from a mean created by

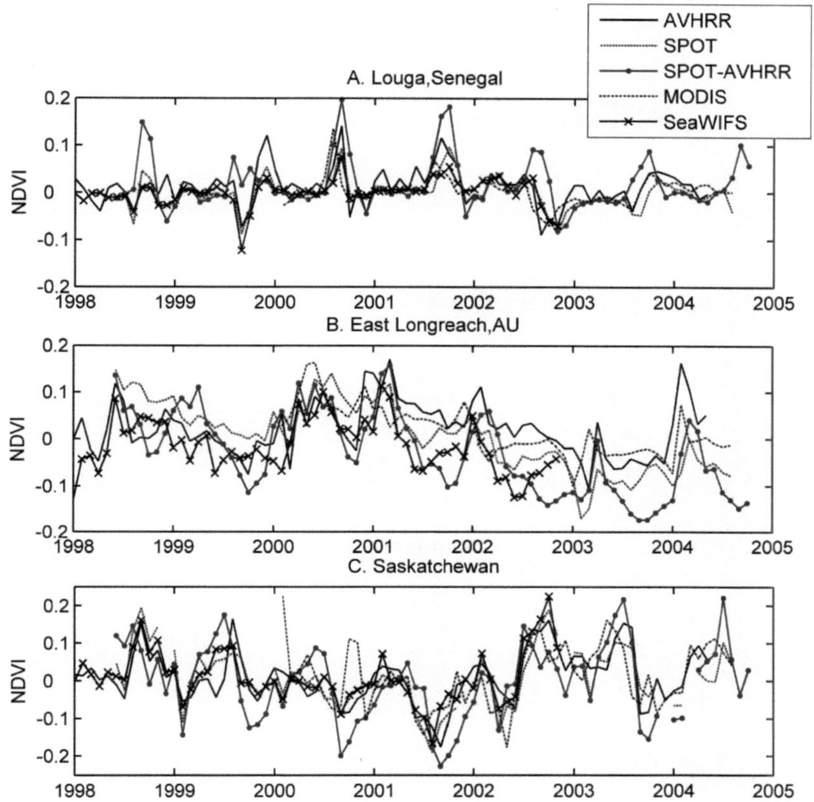

Fig. 6.6 NDVI time series anomalies for three locations that have experienced drought events during the past four years, Louga, Senegal, E.Longreach, Queensland, Australia and Saskatchewan, Canada sites. AVHRR, SPOT VGT, SeaWiFS and MODIS monthly anomalies were created using the mean of all available months subtracted from the actual month. SPOT-AVHRR refers to a SPOT VGT time series anomaly created using a monthly climatology from AVHRR, 1998–2004

averaging the AVHRR data over that location, this approach does not have a lot of stability. Only in the least cloud-contaminated environments can it be done without considerable error. In addition, the variance introduced into the dataset due to the substantial differences between the two datasets may overwhelm the improved characterization of the ground that can be obtained by using 20 more years of data from the AVHRR sensor with the higher resolution SPOT Vegetation data. Thus FEWS NET has, with good reason, a conservative view of innovations in remote sensing and does not use a new product unless it has overwhelming utility or extremely obvious improvements over its predecessors. A mixed-sensor anomaly is not one of these innovations.

For the past decade newer and more sophisticated sensors are becoming operational, providing biophysical measurements that are aimed at addressing various

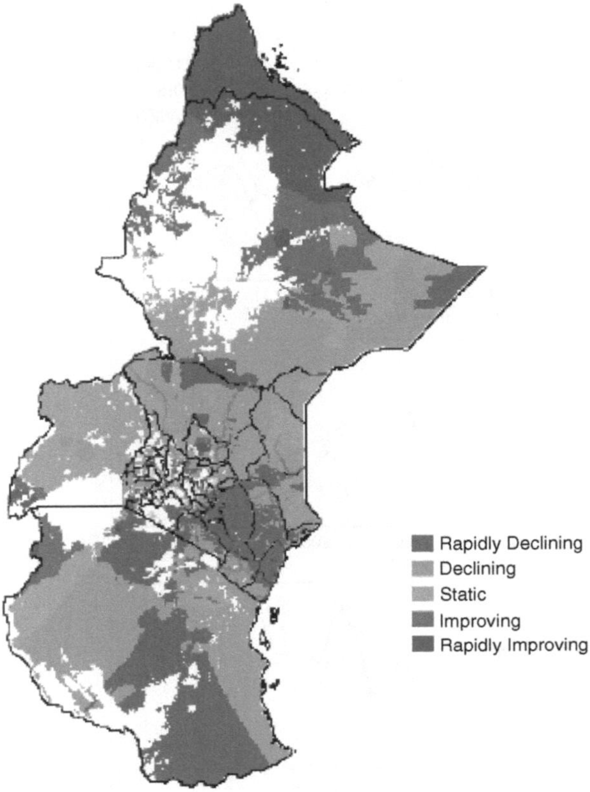

Fig. 6.7 Forage condition map created from NDVI anomaly projections for the horn of Africa

global change related questions. NASA's MODIS sensors on-board Terra and Aqua satellites are providing a series of advanced remote sensing based land products (Justice et al., 1998; MODIS-Web, 2004). FEWS NET is interested in higher resolution vegetation index products that can provide much higher detail, even at the expense of a longer time series. At the moment, three vegetation index products are being provided to analysts, although sometimes they do not agree in their assessment of land surface conditions. The new Global Agriculture Monitoring (GLAM) project will provide a new and much more flexible alternative to any of the current datasets. When this new product is available, new analysis will be conducted to determine its stability and comparability to the other products that FEWS NET uses.

This research suggests that progress can be made toward a unified NDVI dataset given that absolute variances across sensors are relatively similar, especially when seasonality is removed. Benefits of this work include enabling the use of the longer AVHRR time series to calculate the normal trends and any anomalies in combination with other sensors, the successful merging of various data sets from SPOT-VEG,

MODIS, and SeaWiFS, and others quantitatively with the historic 25-year AVHRR NDVI data record, and the potential associated with using multiple NDVI data sources, especially in the event of one dataset's absence (Brown et al., 2006).

6.4 NDVI for Monitoring Forage Conditions

FEWS NET monitors many communities who gain a significant portion of their income from agropastoral and pastoral activities, and NDVI is a primary monitoring tool for pasture (Fig. 6.7). The Livestock Early Warning System (LEWS) is intended to provide an additional 6–8 weeks advanced notice on the current early warning systems in East Africa, and to increase the accuracy of the climatological data for pasture ecosystems. The near complete lack of accurate historical livestock density and distribution statistics in Africa makes developing an effective early warning system very difficult task. In addition, FEWS NET is focused primarily on agricultural systems, and therefore its interventions and monitoring tools are substantially less effective in pastoral regions.

The LEWS project combines predictive and spatial characterization technologies with the formation of a network of collection and measurement sites in East Africa. The system is based on near infrared (NIR) spectral data and fecal profiling technology supported by advanced grazing land and crop models. The foundational technology is comprised of a Geographic Information System data set used by the Spatial Characterisation Tool (Corbett and O'Brien, 1996) which provides spatial analysis of weather, soils, terrain conditions and human and livestock populations. LEWS links several new technologies capable of predicting the current nutritional status of free-ranging animals and the impact of weather on forage supply and crop production among a carefully selected set of households, reflecting a variety of effective environments across diverse landscapes of East Africa. These technologies include the Almanac Characterization Tool, Near Infra Red Spectroscopy (NIRS) fecal profiling technology linked with a livestock nutritional model (NUTBAL PRO) and linkage of satellite weather and NDVI data with point-based biophysical grazing land modeling using the PHYGROW model. GIS co-kriging and kriging techniques are used to extrapolate point-based model output to non-monitored areas. Making these technologies work together in an operational environment is very challenging, and LEWS has never quite attained the status of an operational product in the regions that FEWS NET works in.

FEWS NET uses LEWS data in regions where pastoral systems are dominant, but it never uses them alone or without additional information. Determining the accuracy of the LEWS projections is difficult because there are virtually no reliable statistics in the region on livestock stocking levels. Organizations continue to invest in improved ways of monitoring both pastoral areas and pastoralist communities to improve intervention, response and ongoing monitoring techniques. FEWS NET uses LEWS products in conjunction with data on the condition of pastures from straight NDVI images, from WRSI imagery tuned to pastures and from local reports.

6.5 Desert Locust Monitoring

The Desert Locust (Schistocerca gregaria) is an insect whose distribution area extends from West Africa to India. During invasion periods, adults form swarms that can fly or be carried by wind over great distances. These swarms can completely decimate crops located hundreds of kilometers from their places of origin, resulting in a sudden, drastic decline in food production in these regions. The Desert Locust Information Service (DLIS) from the UN Food and Agriculture Organization (FAO) collaborates with National Locust Units to collate, summarize and analyze field data (e.g., vegetation, rainfall, locust and control information) in order to assess the current situation and forecast the scale, timing and location of locust breeding and migration. The warnings, assessments and forecasts produced by DLIS are used by affected countries to plan survey and control operations and by the international donor community to target assistance, especially during emergencies. FEWS NET contributes to the locust control information dissemination and monitoring through its bulletins and networking operations.

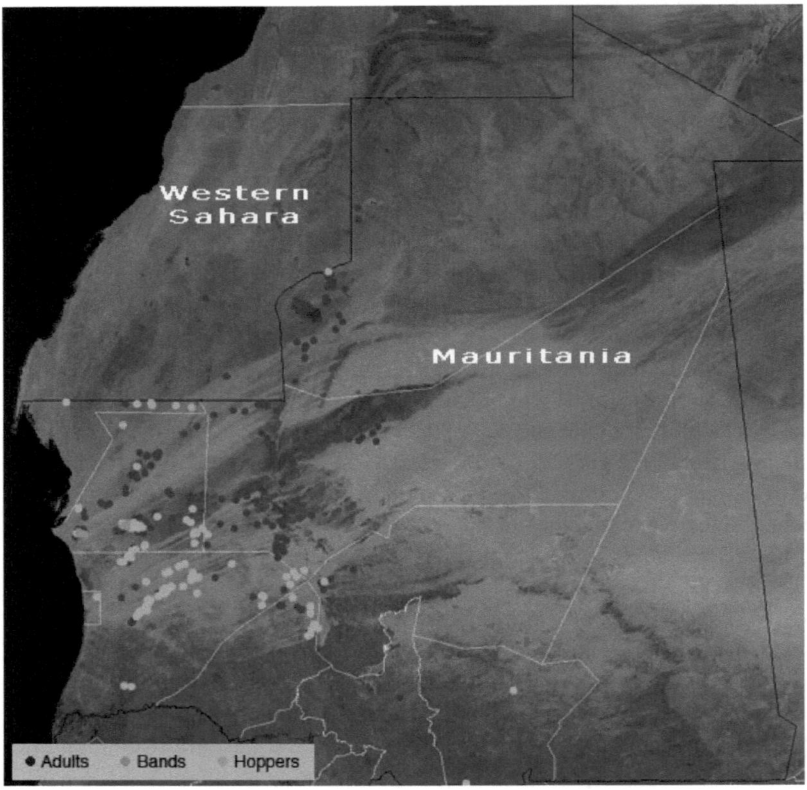

Fig. 6.8 MODIS Vegetation map with locations of locust adults, bands and hoppers in October 2006 from the FAO Desert Locust Information Service

Using maps and analysis products to illustrate recent climate conditions, such as rainfall and vegetation, which provide ideal breeding conditions for the locusts, the DLIS provides information that can be used by FEWS NET to identify and forewarn its partners of impending food security problems. The DLIS uses AVHRR NDVI, SPOT NDVI, RFE and NOAA CMORPH rainfall products to estimate the general location of locust breeding areas. Since 2005, the FAO locust project has been providing locust-affected countries with high resolution (250 m) MODIS satellite imagery to help detect areas of green vegetation. MODIS NDVI is used to guide survey teams to areas where locusts may be present and to reduce the large areas that must be monitored. Higher resolution imagery reduces the occurrence of false positive incidents, that is, where the image suggests that it is dry when in fact there is sufficient vegetation to support the Desert Locusts (Fig. 6.8).

During the past few years, the precision of pesticide applications during locust control operations has improved significantly with the increased reliance on differential geographic positioning systems and track guidance systems used to guide the operator when applying pesticide. Initially, these systems were limited to aircraft but they have recently expanded for use during ground control operations with vehicle-mounted sprayers. During the 2003–2005 upsurge in locust activity corresponding with increased rainfall in the northern desert margin regions of the Sahel, all aircraft contracted by FAO were required to have GPS and track guidance systems. The use of these systems has contributed to a reduction in pesticide wastage and related negative effects on the environment.

6.6 MODIS Snow Product for Afghanistan

Unlike regions in the tropics, Afghanistan has its wet season in the winter, when snow accumulates to become its primary source of water for agriculture during the summer. To measure how much water will be available for growing crops the following year, FEWS NET monitors the rate of snow accumulation and later melting. A new index from MODIS is used to estimate this snow cover. The automated MODIS snow-mapping algorithm uses satellite reflectances in MODIS bands 4 (0.545–0.565 μm) and 6 (1.628–1.652 μm) to calculate the normalized difference snow index (NDSI) (Hall et al., 1995): NDSI $=$ band 4 $-$ band 6 / band 4 $+$ band 6). A pixel in a non-densely forested region will be mapped as snow if the NDSI is ·0.4 and reflectance in MODIS band 2 (0.841–0.876 μm) is $>11\%$. The product is able to distinguish clouds from snow using these reflectance thresholds (Fig. 6.9).

The MODIS snow product uses 500 m MODIS data to define the maximum 8-day snow extent for Afghanistan. Daily snow depth values from the Air Force Weather Agency (AFWA) and NOAA are used to create a maximum 8-day snow depth corresponding to the MODIS composite period (Fig. 6.10). These two data sources are then merged to produce a unique FEWS NET product that identifies snow depth in terms of the snow extent identified using MODIS. The merged product maintains the

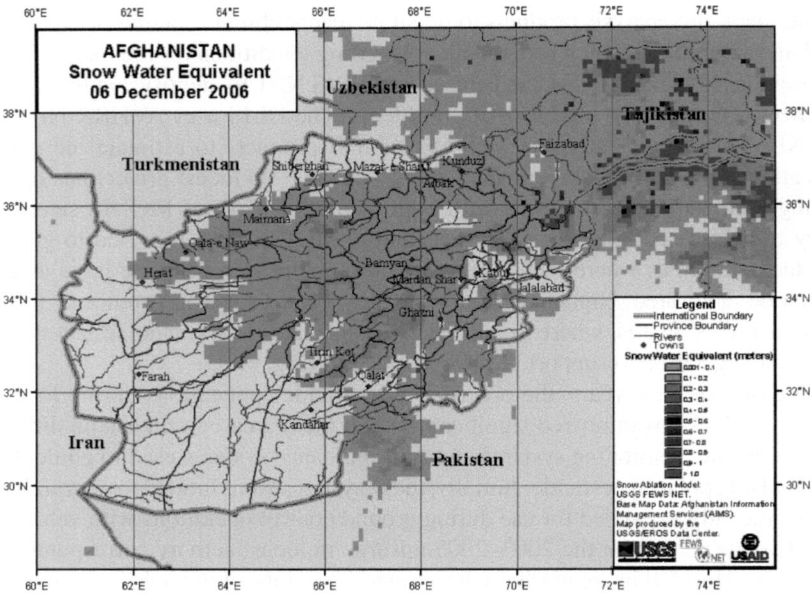

Fig. 6.9 Snow water equivalent image for Afghanistan based on MODIS and the Air Force Weather Agency model

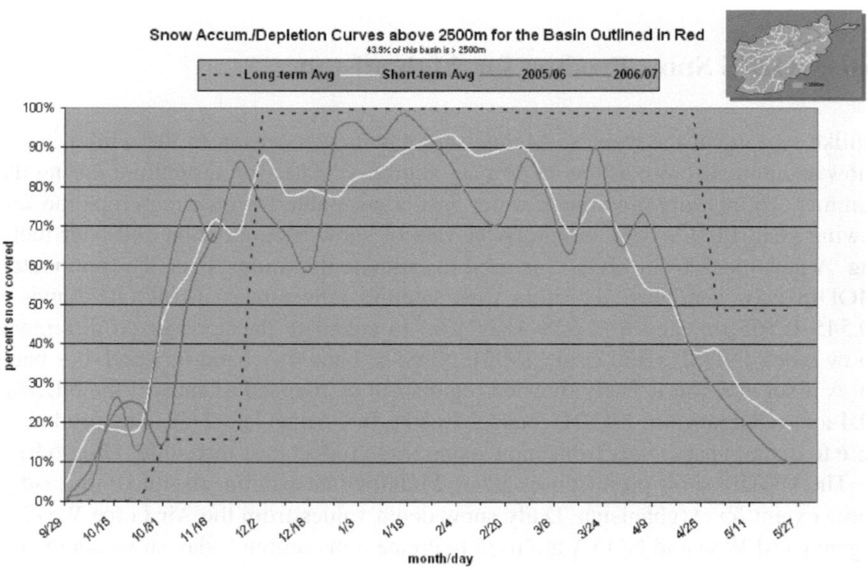

Fig. 6.10 Snow water accumulation/depletion curves for an individual basin

spatial resolution of MODIS and provides snow depth measurements in centimeters. This product helps FEWS NET anticipate water shortages in Afghanistan that will significantly affect food production, and ultimately food security.

6.7 High Resolution Spectral Vegetation Data for Cropped Area Estimates

FEWS NET needs to estimate, as accurately as possible, the amount of food produced in regions at risk. Food production can be measured by multiplying the yield per hectare times the number of hectares in cultivation. Until recently, FEWS NET focused on only estimating the variation in yield, with little attention paid to variations in cultivated area for a particular crop. Year to year variations in yield would be used with previous year's production figures to estimate change in production due to the weather. It has been recognized, however, that large macroeconomic upheaval and political change often result in significant variations in cropped area, which when combined with additional variations due to weather conditions, leave FEWS NET in the dark as to how much food is being produced in a particular region. Thus, FEWS NET has begun to invest in remote sensing systems that can identify fields under cultivation and estimate the amount of area being cropped to determine the level of food production for food security. High resolution visible and near infrared imagery is used by FEWS NET for cropped area estimation.

The 2005 cropped area for Zimbabwe was estimated by FEWS NET partner, the Climate Hazard Group (CHG) at the University of California at Santa Barbara, from Landsat imagery using a dot grid sampling approach. This methodology was developed by Gregory Husak and others at the Climate Hazards Group. The following description was taken from the CHG report on estimating cropped area in Zimbabwe for 2005 (Smith et al., 2006).

The NIR and Red information from Landsat imagery was acquired over the entire country, which required a total of 24 scenes. Cloud free scenes were selected from the harvest period which occurs in the April to May time frame. Although Landsat 5 data were available for that time frame through international stations, Landsat 7 data from after the malfunction of the Scan Line Corrector were used because it was assumed that the sampling method would not be affected by the data gaps and these data were much more cost effective ($250 per scene vs. $810 per scene) than the Landsat 5 data.

SLC-off data has linear gaps in each scene caused by failure of the Scan Line Corrector (SLC), which compensates for the forward motion of the satellite. Without an operating SLC, the Enhanced Thematic Mapper Plus (ETM+) line of sight now traces a zig-zag pattern along the satellite ground track with a resulting duplication of imaged area that increased toward the scene edge. Because the cropped area technique does not require a continuous scene, these data are acceptable for the application. The 2000 GeoCover data were also acquired for the entire country as a supplemental data source to help clear up any questions that may arise due to the data gaps. A few selected ASTER scenes were also available as supplemental data to provide some assessment of potential scale issues.

The sampling points were generated as a regular grid spaced 2 km apart. This created a total of 97,736 points across the country. Of these 16,325 were lost due to the SLC-off data gaps and an additional 748 points were lost due to cloud cover. This left 80663 points which were visually interpreted as either planted crop area or not. Tests were completed to verify that these dropped points would not bias the results. Regularly gridded points spaced 1 km apart were interpreted for one SLC-on Landsat scene. The proportion of planted area was calculated. This set of interpreted points was randomly placed over an actual binary map of the SLC-off data gaps for the 2005 data being used in Zimbabwe. Points falling within the SLC-off gaps were removed and the proportion of cropped points recalculated. This was repeated 10 times to test the variance of proportions related to the spatial variance of the number of points being dropped. This test was repeated using a 2 km spacing of interpreted points derived from the original classified dataset.

The actual interpretation and recording of the sample attributes was greatly aided by a tool called LCMapper (under development) at USGS/EROS by M. Cushing which essentially simplifies the process to a drag and click technique. Because of the large number of points needing to be interpreted the country was divided up into the 3 regions and the task was split among 3 interpreters. The points were split by connected regions rather than randomly because interpretation becomes increasingly more difficult at scene edges where the SLC-off gaps are larger, having overlapping scenes leads to much better interpretation.

All three image interpreters were given expert image interpretation training on the Zimbabwe Landsat data to ensure that all were seeing the same thing in the imagery. The visual interpretation considers color, shape, size, pattern, texture and context as opposed to a spectral signal which can be quite variable. One scene was selected to be interpreted by all 3 to get an assessment of any personal biases between interpreters. This initial test scene underwent comparison, and review and led to additional training and agreement for difficult areas. This scene was reinterpreted by all for another comparison of potential bias. After each interpreter had completed their region, the datasets were assembled for the country. Overlapping areas of disagreement were resolved by reinterpretation by the compiler who had access to all of the available imagery.

Geo-referenced ground photos taken during the USDA Crop Assessment Tour were used to validate and/or modify the classified samples. After closely relating the ground photos to what the imagery was showing a recheck of the entire country classification was done by a single interpreter (Fig. 6.11). This step was important to remove any biases due to different interpreters and also to make sure that the entire classification was in line with the only ground truth available.

The results were compared with the 1992 Landsat derived crop classification from South Africa's Council on Scientific and Industrial Research (CSIR) Regional Land Cover Database and the spatial patterns matched quite well. The total cultivated area for 2005 was calculated for each district from the proportion of cultivated vs. non-cultivated samples. The percentage of samples that were classified as planted was 33.22%. The total area planted in Zimbabwe, as calculated from the 2005 classification by applying this percentage to the total country area, is 12,919,710 hectares. The planted area was also calculated per district from the 1992 classification for comparative purposes, these data produced a figure of 10,730,616 hectares for the country. It should be noted that 1992 was one of the worst drought years on record.

Cropped area estimation from Landsat data using the technique described above results in an overestimated amount of land in cultivation. This is due to the large amount of mixed pixels and that many fields are smaller than the 30 m pixels available from Landsat data, thus if a pixels is considered to be cultivated, the entire 30×30 m area is considered cultivated although in reality only a portion of the pixel may be a field. To rectify this, FEWS NET and its UCSB CHG partners are using 4 m IKONOS and Quickbird imagery for select areas. These high resolution images are used to identify the exact extent of fields in a sampling scheme throughout the country and then correct the Landsat evaluation statistics with a scaling value. With this high resolution analysis, FEWS NET hopes to get at a measured, accurate and substantiated cropped area estimation which can be used to achieve consensus in the humanitarian community regarding the amount of grain produced in any particular year. This would be a huge step forward, because until now, estimates of food production in subsistence agricultural areas varied by nearly two fold depending on the source. Obtaining a repeatable and statistically valid estimation that can be verified would improve humanitarian action by both improving the identification of the location of reduced production, and by improving the estimates of overall need.

Fig. 6.11 Validation of classification with ground photos (from Smith et al., 2006). Thanks to Greg Husak of UCSB for the imagery

6.8 Summary

This chapter has summarized vegetation indices, what they are, how they are measured and has presented several uses of the information in the context of FEWS NET. The biophysical basis of vegetation indices is reviewed and the scalability of the measurement discussed. Low and moderate resolution imagery from AVHRR and MODIS sensors were presented, along with examples and discussion of how images from multiple periods are constructed to create a time series, enabling anomaly images to be made that enable analysis. Anomalies of vegetation images are used in conjunction with rainfall imagery to monitor the progress of the growing season during operational hazard monitoring by FEWS NET.

Two examples of specific uses of high resolution imagery were presented in Sects. 6.4 and 6.5 of this chapter. The first involves detection of small regions of vegetation in desert valleys for desert locust identification and eradication to prevent food insecurity due to swarms. The second example involves extremely high resolution 1 and 2 m data from IKONOS and Quickbird, which when used in conjunction with 30 m data from Landsat, can provide highly precise measurements of the area of cultivation. FEWS NET is interested in determining the amount of food produced in order to estimate more precisely the amount of food aid needed in a particular country.

References

Beard, G. and Trewin, B., 2002. Drought Statement – Issued 1st November 2002: Rainfall deficiencies worsen following dry October National Climate Centre, Bureau of Meteorology, Government of Australia, Melbourne, Victoria, Australia.

Brown, M.E., Pinzon, J.E., Didan, K., Morisette, J.T. and Tucker, C.J., 2006. Evaluation of the consistency of long-term NDVI time series derived from AVHRR, SPOT-Vegetation, SeaWIFS, MODIS and LandSAT ETM+. IEEE Transactions Geoscience and Remote Sensing, 44(1): 1787–1793.

Calera, A., Gonzalez-Piqueras, J. and Melia, J., 2004. Monitoring barley and corn growth from remote sensing data at field scale. International Journal Of Remote Sensing, 25(1): 97–109.

Corbett, J.D. and O'Brien, R.F., 1996. The spatial characterization tool—Africa v1. 0, Blackland Research Center Report.

Cracknell, A.P., 1997. The Advanced Very High Resolution Radiometer. Taylor and Francis, London, 534pp.

Cracknell, A.P., 2001. The exciting and totally unanticipated success of the AVHRR in applications for which it was never intended. Advanced Space Research, 28(1): 233–240.

D'Souza, G., Belward, A.S. and Malingreau, J.-P., 1996. Advances in the Use of NOAA AVHRR Data for Land Applications. Kluwer Academic Publishers, Dordrecht, Netherlands.

DeFries, R.S. and Belward, A.S., 2000. Global and regional land cover characterization from satellite data: and introduction to the special issue. International Journal of Remote Sensing, 21(6–7): 1083–1092.

FAS, 2002. Senegal: Drought Causes Worries in Peanut Producing Region, Production Estimates and Crop Assessment Division, Foreign Agricultural Service, USDA, Washington DC.

Gray, T.I. and McCrary, D.G., 1981. Meteorological Satellite data – A tool to describe the health of the world's Agriculture. EW-NI-04042, Johnson Space Center, Houston, Texas.

Hall, D.K., Riggs, G.A. and Salomonson, V.V., 1995. Development of methods for mapping global snow cover using Moderate Resolution Imaging Spectroradiometer (MODIS) data. Remote Sensing of Environment, 54: 127–140.

Holben, B., 1986. Characteristics of maximum-value composite images from temporal AVHRR data. International Journal of Remote Sensing, 7(11): 1417–1434.

Holben, B.N. and Fraser, R.S., 1984. Red and near infrared sensor response to off-nadir viewing. International Journal of Remote Sensing, 5: 160–166.

Huete, A., Didan, K., Miura, T., Rodriguez, E.P., Gao, X. and Ferreira, L.G., 2002. Overview of the radiometric and biophysical performance of the MODIS vegetation indices. Remote Sensing of Environment, 83(1–2): 195–213.

Huete, A.R., Huemmrich, K.F., Miura, T., Xiao, X., Didan, K., van Leeuwen, W., Hall, F. and Tucker, C.J., 2006. Vegetation Index greenness global data set. College Park, MD.

IGBP, 1992. The international geosphere-biosphere programme: A study of global change, improved global data for land applications. IGPB Report Number 20, IGBP Secretariat, Stockholm, Sweden.

Jacquemoud, S. and Baret, F., 1990. Prospect – A model of leaf optifcal properties spectra. Remote Sensing of Environment, 75–91.

Justice, C.O., Vermote, E., Townshend, J.R.G., Defries, R., Roy, D.P., Hall, D.K., Salomonson, V.V., Privette, J.L., Riggs, G., Strahler, A., Lucht, W., Myneni, R.B., Knyazikhin, Y., Running, S.W., Nemani, R.R., Wan, Z.M., Huete, A.R., van Leeuwen, W., Wolfe, R.E., Giglio, L., Muller, J.P., Lewis, P. and Barnsley, M.J., 1998. The moderate resolution imaging spectroradiometer (MODIS): Land remote sensing for global change research. IEEE Transactions on Geoscience and Remote Sensing, 36(4): 1228–1249.

Kaufmann, R.K., Zhou, L.M., Knyazikhin, Y., Shabanov, N.V., Myneni, R.B. and Tucker, C.J., 2000. Effect of orbital drift and sensor changes on the time series of AVHRR vegetation index data. IEEE Transactions on Geoscience and Remote Sensing, 38(6): 2584–2597.

Kidwell, K.B., 2000. NOAA KLM Users' Guide. National Oceanic and Atmospheric Administration.

Kodani, E., Awaya, Y., Tanaka, K. and Matsumura, N., 2002. Seasonal patterns of canopy structure, biochemistry and spectral reflectance in a broad-leaved deciduous Fagus crenata canopy. Forest Ecology And Management, 167(1–3): 233–249.

Los, S.O., Collatz, G.J., Sellers, P.J., Malmstrom, C.M., Pollack, N.H., DeFries, R.S., Bounoua, L., Parris, M.T., Tucker, C.J. and Dazlich, D.A., 2000. A global 9-yr biophysical land surface dataset from NOAA AVHRR data. Journal of Hydrometeorology, 1(2): 183–199.

MODIS-Web, 2004. MODIS Documentation and Data Source. NASA Goddard Space Flight Center.

Monteith, J.L., 1977. Climate and the efficiency of crop production in Britain. Philosophical Transactions of the Royal Society B, 271: 277–294.

Morisette, J.T., Privette, J.L. and Justice, C.O., 2002. A framework for the validation of MODIS land products. Remote Sensing of Environment, 83(1–2): 77–96.

Myneni, R.B., Hall, F.G., Sellers, P.J. and Marshak, A.L., 1995. The interpretation of spectral vegetation indexes. IEEE Transactions Geoscience and Remote Sensing, 33(2): 481–486.

Pinzon, J., Brown, M.E. and Tucker, C.J., 2005. Satellite time series correction of orbital drift artifacts using empirical mode decomposition. In: N. Huang and S.S.P. Shen (Editors), Hilbert-Huang Transform: Introduction and Applications, World Scientific, Singapore, pp. 167–186.

Running, S.W., 1990. Estimating terrestrial primary productivity by combining remote sensing and ecosystem simulation. In: H. Mooney and R. Hobbs (Editors), Remote Sensing of Biosphere Functioning. Springer-Verlag, pp. 65–86.

Smith, J., Tappan, G.G., Husak, G. and Crane, M., 2006. Zimbabwe 2005 Remotely Sensed Cropped Area Assessment. University of California, Santa Barbara.

Townshend, J.R.G., 1994. Global data sets for land applications from the advanced very high resolution radiometer: An introduction. International Journal of Remote Sensing, 15(17): 3319–3332.

Townshend, J.R.G. and Tucker, C.J., 1981. Utility of AVHRR NOAA-6 and -7 for vegetation mapping. Remote Sensing of Environment, 8: 127–150.

Trenholm, L.E., Carrow, R.N. and Duncan, R.R., 1999. Relationship of multispectral radiometry data to qualitative data in turfgrass research. Crop Science, 39(3): 763–769.

Tucker, C.J., 1977. Use of Near Infrared/Red Radiance Ratios for Estimating Vegetation Biomass and Physiological Status. X-923-77-183, NASA Goddard Space Flight Center, Greenbelt.

Tucker, C.J., 1979. Red and photographic infrared linear combinations for monitoring vegetation. Remote Sensing of Environment, 8: 127–150.

Tucker, C.J., 1980. Remote sensing of leaf water content in the near infrared. Remote Sensing of Environment, 10: 23–32.

Tucker, C.J. and Brown, M.E., 2008. Red and photographic infrared linear combinations For monitoring vegetation: An update. International Journal of Remote Sensing, in revision.

Tucker, C.J. and Garratt, M.W., 1976. 3-Dimensional chlorophyll concentrations in a high biomass blue grama canopy. Journal of Range Management, 29(2): 170–171.

Tucker, C.J., Pinzon, J.E., Brown, M.E., Slayback, D., Pak, E.W., Mahoney, R., Vermote, E. and Saleous, N., 2005. An Extended AVHRR 8-km NDVI Data Set Compatible with MODIS and SPOT Vegetation NDVI Data. International Journal of Remote Sensing, 26: 4485–4498.

Vermote, E.F., Tanre, D., Deuze, J.L., Herman, M. and Morcrette, J.-J., 1997. Second simulation of the satellite signal in the solar spectrum, 6S: An overview. IEEE Transactions on Geoscience and Remote Sensing, 35(3): 675–686.

Wittrock, V., 2004. Preliminary Description of the 2001 Drought in Saskatchewan. SRC Publication No. 11501-1E02, Saskatchewan Research Council, Saskatoon, SK, Canada.

Chapter 7
Climate Forecasts

In order to provide earlier early warning that can effectively be acted upon, FEWS NET must do more than simply monitor current conditions and report on growing season progress. FEWS NET has worked with NOAA's Climate Prediction Center and with its other partners to develop a suite of products that look ahead to future changes. The Weather Hazards Assessments integrate existing current imagery, official medium (3–5 day), extended (6–10 day) and long-range (three month) forecasts in outlooks that use state-of-the-art science and technology in their formulation. To increase the forecast utility for analysis, FEWS NET partners have also developed tools that enable the interpretation of these climate forecast probabilities into probable observed rainfall amounts.

Dynamical forecasts are complemented with one to four month statistical projections of rainfall and vegetation data based on current observed conditions. Projections are much simpler to couple with observations than probabilistic climate model output, and thus will have tremendous utility for FEWS NET's applications. Although there is always uncertainty associated with predictions of biophysical parameters, they can be very useful in estimating quantitatively future changes. This chapter will describe the various kinds of weather forecasts, statistical projections and seasonal outlooks that FEWS NET uses in its operations.

7.1 Numerical Climate Forecasts

Beyond the current climate observations, forecasts are used as analysis tools to analyze possible future changes in the climate that are likely to have food security implications. FEWS NET uses forecasts at a wide variety of scales. Figure 7.1 shows forecasts at different time steps that are used routinely both in the context of the weekly hazard assessment and during analyses of future conditions by FEWS NET climate experts both in the field and in the United States. Each forecast has its strengths, weaknesses, and usefulness in the regions where FEWS NET works. Some regions in sub-Saharan Africa have only limited predictability in climate forecasts

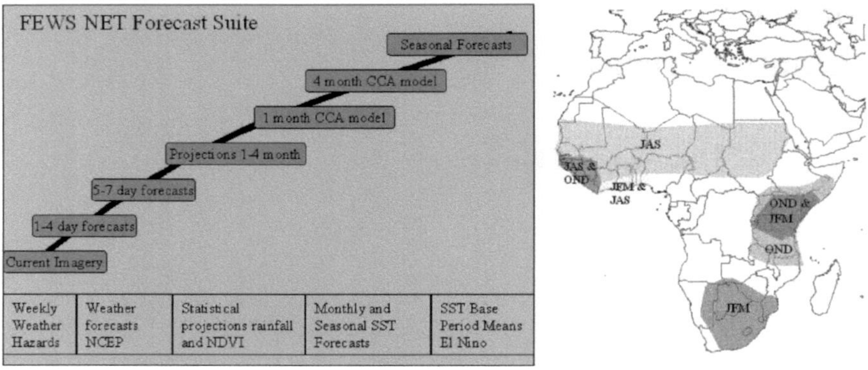

Fig. 7.1 FEWS NET climate forecast products (**left**) and map of Africa regions (**right**) showing areas where seasonal precipitation can be simulated with an acceptable degree of skill based on dynamical model simulations over 1950–1999 period, according to the Columbia University International Research Institute (IRI) (Williams, 2005). The labels indicate seasons when predictability exists: July-August-September (JAS), October-November-December (OND), and January-February-March (JFM)

(Fig. 7.1). The highlands in Ethiopia, for example, are of great interest to FEWS NET because of their extensive food security problems. Unfortunately, climatological forecasts often do not help with monitoring of impending hazards in the mountainous area of the Greater Horn because of its very low predictability in climate models due to their inability to capture orographic rainfall. Additional work is needed to develop new tools that provide improved predictive capability in such regions.

The International Weather and Climate Monitoring Project in NOAA's Climate Prediction Center is an extension of an earlier USAID Famine Early Warning System program which originally covered only sub-Saharan Africa. The project has now grown to encompass all of Africa, Afghanistan, Central America and the Caribbean, the Mekong River Basin, and much of southern Asia. Work is underway to create a global weather and climate monitoring program to address any international region where humanitarian support is needed. The goal of the program is to provide weather and climate related information to users within USAID as well as international partner organizations in regions where FEWS NET is not yet established, so that a greater level of humanitarian assistance may be offered. This goal can only be accomplished through constant interaction with FEWS NET partner groups such as the USGS, NASA, Chemonics, USAID, and local organizations. A more thorough and accurate analysis of conditions is possible via these collaborations.

7.1.1 1–4 and 5–7 Day Forecasts

NCEP Global Forecast System (GFS) is the NOAA's National Weather Service's primary global spectral medium range forecast model, and has a spatial resolution of

0.375° for all of Africa. This numerical weather prediction model is run four times a day and produces forecasts up 16 days in advance, but with decreasing spatial and temporal resolution over time. The model is run in two parts: the first part has a higher resolution and goes out to 180 hours in the future, the second part runs from 180 to 384 hours. The resolution of the model varies in each part of the model: horizontally, it divides the surface of the earth into 0.375° grid squares; vertically, it divides the atmosphere into 64 layers and temporally, it produces a forecast for every 3rd hour for the first 180 hours, after that they are produced for every 12th hour. The GFS is the only national weather forecasting model that is provided free, and is the basis for most of the weather forecasts provided in the United States, particularly those provided by private weather services.

FEWS NET uses the GFS to monitor precipitation 1–4 and 5–7 days into the future. This is particularly useful in the context of the weather hazard discussion, so that the analyst can see if those areas that are receiving adequate rainfall will continue to do so, and those areas that are experiencing dryness might get some relief in the coming week. The GFS model produces hundreds of parameters, some of which have no specific observational analogs. FEWS NET examines only the output from the model for precipitation, minimum and maximum temperature, and the heat index at 2 meter height. The GFS model is much more accurate in the near term (1–4 days) than in later periods (5–7 days), which is why NOAA has two specific time step images instead of one.

Figure 7.2 shows the total 7 day precipitation estimate from the GFS, predicting good rains for the period.

7.1.2 One and Four Month Canonical Correlation Analysis Climate Model

The difference between the forecasts described above and the climate models that can predict one to four months into the future is that the forecast is predicting weather and the Canonical Correlation Analysis model is assessing climate. The most important thing to know for tomorrow's forecast is today's weather (the current temperature, pressure, and winds), in as much detail as possible. The weather forecast uses a wide variety of parameters, some of which have no known observational analog, to forecast whether or not it will rain in the coming week and how much, the temperature, humidity and other parameters. The climate models only say what the average will be, and in far less detail. The temperature of the ocean is the primary input to the model that FEWS NET uses regularly, and it changes only very slowly so that its effects on overall climate can be assessed using historical datasets. Then, when the ocean temperatures are above or below normal, we can say the impact of these changes on the overall amount of rainfall or temperature in a particular region.

The precipitation forecast methodology utilized by NOAA's Climate Prediction Center Africa Desk is a statistical technique entitled Canonical Correlation Analysis

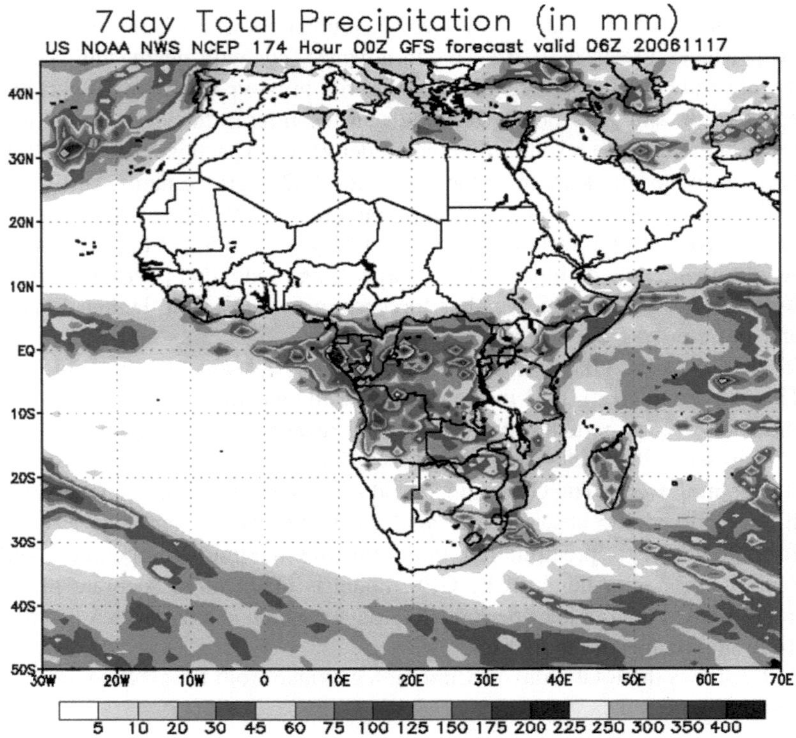

Fig. 7.2 7 day precipitation forecast by the NOAA National Weather Service GFS model

or CCA for short. In this forecast, sea surface temperatures between 40°S and 60°N at a resolution of 10° by 10° latitude-longitude are used to predict rainfall patterns over the area and season of specific interest. The predictions are expressed in terms of tilting of the odds favoring the categories of above, near, or below rainfall. These forecasts are made with a lead time of 4 months, where lead time is defined as the time between the end of the latest predictor or observed sea surface temperatures and the end of the predicted season.

The CCA technique finds the components of the climate system that are specifically related to variations in sea surface temperatures. The prediction design used here is the same as that of the CCA used as one of the tools for operational climate prediction in the U.S. (Barnston et al., 1996), based on earlier work of (Barnett and Preisendorfer, 1987). Four consecutive 3-month predictor periods are followed by a lead time and then a single 3-month predictand, or target, period. Sea surface temperature records from 1955 to 2000 are used as the only predictor in this model. While additional fields such as upper air geopotential height, tropical low-level wind or outgoing longwave radiation might well enhance skill further, data sets of these fields do not extend far enough into the past to satisfy the CCA's need for a long-term (at least 25-year) data record from which to identify the dominant relationships. The predictor and predictand data sets used in the GFS begin in 1955.

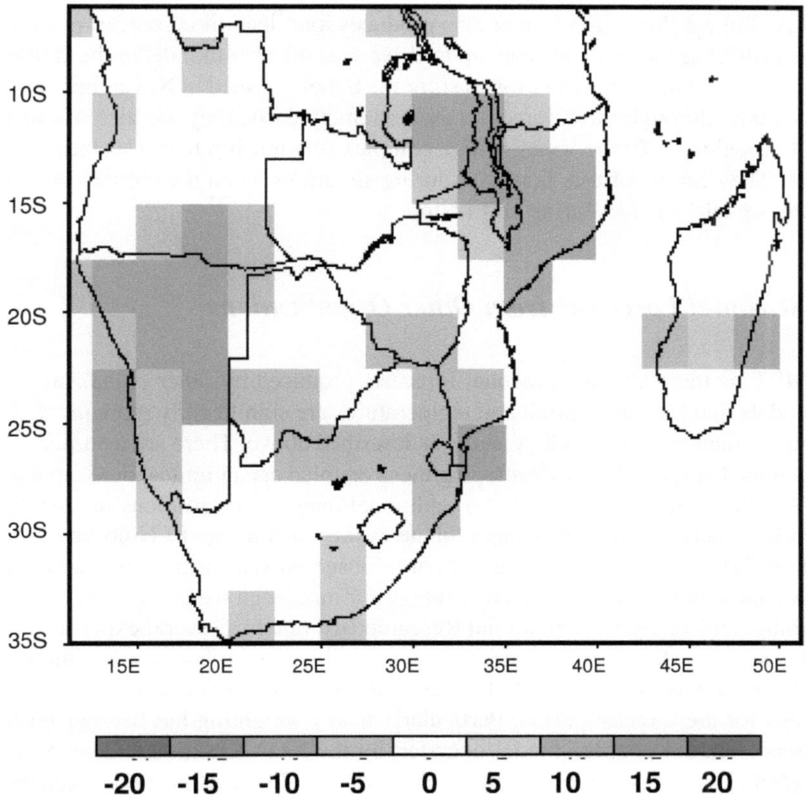

Fig. 7.3 CCA model prediction for Southern Africa one month ahead of the predicted period. The resolution of the model is 10 by 10 degrees

For the 1996 southern Africa rainfall prediction shown in Fig. 7.3, the predictor data are the global SST anomaly field over the four 3-month periods of Nov-Dec-Jan 2006–7. Using data from 1955 to 2004, relationships between the prior year's SST anomaly evolution and the target year's three month seasonal period rainfall anomaly patterns are linearly modeled by the CCA. The predictor SST data for the current forecast are then projected onto the preferred relationships derived from the past years, and a forecast for the target season is developed. Here the lead time is 1 month, as the latest predictor data used are those of period preceding the beginning of the target period by 1 month.

The climatological probability of each of the three categories, above, near, or below average rainfall is 100% divided by 3 or 33%. The probability for the near average category remains fixed at 0.33 because CCA has virtually no skill in predicting this category. The tilting of the odds are presented in units of 5% above or below the average. When forecast skill is low, climatology is suggested and no forecast is made.

NOAA Climate Prediction Center also produces long-lead time forecasts. These forecasts project seasonal conditions up to three seasons or 9 months in the future. For example, the August-October forecast for the Sahel is issued in November in the previous year. Although these forecasts are extremely early, they are also of fairly low skill in regions, like the Sahel, with a complex relationship to sea surface temperatures. They can be of use, however, during situations when the climate system has strong signals, such as during an El Niño.

7.1.3 Seasonal Forecasts from Other Organizations

FEWS NET partners also use seasonal forecasts produced by other organizations. Some models that forecast rainfall and temperatures are significantly more involved and complex than used in the CCA analysis described above. There are a number of organizations that specialize in developing these coupled ocean-atmosphere models, and FEWS NET uses their products for additional insight for variations in climate, particularly in periods without strong climate events such as the El Nino Southern Oscillation. When there is no strong pattern in observed sea surface temperatures, the CCA approach may show only climatology for most regions.

Columbia University's International Research Institute has several experimental climate models for Africa. To extend predictive skill beyond a few months into the future, these models predict future changes in the surface temperature of the tropical oceans for the forecast period. Particularly heavy weighting has been given to predictions from the coupled model operated by the NOAA National Centers for Environmental Prediction, Climate Modeling Branch. Global atmospheric general circulation model (GCM) predictions of the atmospheric response to the present and predicted sea-surface temperature patterns provide additional skill to the rainfall response to these variations. Other sources of information include NASA's Seasonal to Interannual Prediction Project (GSFC-NASA) and also seasonal prediction research at the Center for Ocean-Land Atmosphere (COLA) studies.

The seasonal to interannual model rainfall projections for each period is dependent on the accuracy of the SST predictions (Fig. 7.4). For the tropical Pacific, these predictions can be expected to provide useful information, but there is some uncertainty concerning the evolution of SSTs. Spread (variation) in global SST predictions is a source of uncertainty in the prediction provided. In particular, the forecasts for the tropical Indian and Atlantic oceans have been an important influence on the forecasts over Africa. Even if perfectly accurate SST forecasts were possible, there would still be uncertainty in the climate forecast due to chaotic internal variability of the atmosphere and its response to variations in ocean temperature. These uncertainties are reflected in the probabilities given in the forecast.

The current status of seasonal-to-interannual climate forecasting allows prediction of spatial and temporal averages, and does not fully account for all factors that influence regional and national climate variability. This prediction is relevant only to seasonal time scales and relatively large areas; local variations should be expected, and variations within the 3-month period should also be expected.

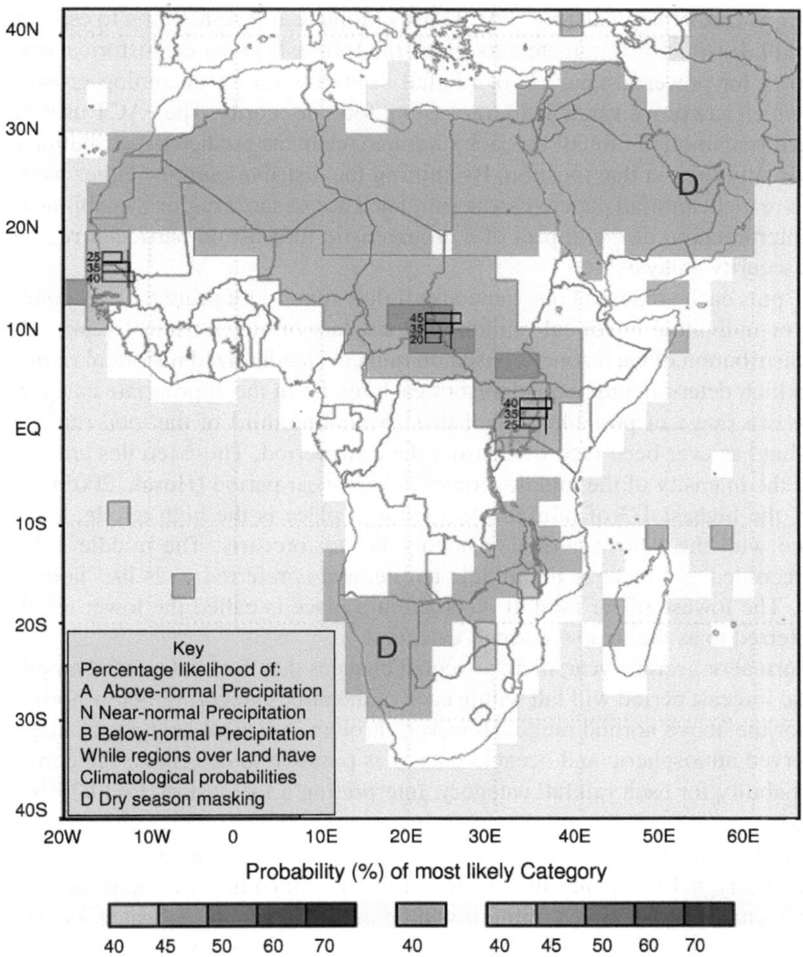

Fig. 7.4 Columbia University forecast four months in advance based on coupled ocean-atmosphere model

7.2 Interpreting Probabilistic Forecasts

Probabilistic seasonal forecasts present the forecast information in terms of the probability of rainfall totals falling within a three months period, as was described above. Interpreting these forecasts can be difficult, as the information is generalized across a very large surface in addition to several months simultaneously. FEWS NET partners at the University of California Santa Barbara have developed a tool called the FEWSNET Agro–Climatological Toolkit, or FACT, that enables the propagation of forecast distributions into actual rainfall values in order to drive its agriculture monitoring products such as the Water Requirement Satisfaction Index (WRSI).

The Agro-Climatological Tool (FACT) uses Monte Carlo resampling to estimate new rainfall distribution parameters to reflect the forecasts given the historical data. This allows for powerful analysis of rainfall scenarios for an upcoming growing season, which can prove valuable in forecasting food insecurity. The FACT uses the observed distribution of rainfall for each station to relate the predicted distribution to what is usually seen in that location. By shifting the distribution to be either wetter or drier, a realistic rainfall pattern can be simulated across large regions, enabling the clearer interpretation of the impact of a probabilistic forecast on particular regions for food security analysis.

FACT puts each area into homogeneous rainfall zones with principal component analysis of individual historical rainfall stations data of at least thirty years. The rainfall distribution of each zone is based on ranked, standardized historical rainfall indices which determine the boundaries of each tercile in the appropriate category. A tercile is a range of possible rainfall that covers one third of the total range of rainfall that has ever been recorded during the time period. These terciles are used to define the intensity of the middle tercile for a 30-year period (Husak, 2005). For example, the highest 1/3 of rainfall recorded at a place is the high tercile, and is associated with the 'above normal' category in the forecasts. The middle 1/3 of rainfall recorded at a place is the middle tercile, and is referred to as the 'normal' category. The lowest 1/3 of rainfall recorded at a place is called the lower tercile, and is referred to as the 'below normal' category.

In a normal or average year, there are equal chances that the total rainfall amount during the forecast period will fall within each of the three categories: below normal, normal, or the above normal range. However, through statistical analysis of historical observed atmospheric and oceanic data, it is possible to determine the current year probability for each rainfall category. Interpreting a forecast of 35/40/25 over the period.

March-May, there is a 35% chance that the total rainfall will be in the above normal category, a 40% chance that the total rainfall will be in the normal category, and a 25% chance that the total rainfall will be in the below normal category. This means that there is a higher probability that total rainfall amount will fall within the normal biased towards the above normal category, and as a result expected rainfall during this period is likely to be slightly higher than average.

Embedded in FACT is the Forecast Interpretation Tool (FIT), which was developed by Dr. Greg Husak. The FIT is a module that warps the climatological cumulative distribution of the rainfall, and creates a new distribution that matches the forecast probabilities presented in the probabilistic seasonal forecast. This distribution warping is achieved by drawing random samples from each of the terciles in the climatological distribution (Fig. 7.5). The number of times samples are drawn from each tercile is proportional to the probabilities presented in the probabilistic forecast. The gamma distribution is then re-fit using the samples. Once the FIT has been applied to a probabilistic forecast, the new probabilities associated with a specific rainfall amount, as well as the new rainfall amounts associated with a specific probability of occurrence can be computed. These new probabilities and new rainfall amounts are conditional on the probabilistic forecast (Husak, 2005).

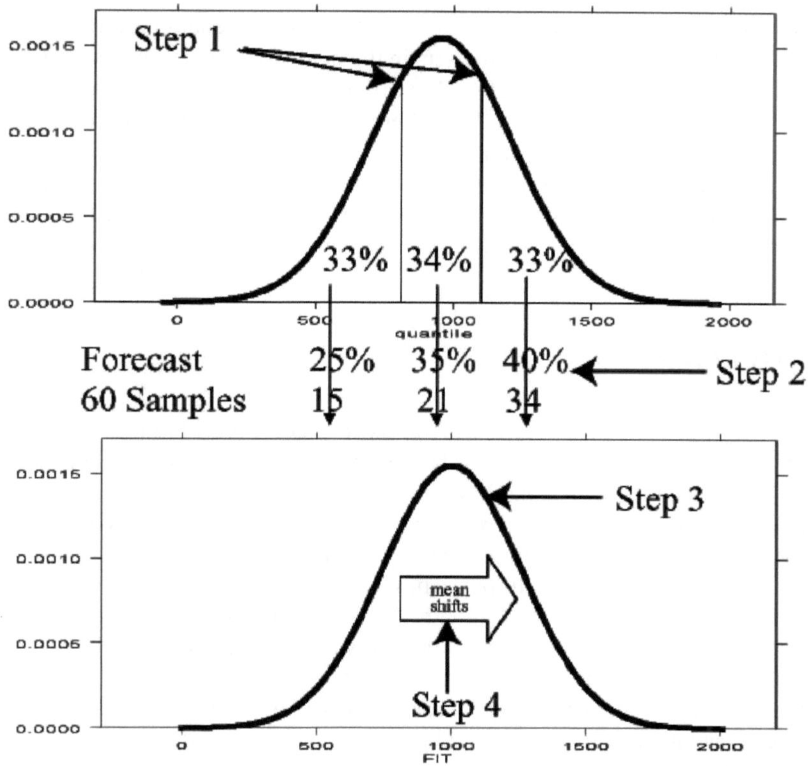

Fig. 7.5 Forecast Interpretation Tool example. Probability distribution functions before (**top panel**) and after (**bottom panel**) the FIT is applied (Husak, 2005)

The probability of meeting the water required by a crop during the forecast period at each point is calculated based on the seasonal forecast. Climatological records were calculated and the climatological probabilities were then compared with the conditional probabilities, so that the increase or the decrease in the chance of obtaining sufficient rainfall to satisfy the maize crop was computed. This entire process was also carried out for 75% of the water required by the crop, so that chances of the crop obtaining 75% of the water it required were analyzed. In this way, the forecast can be translated into probable variations of food production, the parameter of interest for FEWS NET.

7.3 NDVI and Rainfall Projections

FEWS NET partners have recently developed an alternate way of forecasting variations in climate for food security estimation that does not involve sea surface temperatures or climate models. The lagged relationship between rainfall and NDVI can be used to estimate vegetation response to current conditions, enabling forecasts of changes in vegetation to be available to users in the field. We refer to these statistical estimates of NDVI change as 'projections' to differentiate them from climate-based forecasts of NDVI (Anyamba et al., 2002; Verdin et al., 1999). The approach described here is distinct from and compatible with climate-modeling approaches. Using satellite-observed precipitation as a basis for forecasting NDVI, no interpretation is required, as the data is already in the same format and is completely interchangeable with current observations.

7.3.1 NDVI Projections

The objective of the NDVI projection model is to predict NDVI values over relatively short time scales (1–4 months), based on observed rainfall and humidity. The method is based on the idea that if it is raining today, it is likely to either remain green or become greener tomorrow. By using observed precipitation, relative humidity and NDVI values, we can estimate the change in NDVI from month-to-month in response to environmental conditions. These values can then be used to estimate future values of NDVI. A simple 1-month-ahead implementation at quasi-global resolution shows the utility of using a simple statistical model to project NDVI.

The NDVI projections are based on estimating the change in NDVI at a particular time and location, which can be approximated by NDVI growth and loss terms. Since vegetation is mostly water, this can be grossly considered as a balance between precipitation/soil moisture uptake and a transpiration related loss component. NDVI state is important. The growth associated with a given precipitation value is greater when the NDVI is far below its maximum value, and zero when this NDVI value is achieved. Similarly the NDVI loss term associated with low relativity humidity values will be greatest when the NDVI value is highest, and zero when the NDVI value is at its historic minimum. Early in the rainy season, precipitation tends to produce rapid green-up and water absorption. Late in the season, increased leaf area can produce higher rates of evapotranspiration and brown-down or vegetation senescence. Constrained NDVI growth estimates are derived from the log of monthly precipitation values. Constrained NDVI loss estimates are based on a transform of relative humidity. We refer to this latter term as relative humidity demand.

This modeling approach automatically limits the range of the NDVI to the historically observed extremes, and provides an unsophisticated water-balance approach to modeling NDVI response. When NDVI values are low, leaf area and soil moisture values are typically low, enabling the plant/soil system to capture and use a large fraction of the available water. When NDVI values are high, soil moisture and vegetation cover are typically near their maximum values, and a larger fraction of rainfall will be lost from the soil/vegetation system through runoff and evapotranspiration from the land and plant surfaces. When NDVI values are low, both plant leaf area and stomatal conductance are low, reducing moisture loss to atmosphere.

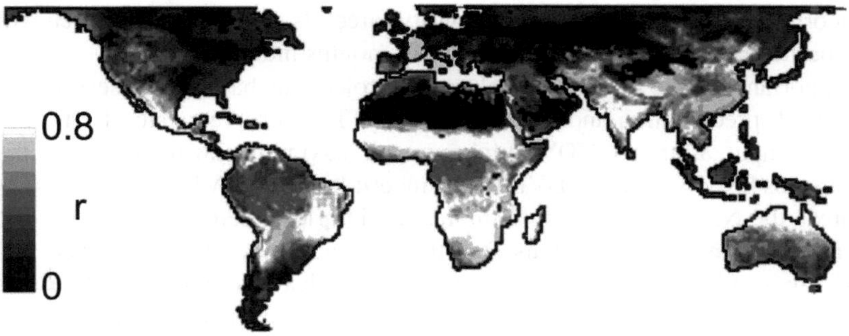

Fig. 7.6 Multiple correlation coefficient from regression of simple model applied quasi-globally

Figure 7.6 shows the shows the simple 1-month ahead forecast model multiple r on a quasi-global scale. The simple model works well across large stretches of the semiarid and semi-humid sub-tropics within ±40°N/S. More details on this model can be found in Funk and Brown, (2006).

NDVI is related to the past several months of rainfall in semi-arid regions (Nicholson et al., 1990; Richard and Poccard, 1998) and NDVI maximum and minimum values are heavily constrained by environmental conditions and can therefore be assumed as stationary for short range forecasting applications. The model, as formulated, is only applicable to semi-arid and semi-humid regions of the tropics. Both Fig. 7.6 and previous research (Potter and Brooks, 1998) suggest a transition between 'cold' and 'warm' regions at about 40° N/S. The semi-arid regions of Africa, Asia and North and South America all stand out in Fig. 7.6 as regions that are controlled by precipitation and humidity. In these regions NDVI is closely coupled to net primary productivity (NPP) and canopy structure (Tucker and Sellers, 1986). The spatial distribution of regions phenologically limited by available water vapor corresponds well with regions showing strong correlations in Fig. 7.6. Alternately, areas with low correlations tend to be either not water limited (tropics) or limited by photoperiod and minimum temperature (mid-latitudes). Thus, in water limited regions precipitation is tightly coupled to inter-annual variability in ecosystem dynamics (Lotsch et al., 2003; Trenberth, 1998), and therefore useful for predicting NDVI.

The equation for the model is:

$$\Delta N_t = b_1(\mathrm{N}_{\max} - \mathrm{N}_{t-1})\ln(1 + \mathrm{P}_{t-1}) - b_2(\mathrm{N}_{t-1} - \mathrm{N}_{\min})(100 - RH_{t-1}) \qquad (7.1)$$

where N = NDVI historical time series, P is precipitation, and RH is the relative humidity for that place in time. The coefficients permit fitting of the model across all observed values for each pixel. The precipitation term is related to green-up of vegetation, so as the precipitation increases the NDVI increases until the maximum historical NDVI value for that location. Likewise the relative humidity term is related to drying of vegetation in humid systems where temperature does not limit growth. Thus as the humidity grows and precipitation drops, the vegetation index gets smaller until it reaches its historical minimum. The model can easily incorporate rainfall and relative humidity forecasts or projections, which will improve performance of the model three to four months into the future.

It is important to note that the model makes projections based on observed values for NDVI, precipitation and relative humidity. The observed state of NDVI is an important input, as are the NDVI minimum and maximum constraints. Since the model depends heavily on the NDVI, it will inherit known NDVI shortcomings. In humid areas the NDVI will tend to saturate, and in arid or semi-arid regions soil, surface moisture and aerosol signals may limit the accuracy of the observed signal. Dark areas in Figure 7.7 shows the regions in Africa where the model has the most skill. Extensions of the model, such as including more environmental parameters like temperature, would likely improve performance, especially in mid-latitude climes.

7.3.2 Rainfall Projections

FEWS NET's team members at the University of California Santa Barbara have been developing methods to project rainfall one to three months into the future in a similar way to NDVI projections. The forecasts are based on matched filter regression based short-lag forecast system. This uses monthly fields of National Centers for Environmental Prediction (NCEP)/ National Center for Atmospheric Research (NCAR) reanalysis data (Kalnay et al., 1996) as the basis of statistical Standardized Precipitation Index (SPI) forecasts. Pacific and Indian Ocean sea surface temperatures (SSTs), Indian Ocean SSTs, and 200 and 500 hPa zonal and meridional winds over eastern and southern Africa are used as inputs (Funk et al., 2007). This statistical framework was used successfully to predict the poor late season rains in southern Africa during the 2002/03 El Niño (Funk et al., 2003), as well as the switch to positive rainfall anomalies during 2003/04. The model is still under development at the time of this writing. The same NASA grant that has funded the implementation of NDVI projections will also enable the development of rainfall projections. The two products will be coupled to further increase the skill of the NDVI projection model.

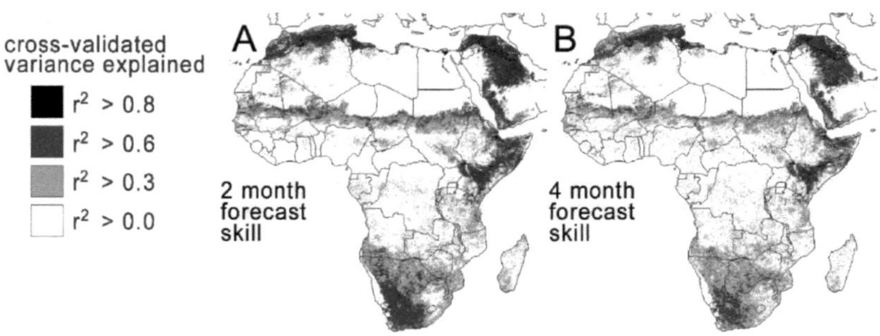

Fig. 7.7 Model skill at 0.1° in Africa from Funk and Brown 2006. Figure used with permission

7.4 Summary

This chapter summarizes the various models and methods that FEWS NET uses to obtain forecasts of climate conditions for use in its monitoring capacity. Numerical weather forecasts for predicting rainfall conditions a week in advance were summarized. To extend these short-range forecasts with those that go out several months, FEWS NET partners have developed models that use canonical correlation analysis to relate variations in sea surface temperature with rainfall in Africa. Global forecasts from the International Research Institute at Columbia University are also used routinely to understand future variations in climate. To help analysts understand these forecasts, a tool was developed that translates forecasts into actual rainfall amounts that may result. Finally, FEWS NET is developing statistical projections of its two most commonly used indices; rainfall and vegetation data. These projections are based on models that use observed conditions to estimate future variations due to persistence in the atmosphere.

References

Anyamba, A., Linthicum, K.J., Mahoney, R., Tucker, C.J. and Kelley, P.W., 2002. Mapping potential risk of rift valley fever outbreaks in african savannas using vegetation time series data. Photogrammetric Engineering and Remote Sensing, 68: 137–145.

Barnett, T.P. and Preisendorfer, R., 1987. Origins and levels of monthly and seasonal forecast skill for united states surface air temperatures determined by canonical correlation analysis. Monthly Weather Review, 115: 1825–1850.

Barnston, A.G., Thiaw, W.M. and Kumar, V., 1996. Long-lead forecasts of seasonal precipitation in africa using cca. Weather Forecasting, 11: 506–520.

Funk, C.C. and Brown, M.E. 2006. Intra-seasonal NDVI change projections in semi-arid Africa. Remote Sensing of Environment, 101: 249–256.

Funk, C., Verdin, J. and Husak, G. 2007. Integrating observation and statistical forecasts over sub-Saharan Africa to support Famine Early Warning. American Meteorological Society. San Antonio TX.

Funk, C., Pedreros, D., Husak, G., Eilerts, G., Verdin, J. and Rowland, J. 2003. June-july-august 2003 rainfall forecast interpretation for central america, FEWS NET – UCSB Climate Hazards Group, Santa Barbara, CA.

Husak, G., 2005. Methods for the Statistical Evaluation of African Precipitation, University of California, Santa Barbara, 221pp.

Kalnay, E., Kanamitsu, M., Kistler, R., Collins, W., Deaven, D., Gandin, L., Iredell, M., Saha, S., White, G., Woollen, J., Zhu, Y., Leetmaa, A., Reynolds, B., Chelliah, M., Ebisuzaki, W., Higgins, W., Janowiak, J., Mo, K.C., Ropelewski, C., Wang, J. and Jenne, R., 1996. The ncep/ncar 40-year reanalysis project. Bulletin American Meteorological Society, 77: 437–471.

Lotsch, A., Friedl, M.A., Anderson, B.T. and Tucker, C.J., 2003. Coupled vegetation-precipitation variability observed from satellite and climate records. Geophysical Research Letters, 30.

Nicholson, S.E., Davenport, M.L. and Malo, A.R., 1990. A comparison of the vegetation response to rainfall in the sahel and east africa, using normalized difference vegetative index from NOAA avhrr. Climatic Change, 17: 209–241.

Potter, C.S. and Brooks, V., 1998. Global analysis of empirical relations between annual climate and seasonality of NDVI. International Journal of Remote Sensing, 19: 2921–2948.

Richard, Y. and Poccard, I., 1998. A statistical study of NDVI sensitivity to seasonal and interannual rainfall variations in southern africa. International Journal of Remote Sensing, 19: 2907–2920.

Trenberth, K.E., 1998. Atmospheric moisture residence times and cycling: Implications for rainfall rates and climate change. Climatic Change, 39: 667–694.

Tucker, C.J. and Sellers, P.J., 1986. Satellite remote sensing of primary production. International Journal of Remote Sensing, 7: 1395–1416.

Verdin, J., Funk, C., Kalver, R. and Roberts, D., 1999. Exploring the correlation between southern africa NDVI and pacific sea surface temperatures: Results for the 1998 maize growing season. International Journal of Remote Sensing, 20: 2117–2124.

Williams, J., 2005. Sustainable Development in Africa: Is the Climate Right? Columbia University Palisades, NY.

Section III
Food Security Analysis

Chapter 8
FEWS NET's Integrated Analytical Areas

FEWS NET uses specific analytical areas of expertise to determine what the impact of varying food production will have on food security. Remote sensing data plays an important role in this analysis, but it is only a small part of the overall picture of what FEWS NET does to effectively warn of current and impending problems. In this chapter, the social and economic datasets and frameworks that are used to monitor livelihoods and social processes will be described. These will include the analyses and tools that are used by FEWS NET food security experts, and the integrated analytical areas that are used to estimate the effects of biophysical variability on household health and well being.

Figure 8.1 shows FEWS NET's analysis and decision support framework which connects livelihoods-based food security and early warning analysis to decision support. The objective is to deliver situation analysis, core messages and recommendations to decision makers who can coordinate short and long-term actions to prevent and mitigate food insecurity and famine. Decision makers at the country, regional and global level are contacted through a variety of information products delivered through email, web site, hard copies, and through briefings, networking and dialog. Decision makers at two levels are targeted:

- Global, strategic decision makers, including those at USAID headquarters, decision makers of donor national governments, United Nations agency headquarters, headquarters of key non-governmental organizations (NGOs);
- National and regional decision makers, such as those of regional offices of USAID, donor NGOs regional offices, UN agencies regional offices, affected national governments, and communities.

Figure 8.1 shows two interlinking systems. On the left are the processes and knowledge systems that will be described here and in subsequent chapters, which include baseline knowledge about the human systems. The four key analyses that are conducted by FEWS NET staff are in the four boxes: analysis on the effects of livelihood systems, food security outlook, baseline livelihood analyses and current and projected hazard and analyses. On the bottom, supporting these systems are key implications of livelihood capital, including financial, natural, social, physical, human

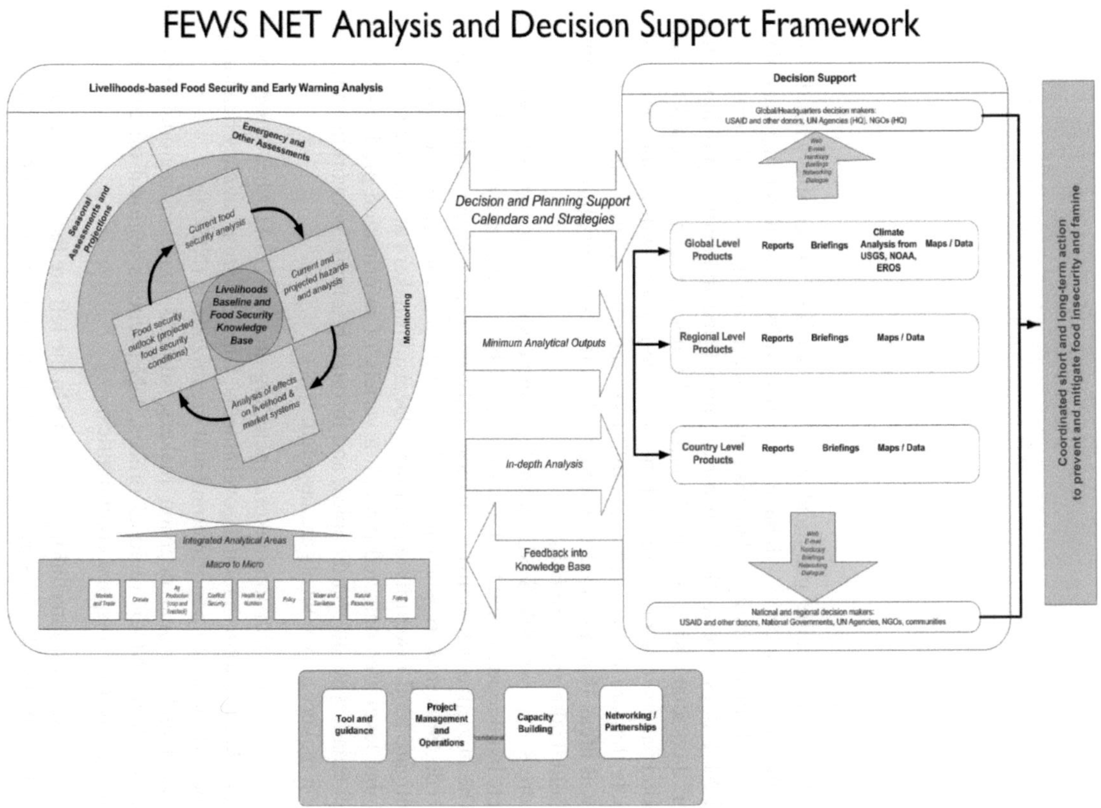

Fig. 8.1 FEWS NET Analysis and Decision Support Framework, describing how food security analysis interacts with decision support and reporting

and political. The box on the right shows how FEWS NET communicates its conclusions with decision makers, including country level, regional and global products to both national and local decision makers in the field as well as USAID and other global-level groups. Connecting these two boxes are decision support and planning strategies through analyses and feedback, along with the planning and decision support strategies.

Through analysis of the current situation with core early warning and food security messages, FEWS NET seeks to impact the budget, the actions and the focus of these decision makers in order to achieve consensus on required short and long-term action to prevent and mitigate food insecurity and famine. The next chapters will explore further who these decision makers are and how they make decisions regarding food security. This chapter will describe the analysis and knowledge generated by FEWS NET regarding food security, decision support mechanisms, and how remote sensing data is used during the analysis.

In order to provide locally relevant data and analysis in a particular food security situation, FEWS NET employs in-country analysts who are primarily responsible for collecting primary information, conducting analysis and writing monthly reports that describe the current situation and the likelihood that intervention would be required. They use a wide variety of information sources in order to write their reports. This knowledge base includes livelihood analysis, markets and trade, climate, crop production, livestock, conflict/security, health and nutrition, policy, water and sanitation, natural resources and other proximate causes of food insecurity. Each of these specific areas will be described below.

8.1 Analysis of Entitlement Decline

The primary analytical method that FEWS NET uses to relate variations in the environment as measured through remote sensing to how these variations affect people's household budgets is entitlement decline through livelihood analysis. The approach is a participatory, rapid community assessment technique that identifies the constraints to peoples' well being as well as their assets and opportunities (Lindenberg, 2002). The method combines features of participatory rural appraisal with local expert information to develop a profile of these constraints which forms a baseline of how the community functions day in and day out. The household economy approach (HEA) was piloted by CARE, one of the world's largest international relief and development not-for-profit organizations. Many of FEWS NET's personnel were involved in the development of the approach, which provides a systematic and integrated way of assessing the impact of a hazard on a particular community (Frankenberger et al., 2000).

Lindenberg (2002) defines household livelihood security as a family's or community's ability to maintain and improve its income, assets and social well-being from year to year. Many families in the developing world can easily be pushed from a state of equal income and expenditures into a crisis where assets are depleted in an effort to cope with a sudden shock. Household livelihood frameworks, such as the one that FEWS NET uses, are useful for community and family assessments because they allow for unique definitions of both family and community that fit the context and permit grouping across diverse geographic and cultural landscapes to enable analysis of the impact of a shock to be more realistic. As the Lindenberg paper states, before the advent of livelihood analysis, CARE's programs were structured in a way so that relief, rehabilitation and development programs were separate, and were organized and delivered by sector (microenterprise, agriculture, population, for example) without regard to what the other sectors were doing. This led to fragmented and redundant programs that were extremely hard to access from any one particular community's point of view. Livelihood security analysis permitted the community or family to be at the center of assistance provision, with all programs being reoriented so that they may work together to provide for the complex needs along a continuum of relief, rehabilitation and development. FEWS NET has drawn from CARE's experience in its implementation of livelihood analysis.

Livelihood analysis provides a glimpse into the diversity of rural household incomes. It offers a holistic view of the survival strategies of the rural poor, and emphasizes diversity and adaptability as positive attributes which exhibit resilience in the short term and viability in the long term. Rural families in developing countries engage in diverse activities, ranging from rural trade, migration to distant cities, casual employment, and, of course, farming and livestock rearing (Ellis, 2000). Studies show that 30 to 50% of rural household income in sub-Saharan Africa is derived from non-farm sources (Reardon et al., 1992). Understanding what these sources are and determining how a particular shock will impact a community's ability to meet its daily needs is the focus of livelihoods analysis and the household economy approach.

8.1.1 FEWS NET and Livelihoods

FEWS NET defines livelihood as the sum of ways in which households make ends meet from year to year, and how they survive (or fail to survive) through difficult times. In the academic literature, livelihoods are defined as follows:

$$\text{Livelihoods} = (\text{subsistence} + \text{accumulation}) - (\text{vulnerability} + \text{shocks}) + \text{coping}$$

In FEWS NET's definition above, 'difficult times' can be read as vulnerability and shocks (Ellis, 2000). The livelihoods definition above includes profit and capital accumulation as well as loss due to shocks. FEWS NET, being an operational agency with an extremely short time horizon, does not recognize these longer term aspects of livelihoods.

FEWS NET uses livelihoods analysis as the lens through which it views a variety of problems, including the effects of interannual climate variability, and it greatly improves its ability to determine their impact. This interest rests upon two basic observations:

1) Information about a given area or community can only be properly interpreted if it is put into the context of how people live.
2) Interventions can only be designed in ways appropriate to local circumstances if the planner knows about local livelihoods and whether or not a proposed intervention will build upon or undermine existing strategies.

Each FEWS NET country that participates in the zoning activity will have a livelihood profile that includes two main products: the National Livelihood Zone Map, and Livelihood Zone Profiles. The profiles describe the major characteristics of each zone, including a brief differentiation of the food security status of different wealth groups. There is some emphasis on hazards and the relative capacity of different types of households in different places to withstand them. In compiling the profiles, a balance has been struck between accessibility and level of detail. The aim has been to present sufficient information to allow a rounded and balanced view of livelihoods nationally, but not so much detail as to require weeks of study before conclusions can be made. The profiles provide a rapid introduction to livelihoods in the country but do not offer localized detail.

The preparation of these profiles is usually a joint activity between FEWS NET, the country of interest and key regional food security organizations such as the Permanent Interstate Committee for Drought Control in the Sahel (CILSS) in West Africa, the Southern African Development Community (SDAC) in southern Africa, or the Inter-Governmental Authority on Development (IGAD) in east Africa. The main focus of FEWS NET's work is early warning, food security monitoring and emergency assessment. The livelihood profiles that FEWS NET produces have been structured primarily with these types of activity in mind. However, it is hoped that they will also prove useful to the wider development community. Figure 8.2 lists countries have had livelihood analysis conducted that can be used for analysis.

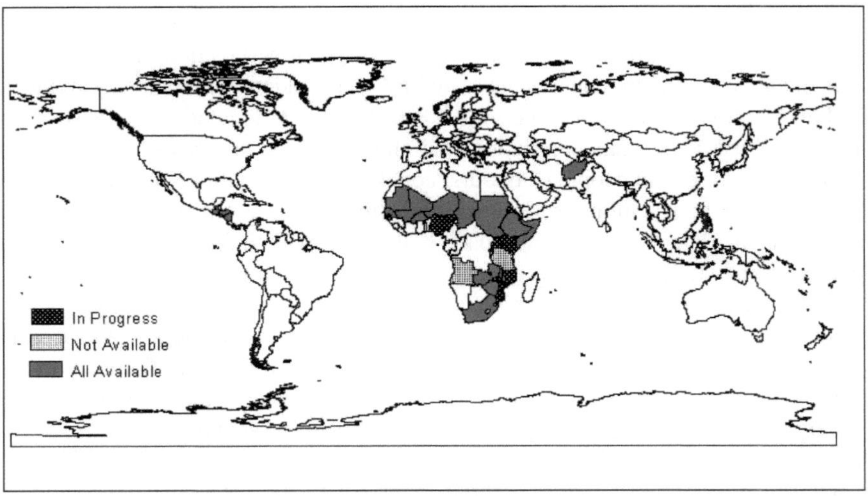

Fig. 8.2 Map showing the availability of livelihood profiles by FEWS NET country as of 2007

The livelihood zoning and profiles offer an analysis of urban and rural food security on a geographical basis. The zones will be useful on three levels: as an introductory guide to food security in the country, as a guide for early warning and response planning, and for policy development. The country is divided into relatively homogeneous zones defined according to FEWS NET's livelihoods framework. The geographical divisions are relatively small, as far as this is consistent with ground realities, so that the food security analyst can take in the general pattern and the basic differences between areas and populations without being overwhelmed by too much detail.

Local food security is often equated with agricultural production outcomes. Hence, a chronic or temporary production deficit against local food requirement is immediately translated into chronic or temporary food insecurity. Consequently early warning and food security monitoring systems draw heavily from two information sources: (i) crop and/or livestock production data which is often derived from remote sensing information; and (ii) market price information. This is never the whole story. A full account of 'food economy' addresses both food availability, or what food people produce, and food access, or what cash people earn to purchase food. Data on casual employment, collection of wild foods, and receiving charity from relatives or the sale of handicrafts may be equally important to the livelihood story as data on crop and livestock production, and knowledge of the relative importance of these can guide the design of more appropriate monitoring systems and better rapid emergency assessments. Using a livelihoods framework allows a more complete view into household capacity to cope with stress, especially failed crop or livestock production and permits an understanding of household activities at different periods in the yearly cycle.

All these sources of information feed directly into FEWS NET's analysis of need, helping to answer key questions. These questions include which areas and what types of household are likely to cope should a hazard strike and which will need assistance? What types of intervention will be most appropriate, and when and for how long should they be implemented? One could point to the position of poor households in a given geographical area who are highly dependent on urban employment. If urban employment declines, their labor will be less in demand. The question an analyst will ask is can this population find alternative income elsewhere, and will they be competing with people from other zones in these activities? National officers working within their national early warning system have an immense knowledge of their countries. The livelihoods approach helps to provide a framework for the full use of that knowledge, as well as adding a new level of information to it.

Finally, the livelihood information may provide a framework for policy development. Disaster management has been the main impetus to the spread of early warning systems. The rationale in early warning is to improve the efficiency in the scale and timing of emergency food aid. However, increasingly planners are looking at alternatives to food aid in early emergency intervention – and this often requires changes in policy and practice. Livelihoods analysis can expose the likely effects of such interventions on different households' capacity to survive a crisis. The analysis can also recommend the optimum timing for intervention.

Livelihood analysis can be applied to other policy changes. For example, if government taxes on kerosene were reduced, or charges made for government veterinary drugs, what would be the impact on households? FEWS NET is increasingly conducting policy analysis and providing recommendations for changes that may improve food security on a national level. More generally, the household viewpoint offers a more secure footing for looking at the increasingly voluminous discussion of poverty alleviation. It allows one to look at the story which lies behind national statistics.

Figure 8.2 shows those countries have completed livelihood profiles, those are in progress, and those that are currently not planned to be conducted. Because the development of the profiles is a collaborative process that requires both the participation and involvement of local and national government officials, regional and international organizations, profiles cannot be developed without the approval and involvement of the national government. Thus there are some countries that may never have livelihood profiles, but use alternative and parallel analysis mechanisms.

8.2 Markets and Trade

In many FEWS NET regions, a significant proportion of the population grow food in subsistence rain-fed agricultural systems as part of their income generating activities (Galvin and Ellis, 1997). Many of these families can grow only a portion of their family's needs, and some have no ability at all to grow crops. Access to food for both those that grow their own food as well as those who cannot involves markets where locally grown and imported grain is bought to be consumed throughout the year. Food security is therefore influenced both and by the price and the availability of food in the market that was produced elsewhere.

Farmers typically sell a portion of their crop in the market after harvest, save a portion for consumption, and purchase food from the market as their own supplies diminish later in the year. This interaction tends to amplify the response of market prices to the production of low-cost, locally grown grains such as millet, because higher prices will yield larger rewards for those who have excess grain to sell, and cost more for those who have food deficits for large portions of the year (Brown et al., 2006). Higher prices can cause food insecurity among the most vulnerable in a population even in times with adequate or even abundant food supplies (Sen, 1981). Early warning of these price increases enables organizations to increase food or income assistance at the appropriate time and to the appropriate households.

Countries that rely on imported food from outside the region are vulnerable to regional scarcity, global pricing and exchange rate volatility. Local cereal prices can double in a short period of time, leaving households that rely on the markets for their food with sharply reduced food access. FEWS NET therefore monitors food prices as a key component of food security monitoring. FEWS NET has worked to gather and analyze prices of both cereals and livestock during the past two decades. Although these databases have in the past five years mostly remained at the country level, most country reports involve some analysis of food prices.

In Eastern Africa, the Regional Agricultural Trade Intelligence Network (RATIN) is a key FEWS NET collaborator, working to provide good market information, enhancing cross border food movements in the East African countries. The RATIN Network was originally developed in 2003 as a three-way collaborative effort among FoodNet, FEWSNET and the Regional Agriculture Trade Expansion Support (RATES) program. The Foodborne Diseases Active Surveillance Network (FoodNet) is the principal food-borne disease component of the Centers for Disease and Control's Emerging Infections Program. The Regional Agriculture Trade Expansion Support (RATES) program was designed to increase value and volume of agricultural trade within the East and Southern Africa region and between the region and the rest of the world. Although very successful, RATES is no longer active although many of its activities are being continued by various organizations in the region. It focused on developing commodity-specific regional trade initiatives through innovative private sector/public sector alliances and partnerships and works primarily through regional trade flow leaders such as regional trade associations, national-level trade organizations, private companies and individual entrepreneurs. RATES supported activities in specialty coffee, maize and pulses, cotton/textiles, livestock and dairy sectors. The aim was to create a reliable source of market and price information that could be used by traders and policy makers to make more informed decisions about trading and regulating grain trade in the region. Its primary decision support tool, RATIN, has been taken on by the regional grain council and thus the information is still available to FEWS NET.

Analysis of grain movements between countries can be a critical component of food security in a region. For example, RATIN has established that between July 2004 and April 2005, Uganda exported maize estimated at 71,000 MT and 17,000 MT to Kenya and Rwanda, respectively. Similarly, Kenya imported some 70,000 MT of maize from Tanzania during the same period. This kind of information is a key component for understanding the likelihood of food shortages and upward spikes in prices that would have direct food security consequences. Figure 8.3 shows the price of maize per kilogram, the price of goats and terms of trade (kilograms of maize that can be purchased for the sale of one goat) at the Garissa Market in Kenya during 2004 and 2005 compared to average. Price analysis

Fig. 8.3 Maize prices, goat prices and the terms of trade in Garissa Market, Kenya, 2000–2005

is strongly regional, as production deficits in one region can result in price increases in areas that had adequate grain production nearby.

Identifying the most appropriate grain or livestock price that should be monitored requires local expertise, as they vary from one region to another. Monitoring the cost of transportation of grain and the exchange rate from production to consumption areas is also very important for effective food access monitoring and understanding of the underlying causes of food price movements. FEWS NET has begun to look into using models to ensure that long-term trends in prices are captured, and to enable their decision support products can look forward into the future. These models can use remote sensing information such as vegetation and rainfall dynamics as a constituent of these price models. This is based on research linking prices and vegetation anomalies as a proxy of agricultural productivity across a region. Figure 8.4 shows the relationship found in West Africa. Although price modeling is, at the time of this writing, only just beginning to be explored by FEWS NET, it has the potential to greatly improve its ability to analyze food prices across large regions, and demonstrate the importance of proximate causes of volatility in food prices in locations that it has little information about. More information about economic modeling can be found in Chap. 12.

Integrating livelihoods information with market price information is a key activity of FEWS NET analysts. Because market prices can reduce income for agriculturalists during both high and low price periods, they are central to much analysis

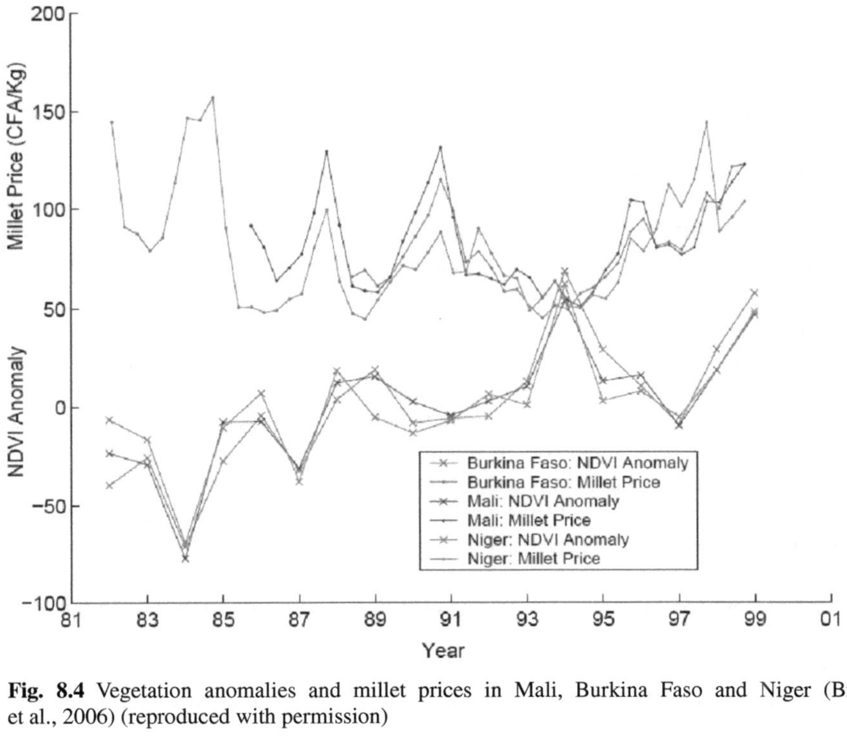

Fig. 8.4 Vegetation anomalies and millet prices in Mali, Burkina Faso and Niger (Brown et al., 2006) (reproduced with permission)

that couple food production and the ways households make a living. Many of the vulnerable across FEWS NET's countries are net buyers of cereal staples. These households rely either partially or entirely on the market to purchase food, and thus food prices are particularly important for them. As food prices fluctuate so does access to food for these households.

8.3 Climate

The impact of droughts, floods and other environmental disasters on economies in sub-Saharan Africa is complex. Early intervention when hazards are detected is critical for maintaining whatever development progress has been made and for preserving the livelihoods of the most vulnerable. As economies develop, their sensitivity to drought increases until they diversify to the point where agriculture is no longer a primary source of export income. Climate variability often threatens to disrupt these countries' ability to maintain their growth trajectory, which is critical for reducing the number of people in extreme poverty. The World Bank has found that the relatively more developed or complex economies in sub-Saharan Africa, such as Senegal, Zambia and Zimbabwe, are more vulnerable economically to drought shocks than less developed and more arid countries, such as Burkina Faso, or countries undergoing conflict emergencies, such as Somalia (Benson and Clay, 1998).

From the food security analyst's perspective, climate data revolves around the quality of the agricultural economy and hazards. If there have been abundant rains which have permitted investment and expansion of agricultural activity, then there is likely to be adequate employment for immigrants, the landless and the under-employed. Supply of grain will be sufficient and investment in export opportunities increased. Prices for food will be moderate or low, which has the impact of increased access to food for the poorest in the community. When rainfall variability creates an unfavorable environment for agriculture, it affects a large portion of the economy, as it reduces the total amount of goods and services being bought and sold throughout the affected region, extending from farmers to the transportation sector, the markets, shops and education. Environmental variability also poses physical hazards, including cyclones, floods and droughts for their overall effect on the economy as well as their specific impacts on households and individuals, particularly those involved in agriculture.

Because FEWS NET takes this broader view, it can be difficult to interest national FEWS NET representatives in specific remote sensing products that address just one part of the food production equation. Coarse resolution vegetation indices, for example, can only speak to the variation in crop yield from year to year, but do not show cropped area or production variations, and do not speak to how these variations are related to population centers. If remote sensing data can be transformed to show spatial variations in per capital cereal production as compared to the norm (or compared to recent years), then these data can be much more easily integrated into the suite of socio-economic information products and analyses. At the moment, however, there is a great deal of separation of analysis, where the socio-economic aspects of environmental variability are analyzed apart from the environmental causes, with only a summary map of the affected area used as input instead of the actual datasets themselves. This is an area where the scientists associated with FEWS NET are working to provide both the raw data as well as much more focused analytical products for decision support.

The following is an example from the Kenya Food Security Update for November 8, 2006 and shows the various ways climate is incorporated into food security analysis:

> The eagerly anticipated 2006 short rains have begun in all pastoral districts, marking the end of another difficult dry season. However, rains have been exceptionally heavy in most pastoral areas and caused flooding in localized areas of Turkana, Moyale, Wajir, Isiolo and Mandera districts. Six pastoralists in Turkana and Isiolo districts were swept away by flash floods and lost their lives. Key supply routes in Wajir, Mandera, Garissa, Moyale and Marsabit were also cut off, rendering several major routes impassable. Consequently, food prices have risen steeply in markets supplied by these roads, while pastoralists in affected areas are unable to access markets to trade livestock. In response to severe local food shortages, the Government of Kenya's National Disaster Centre, in collaboration with the Kenya Air Force, has delivered food to some affected areas.

Thus, although good rains are a prerequisite to a good agricultural season, there are consequences to severe rainfall that can also be negative. Maps showing areas where rainfall was intense were used to analyze the adverse impacts on several sectors, including trade, prices, transportation, and food availability. The resulting distribution of food aid is a key response to these disruptions due to climate.

8.4 Crop Production

Variations in production of specific crops are monitored by FEWS NET staff members for their impacts on vulnerable segments of the population. Livelihood zones are used to interpret specific variations in climate to determine the impacts of production variations, along with the ancillary problems that may occur. Movement of cereals from exporting regions to regions that do not produce enough cereals to support the population is a critical part of food security analysis, and is carefully monitored by FEWS NET. Thus the production of cereals in these exporting regions is carefully monitored as the consequences of reduced availability of cereals are large to both the local economy and to regional food security.

Accurate production figures are a key part of estimating how much food aid may be required in a particular year. Unfortunately, accurate crop production figures are very hard to find in Africa. Production amounts vary between the national statistics and the United Nation's Food and Agriculture Organization (FAO) figures that are based on local assessments and export statistics. Disagreements can be significant. For example, the Ethiopian statistical authority's production figures are typically lower than FAO/WFP figures (e.g. 20% lower for 2002 Meher production). Figures are reported by administrative units or districts, whose borders do not remain static, with the amount of agricultural land unevenly distributed within it. These issues vary from country to country, and FEWS NET often works to build consensus among various parties, including the local government, regional authority and international organizations such as the FAO and NGOs that provide food aid. Consensus figures are useful politically but can have significant bias through time and thus are not very useful for quantitative estimates based on remote sensing.

FEWS NET has recently begun to develop remote sensing-based production estimates through an analysis of country-level area, calculated from extremely high resolution remote sensing data combined with yields to come to a completely independent estimate of agricultural production in countries such as Ethiopia and Zimbabwe that have particularly poor agricultural statistics. This is a new effort, but it is hoped in the next few years that physical-science based production figures can begun to be an operational product used to correct and counter the bias that a consensus approach to per capita food production introduces into the system.

8.5 Livestock

Livestock density statistics are even more poorly specified than crop production figures. Whereas agricultural fields change location only slowly and have well specified and known distribution across the landscape, locating livestock and the people who tend them can be extraordinarily challenging. There are very few reliable figures regarding the numbers of cattle in pastoral areas in East Africa, for example, for several reasons. First, pastoral herders do not raise the animals for sale but to harvest renewable products from them such as milk and blood for consumption and sale. Thus sale of animals in the market is related far more to scarcity of fodder than to actual numbers on the ground. Second, there are very few reliable statistics throughout the pastoral regions in Africa regarding carrying capacity of the land for traditional cattle and camel breeds, which coexist with varying numbers of goats and sheep. Thus a highly detailed survey producing the numbers of animals in one region cannot be used to extrapolate numbers in nearby regions.

Finally, pastoral herders move from one region to another in search of food for their animals without regard to national and regional boundaries. This makes a particular group of vulnerable people very hard to target, as humanitarian assistance is invariably organized through these national and district boundaries. These issues are all very difficult and thwart the traditional kind of quantitative analysis conducted for estimating needs, and thus FEWS NET tends to rely on local knowledge coupled with remote sensing of pastoral lands to determine quality of fodder resources instead of quantitative analysis to estimate the numbers of pastoral households at risk for food insecurity.

8.5.1 Example of Pastoral Crisis Caused by Variable Rainfall: Kenya

Kenya has over a third of its population of about 31 million located in pastoral, agropastoral, and marginal agricultural areas. Over the past decade and a half, the extent of food insecurity in Kenya has increased, with a large portion of the population at risk being in these areas. While drought is often thought to be the principal catalyst to chronic food insecurity, research suggests that household livelihood characteristics and the presence of appropriate interventions are critical in determining the ability of households to mitigate and recover from shocks and natural hazards, including drought. Food insecurity among pastoralists is severe and increasingly chronic in nature, attributed to distinct characteristics inherent in the pastoral livelihood, such as:

- Highly variable agroclimatic conditions;
- Agroecology characterized by frequent droughts.
- Dependence on livestock as the sole source of food and income.
- Reliance on volatile markets for food when the productivity of livestock declines.
- Debilitating conflict/livestock raiding is increasingly common.
- Rising numbers of destitute pastoralists.
- Fragile and rapidly degrading physical environment.
- Historically, both chronic and acute food insecurity that has largely only been addressed through food aid.

The Kenya pastoral livelihood has with increased frequency experienced several seasons of unfavorable agroclimatic conditions, particularly during the past 10 years. These include droughts in 1992/3; 1996–1997; October 1998–October 2001; from October 2003-March 2006; and floods from October 1997 to February 1998. Figure 8.5 shows a timeline with the continuous period of poor seasons over the past 10 years. The impacts of poor agroclimatic conditions have heightened food insecurity in the pastoral districts. The figure shows how these conditions are captured with rainfall anomaly images, showing the anomalous conditions that have resulted in food security crises.

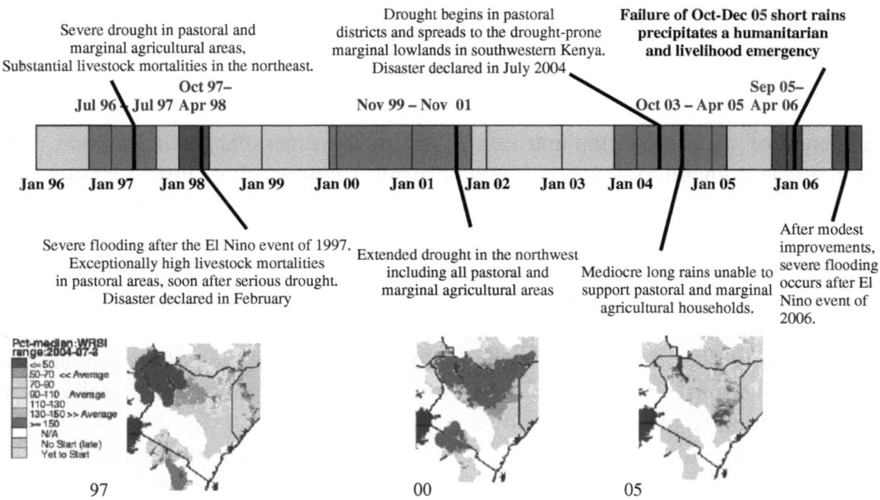

Fig. 8.5 Major events timeline from FEWS NET Kenya pastoral livelihoods analysis shows the succession of droughts and floods over the past 10 years that precipitates humanitarian emergency

8.6 Conflict/Security

FEWS NET's role as an information gathering organization extends to areas that have seen civil and political conflict. Populations in conflict zones are particularly vulnerable to food insecurity due to the restriction in movements of people and goods, increased likelihood of raiding and theft of basic food and livestock, and the inability of farming families to sow, tend and harvest their fields over the course of a season.

FEWS NET works in Southern Sudan, for example, and in 2006 the region has been in a post-conflict phase for two years. In the December 2006 country report for the region, civil insecurity caused by organized armed groups in Juba County was reported, which continued to limit movement of people and the trade of goods. In Gorgial County, food security improvements that were made during the previous four months are unlikely to last beyond February due to reduced cultivation caused by inter-ethnic conflict which limited access to land during the June-September season. The disarmament process conducted by the Government of South Sudan earlier this year in Diror, Pulchol and Nyirol counties has left households more vulnerable to cattle raiding by their armed neighbors in Pibor County. Tension persists in Malakal Town in Shilluk County where fighting between two security forces erupted towards the end of November.

FEWS NET's mandate does not extend into areas in full-out conflict, however. When a conflict becomes serious enough to require evacuation of US supported personnel, FEWS NET's role of information gathering and analysis falls to UN agencies or to private non-profit organizations. FEWS NET will continue to publish information and provide analysis on the country from the regional center and from the central office in Washington DC during and will resume a presence after the conflict, similar to what was done in Southern Sudan.

8.7 Health and Nutrition

Evidence of widespread malnutrition in the most vulnerable has long been used to motivate humanitarian action. Food security is composed of three components: food availability, food access and food utilization. Health and nutrition addresses the food utilization portion of food security. During times of stress, contaminated drinking water, poor environmental hygiene and poor health infrastructure lead to poor assimilation of the food that is consumed. Thus, a person cannot be said to be food secure without environmental hygiene, primary health care and clean drinking water security. Nutritional status is the outcome of a combination of household food security, health service and provision of care, and therefore links several critical aspects of food security and individual outcome.

Nutritional evidence, however, is a trailing indicator of a food security crisis. Because FEWS NET is interested in providing early warning of food security problems, it must not wait until large portions of the population are suffering from malnutrition; must find and ameliorate the causes of the malnutrition before they become systemic. Therefore, FEWS NET does not invest directly in nutritional surveys as they do other aspects of humanitarian work.

FEWS NET does evaluate and report on the likeliness of wide-spread communicable disease and other health problems that may affect nutrition. FEWS NET uses as secondary sources other organization's nutritional surveys to determine the degree of malnutrition present in a particular community. Although FEWS NET does not directly collect these statistics themselves, staff members are often very involved in the humanitarian community that does collect them, with the local FEWS NET representative knowing the parties and actors involved, often working with them to set up and implement the surveys. FEWS NET is in a very good position to know the quality and reliability of the statistics produced.

8.7.1 Nutrition and Food Consumption

Many other factors aside from food production impact the nutritional status of children. For example, despite the large international emergency response to the most vulnerable areas in southern Niger in 2005, including nearly $100 million in food aid, malnutrition rates failed to fall below emergency levels. In Niger (as in many countries) the reasons for persistently high rates malnutrition are multi-faceted and multi-causal. The standard reference tool for understanding and analyzing the causes (direct, indirect and underlying) is the Malnutrition Causal Framework (Fig. 8.6). This framework has generally been adopted by international agencies and should form the basis for nutritional assessments in emergencies. The framework cites food intake and disease as the immediate cause of malnutrition with three underlying and overlapping causes (inadequate household food security; inadequate care and feeding practices; and poor public health access and environment). The third tier of causes includes socio-political and economic causes. The framework is recommended for investigation of the relative importance of the different causes of malnutrition and mortality, and thereby prioritization of interventions and coordination of different sectoral responses.

There is a general consensus that it is not possible to use nutrition data alone for decision making and that information on the underlying causes of malnutrition is necessary. The following factors need to be considered in any interpretation: underlying causes; mortality; and seasonality. Putting this together in practice and translating it to action however is complex; data are not always available, analysis is poor and current guidance is unavailable or contradictory. For example there is little guidance on how to analyze malnutrition and mortality together.

It is highly likely that malnutrition levels in the 2005 in Niger, for example, increased as a direct result of declines in food availability and household purchasing power due to poor crop yields for the 2003/4 agricultural year and rising food prices in 2004/5. While access to food was a key problem, food availability and consumption patterns are also constraints to food security. From a nutritional (as well as psychological) viewpoint, a monotonous diet is the hallmark of chronic under-nutrition.

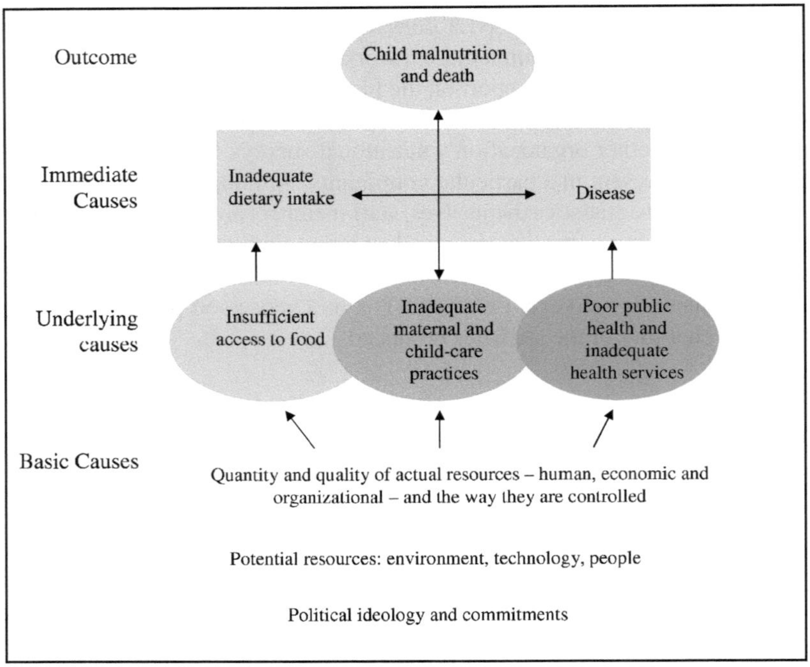

Fig. 8.6 Causal framework linking access, nutrition and health, accepted by WHO, UN, and a wide variety of other organizations

Increasing poverty tends to be associated with lower dietary diversity. In many cases in Niger, malnutrition in children is not principally caused by poor availability of food in the household. Parents may eat a variety of foods, while children are fed predominantly a millet-based porridge every day with very little dietary diversity. This results in a lack of quantity (energy) and quality (micronutrients) in the diet, which often leads to a lack of appetite and concurrent weight loss. Inadequate nutrient intake often precedes weight loss and the anthropometric changes used to define malnutrition. Thus careful monitoring of food consumption patterns, particularly for mothers and children, is essential for understanding the underlying food security issues in a region (Grobler-Tanner, 2006).

8.7.2 Nutrition and Disease

FEWS NET actively monitors HIV/AIDS, cholera, malaria, Rift Valley Fever, and other eco-climatically coupled diseases for their impact on food security. These diseases affect directly both the ability of an affected person to absorb available nutrients, and the economy of the region. Measures to control the spread of cholera, such as a ban on the sale of meat, fish and vegetables in local markets to prevent the spread of the disease, can negatively affect the purchasing power of households.

Rift Valley Fever is an acute fever causing viral disease that affects domestic animals (such as cattle, sheep, goats and camels) and humans. It is generally found in regions of southern and eastern Africa where sheep and cattle are raised, but the virus also exists in most countries of sub-Saharan Africa (FEWSNET, 2000). RVF is most commonly associated with mosquito borne epidemics during years of heavy rainfall and localized flooding (Linthicum et al., 1999; Woods et al., 2002). Humans can get Rift Valley Fever from mosquito bites, handling infected meat at slaughter or preparing infected food. Approximately 1% of humans infected die from the disease. There are no established courses of treatment or human vaccines. Prevention and control focus on mosquito control and limiting contact with infected animals. The closure of the sale of meat from pastoral areas of Eritrea, Kenya and Ethiopia to the Aden Gulf Coast in an effort to stop the spread of Rift Valley Fever has severely reduced the price of cattle in markets in the region, undermining the ability of the pastoral economy to provide for its participants (FEWSNET, 2001) (Fig. 8.7).

During times of emergency, efforts to reduce the spread of disease can save many lives. The threat of an outbreak of water borne diseases in areas experiencing flooding conditions can be extreme, with large numbers of people at risk. Unusual concentrations of people in camps and feeding centers puts many people

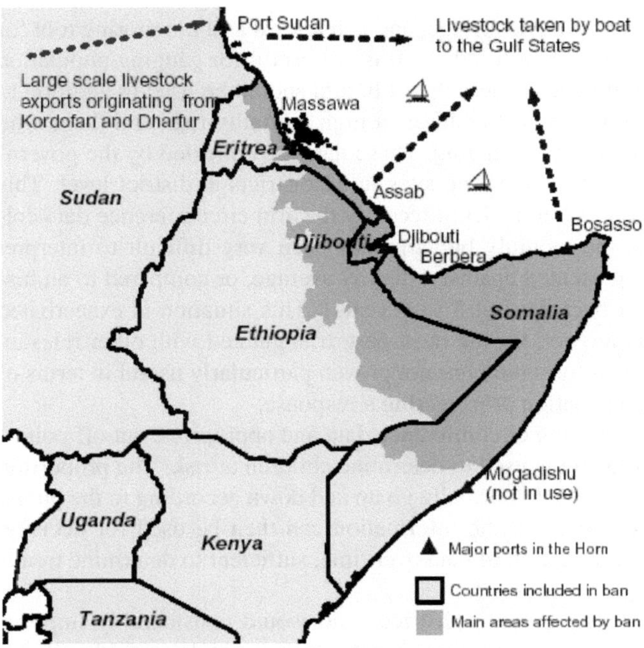

Fig. 8.7 Main livestock production areas in the Horn affected by the Rift Valley Fever export ban in 2006–2007

at risk for diseases that can spread rapidly and have a high mortality rate. In its recommendations, FEWS NET attempts to ensure that these issues are as high on the priority list as food assistance and logistics as is possible. Efforts at vaccination of the most vulnerable and provision of clean water supplies can significantly reduce mortality during times of crisis. Although FEWS NET uses the nutrition information available, it does not collect it directly. The assessment was conducted by other organizations, but are reported by FEWS NET as primary evidence as to the health and nutrition level of a particular community in which it works.

8.7.3 Example of Nutrition Surveillance and Monitoring in Kenya

There are two types of nutrition information routinely collected in Kenya: Middle Upper Arm Circumference (MUAC) data gathered by government monitors on a monthly basis at selected 'sampling' sites in a given district, and routine growth monitoring of children under five, collected by the Child Health and Nutrition Information System (CHANIS). This dataset is extremely limited as a monitoring tool. It is based on Growth Monitoring and collected only from health clinics (facility-based). Thus, coverage is extremely poor, particularly in the northeast. After the child is one year of age, visits become even more infrequent. In the northeast, growth monitoring is extremely erratic, rendering it redundant as a tool for determining trends.

Growth monitoring uses weight-for-age, and is intended as a monitoring tool for individual children to detect growth failure. It is not useful for gauging population level trends. Arm circumference of less than 110 mm should be used in addition to routine growth monitoring to detect children at high mortality risk, and those who should be referred for therapeutic feeding. These data are collected by the government on a monthly basis from sampling sites in 22 districts at district level. This is reported in the monthly bulletins. Until recently the arm circumference data collected and presented in the monthly bulletins has been very difficult to interpret because it has not been presented against a district average, or compared to an historical trend (mean data over the last 5 to 6 years). This situation is exacerbated by the fact that the data are not, for the most part, triangulated with other relevant information. As such, the information has not proven particularly useful in terms of gauging the severity of a situation or triggering a response.

There are also issues in using circumference data and appropriate cut off points. Currently a cut-off of 135 mm is used to determine children at risk. The proportion of children below this level would normally go up and down according to the public health and nutrition situation, and the information can then be used for decision making, if there is an accumulation of data over time, sufficient to determine trends (five years).

This historical data exists in many districts, and would considerably improve the surveillance system by reporting current data against trends, and identifying significant changes from normal condition. However the problem is not only with the limited ability to analyze the data. These institutions and their personnel using this data do not understand or make effective use of it. OXFAM has provided some useful training and analytical tools for collecting and analyzing arm circumference data in the surveillance system for use by the government. This has made some difference, but has not yet trickled down to the district level where the understanding of basic nutrition concepts and indicators is minimal. Similarly while nutrition partners at the district level were very familiar with the survey instruments, they were often unaware that nutritional data was being collected by government, suggesting the need for better coordination and collaboration with technical experts at district level in the interpretation and analysis of the data collected.

Quantitative data and information (food security, malnutrition prevalence rates, surveillance and admissions to therapeutic feeding programs) are not alone sufficient to determine the severity of a crisis, or to design a response. An understanding and analysis of the causes of malnutrition, and an interpretation of the data in context are required. For the most part, analysis of high global acute malnutrition rates in Kenya has tended to over-emphasize the absence of food, and thus the response tends to be overwhelmingly food-related. This is exacerbated by donor 'preferences' for food, versus non-food responses. The over-emphasis on food obscures the contribution of other causal factors that contribute to malnutrition, such as poor hygiene, morbidity and care practices.

8.8 Water and Sanitation

Strongly related to nutritional issues described in the previous section are water availability and sanitation. FEWS NET monitors water availability, particularly in areas that rely on shallow, hand dug wells that can dry up during times of severe drought. Lack of clean potable water and poor sanitation can lead to frequent illness and large outbreaks of water-borne diseases such as Cholera. In Niger, access to clean water is found in 36 percent and adequate sanitation is found in only 4 percent of rural areas. Highly infectious excreta-related diseases such as cholera still affect communities. Improvements in safe water supply, and in particular hygiene (including use of soap for hand-washing) could reduce the incidence of diarrhea and the number of deaths due to diarrhea by more than half. In regions experiencing food crises, the provision of clean, potable water along with food can save lives, but is often expensive and difficult to maintain.

8.9 Challenges for FEWS NET

This chapter has presented a wide range of information sources that FEWS NET uses to evaluate the food security status of a locality. A major challenge for FEWS NET is to ensure the same quality of analysis and information is available for each country and region in which it works. For example, it has mandated that some analysis of the local food market is put in each monthly report. This has been facilitated by providing an expert at the regional level who can assist the national representatives in the collection, archiving and analysis of necessary price datasets. Every FEWS NET representative is an individual with different skills, however, so some are more interested in market price data then others, which results in variations in the skill to which these analyses are conducted.

Similarly, although FEWS NET officially uses the livelihoods analytical methodology described above, implementation of the system is spotty. Some country representatives provide analysis based on livelihood zones, but many others simply state facts about crop conditions, livestock health, market prices and government policies, without providing any analysis as to the food security consequences of these facts. This means that if these analyses are to be done, they must either be provided by the central FEWS NET office or the national representative must be forced to do it. The time, energy, and attention of the national staff may be either focused elsewhere or lacking and thus the reports get posted as they were originally written. This challenge is part of the difficulty of moving new concepts and methodologies from the center to the periphery, particularly when working in the developing world.

Another challenge for FEWS NET is the divergence between advancements in understanding of livelihoods, food security and poverty in the academic literature and the realities of implementing these new understandings in crisis situations. Innovations occur in the field, but it is an ongoing challenge to learn from and integrate conclusions derived from academic studies. This is not unique to FEWS NET, but exists across many development agencies and organiations.

The complexity of the indicators food insecurity has been increasing in recent years. Many in the humanitarian community now accept the relevance and centrality of social, political and economic factors, local context and geopolitical pressures on the food security situation in a locality (Maxwell, 1996). Government officials and local community activists, however, may not have the same perspective as these international experts as they were educated in an era when food security was equated with food availability. This may result in a strange tension between those using a wide variety of information sources and a complex methodology to quantify limited assistance for specific vulnerable groups in particular livelihood zones in a country and those who would like to use remote sensing data alone to justify massive food aid influx due to a biophysical hazard. Thus the diverse information sources listed in this chapter should be seen as what FEWS NET strives to do in every place that it works. Its success depends as much on the resources it has available to it as it does the local expertise, government cooperation and the nature of the problem at hand.

8.10 Summary

In this chapter, the primary analytical methods that FEWS NET uses to identify and analyze the proximate causes of food insecurity are described. This knowledge base includes livelihood analysis, markets and trade, climate, crop production, livestock, conflict/security, health and nutrition, policy, water and sanitation, natural resources and other proximate causes of food insecurity. Each of these specific areas is mentioned in reporting from each region or country in the monthly report in order to identify possible sources of food insecurity. Remote sensing data is used in the context of these analytical areas, and thus remains an important component to FEWS NET's work.

References

Benson, C. and Clay, E., 1998. The Impact of Drought on Sub-Saharan African Economies. The World Bank, Washington, DC.

Brown, ME., Pinzon, J.E. and Prince, S.D., 2006. The sensitivity of millet prices to vegetation dynamics in the informal markets of Mali, Burkina Faso and Niger. Climatic Change, 78: 181–202.

Ellis, F., 2000. Rural Livelihoods and Diversity in Developing Countries. Oxford University Press, New York, 296pp.

Galvin, K. and Ellis, J., 1997. Climate Patterns and Human Socio-Ecological Strategies in the Rangelands of Sub-Saharan Africa. In: O.T.E. Odada, M.S. Smith, and J. Ingram (Editors), Global Change and Subsistence Rangelands in Southern Africa. IGBP Committee, Gaborone, Botswana.

FEWSNET, 2000. Horn of Africa Food Security Update: October 20, 2000, Chemonics International and USAID, Washington DC.

FEWSNET, 2001. Ethiopia Country Report December 2001: Poor pasture, water availability for livestock, Chemonics International and USAID, Washington DC.

Frankenberger, T., Drinkwater, M. and Maxwell, D.G., 2000. Operationalizing Household Food Security: A Holistic Approach for Addressing Poverty and Vulnerability, CARE USA, Atlanta.

Grobler-Tanner, C., 2006. Understanding nutrition data and the causes of malnutrition in Niger: A special report by the Famine Early Warning Systems Network (FEWS NET), Chemonics International, Famine Early Warning System Network, Washington DC.

Lindenberg, M., 2002. Measuring household livelihood security at the family and community level in the developing world. World Development, 20: 301–318.

Linthicum, K.J., Anyamba, A., Tucker, C.J., Kelley, P.W., Myers, M.F. and Peters, C.J., 1999. Climate and satellite indicators to forecast rift valley fever epidemics in Kenya. Science, 285: 397–400.

Maxwell, S., 1996. Food Security: a post-modern perspective. Food Policy, 21: 155–170.

Reardon, T., Delgado, C. and Matlon, P., 1992. Determinants and effects of income diversification amongst farm households in Burkina Faso. The Journal of Development Studies, 28: 264–296.

Sen, A.K., 1981. Poverty and Famines: An Essay on Entitlements and Deprivation, Oxford, Clarendon Press.

Woods, C.W., Karpati, A.M,, Grein, T., McCarthy, N., Gaturuku, P., Muchiri, E., Dunster, L., Henderson, A., Khan, A.S., Swanepoel, R., Bonmarin, I., Martin, L., Mann, P., Smoak, B.L., Ryan, M., Ksiazek, T.G., Arthur, R.R., Ndikuyeze, A., Agata, N.N. and Peters, C.J., 2002. An outbreak of Rift Valley fever in northeastern Kenya, 1997–98. Emerging Infectious Diseases, 8: 138–144.

Chapter 9
Dynamic Communication and Decision Support

FEWS NET's primary function is to communicate the food security status of the communities in which it works to decision makers. A clear understanding of who these decision makers are and what their needs are is a key aspect of what FEWS NET does. By working to define the needs of these decision makers, they influence the types of decisions that can be made by providing information and analysis that may never have been available previously. There is considerable flow of personnel and expertise between FEWS NET staff and the staff of organizations that use the information produced. Integrating remote sensing information of biophysical hazards into the analysis of food security for decision makers is one of FEWS NET's key contributions.

This chapter will describe the process by which remote sensing data becomes actionable information that is then responded to by the humanitarian community with food aid and other assistance. By examining the entire process, the role of remote sensing information and how it can lead to early and effective intervention can be illuminated. Early warning indicators of an impending crisis, reported in FEWS NET monthly country and regional reports, are followed up on with three rounds of assessments funded by the UN's World Food Program (WFP) and local NGOs: an initial assessment during the growing season, a provisional assessment in the post-harvest period, and a final assessment during the peak hungry period, usually a the start of the next growing season. During each assessment, FEWS NET works to analyze the underlying causes of food insecurity, provide spatial and temporal information to the assessment team, and often works directly with the team to complete and analyze the results. Remote sensing information is used to target these activities and illustrate their results, and the assessment results motivate concrete budgetary response by aid organizations.

FEWS NET seeks to include information regarding the situation to its stakeholders at a variety of levels, including comparative global analysis that focuses resources to the locations with the worst problems, instead of just the most visible. A variety of products are delivered in a number of ways to ensure that the proper people get the information that they need. This chapter will describe the wide variety of decision makers and stakeholders that use its information, the types of decisions they make, describe the suite of products and analysis that have been developed to meet the information requirements of these decision makers, and demonstrate how remote sensing information fits into each part of the picture.

9.1 Decisions and Decision Makers

Who are the decision makers who read and act upon FEWS NET's products? They are a very diverse group, with a wide range of interests, expertise levels and information requirements. From international program managers at USAID and the United Nations to local community organizations whose only interest is in their local markets, FEWS NET's stake holders are all interested in information that will help them negotiate a complex world. They range from local non-governmental and non-profit groups to large international donor organizations. Not everyone is involved in every food security process, nor are their efforts at planning for and responding to food security crises necessarily coordinated or complementary. There are, however, a broad array of stakeholders that need to be communicated with during times of food crises and famine (Choularton, 2005), shown in Table 9.1.

Despite this diversity, FEWS NET has as its focus informing and providing products to its funding agency, the US Agency for International Development Food for Peace (FFP) division, and its direct partners in the regions where it works. In its

Table 9.1 Decision makers at a different scales (from (Choularton, 2005)). Acronyms: NGO: non-profit organization, ECOWAS: Economic Community Of West African States, SADC: Southern African Development Community, CILSS: Comité Permanent Inter-Etats de Lutte contre la Sécheresse dans le Sahel (Permanent Interstate Committee for Drought Control in the Sahel), FAO: United Nation's Food and Agriculture Organization, USAID: US Agency for International Development, DFID: United Kingdom's Department for International Development, EU: European Union

	Local	National	Regional	International
Community	Individuals with access	–	–	–
Civil Society	Local NGOs	National NGOs	Regional NGOs	International NGOs Save the Children Oxfam
Government or intra-governmental	Municipal or Departmental government	Ministry of Agriculture or Parliament	ECOWAS SADC African Union	United Nations General Assembly
Private sector	Local shop owner or trader	National companies	Regional companies	Transnational corporations
International Orgs	–	National government Private sector	CILSS	FAO WFP
Donors	–	–	African Development Bank	USAID DFID EU

statement of work for its primary contractor, FEWS NET keeps the focus on the DC national decision makers. It is stated explicitly in FEWS NET's authorization documents that FEWS NET and its Washington DC office personnel will produce decision support products and activities for AID/Washington offices, which will also be posted at the FEWS NET website, to be found at www.fews.net. These reports include:

- Bi-weekly Executive Overview of Sub-Saharan Africa Humanitarian Crises bulletin (EOB);
- Monthly Food Security Implications Briefings (FSIB) in Washington where the material from the Overview is presented;
- Graphs, maps, briefings and short situation reports for USAID/Washington offices, and,
- Household livelihood maps, profiles, and analyses.

The objective of these reports is to support decision making at the level where decisions regarding food aid and other humanitarian interventions are made in order to improve the linkage between early warning and assessments and the response. The time between deciding that there is a problem and the arrival of actual assistance can be nine months or more, thus early warning information needs to be clear and easily interpreted, ensuring that the decision makers have the information that they need when they need to make budgetary decisions. Figure 9.1 shows the timeline

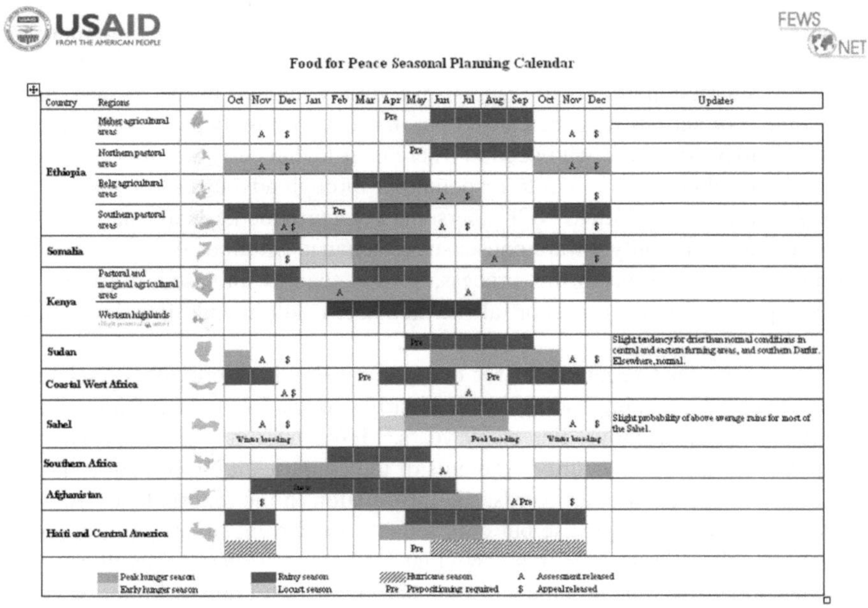

Fig. 9.1 Seasonal planning calendar for USAID Food for Peace that identifies decision making deadlines ($ symbol) simultaneously with agricultural calendars and deadlines for food aid prepositioning in various regions of Africa

for all FEWS NET countries, identifying the agricultural seasons and corresponding periods when access to food tends to be lowest ('hungry season'), along with the periods when nutritional assessments are conducted and when budget decisions are made ($ symbol). The 'Pre' on the calendar shows when the food aid must be acquired and delivered to locations near local communities in order to be available for distribution during the hungry season. In many locations, the roads become increasingly impassable during the rainy season and trucks become scarce, being needed for agricultural use. Thus acting upon early warning information provided by FEWS NET in a timely manner is critical (Choularton, 2005).

Decision makers at the local level have very different information requirements to national or international decision making processes. Decision makers at the local level may need to make comparisons among the emergency-affected populations for targeting purposes or make decisions about the need for feeding programs. In contrast, at the national or international level there may be a need to compare different emergency-affected populations across a border or measure the overall effectiveness of the humanitarian operation. A wide variety of assessment and other information gathering strategies and sources are required in order to meet the needs of all these different decision makers (Young and Jaspars, 2006).

For a timely response from the humanitarian community to be possible, a consensus between all the primary actors must be achieved. The national government must be aware of the problem and must have the political will to respond (Buchanan-Smith, 2000). Assessments of the nutritional deficiencies of a community will be carried out to determine the level of need and the most appropriate response. The national government has to issue a formal request to the international community for assistance before action can be taken. There must be the political will at the international level to respond in a timely manner to the request, and a huge number of actions must be carried out in order for timely and appropriate assistance to arrive for those who need it. There is little that early warning practitioners can do to influence this will to respond, but through advocacy, networking and personal relationships, FEWS NET's personnel can provide key information which will move the community toward consensus (Buchanan-Smith et al., 1994). Remote sensing data is important in this process, as it is seen as unbiased, objective and definitive proof of production shortfalls, which may result in widespread nutritional impact.

Another important requirement for decision makers is to provide sufficient information regarding an impending problem early enough for the large humanitarian apparatus to begin to move. For many years, FEWS NET has provided excellent information on the food security problems of the countries it covers, but it has not been future oriented. In other words, the information in the reports listed above is accurate, actionable and precise but only reflects the current situation. In order for USAID to begin to move food to an area that requires assistance, many things need to happen and six to nine months will pass before the first grain of imported cereal arrive unless heroic measures are put into place (including fantastically expensive helicopter air lifts of food aid to Niamey from coastal zones). In order to more effectively meet these information demands, FEWS NET is beginning to integrate into its reporting structure a monthly requirement that its food security analysts speculate about the aid needs of the people of the area two to three seasons ahead.

FEWS NET can maintain its excellent reputation regarding the quality of its information while simultaneously providing information about potential future food insecurity because the precision of information required about problems in the future is very low. If a problem is anticipated that is clear and actionable, then USAID and other international aid organizations will request assessments to be conducted in the region in order to prepare for humanitarian assistance. Other steps will be taken which are preparatory and necessary but of little expense, thus if by the next FEWS NET monthly report the threat is lessened, little is lost. However, if no warning is given, then the response to a humanitarian crisis will likely be delayed until all the planning can be completed. More discussion of decision support and how politics influences FEWS NET's activities can be found in the last two chapters of this book.

9.2 Nutritional Assessments

Vulnerability to climate hazards, economic shocks and other threats to food security is often uneven in a community. Some individuals and households are more susceptible to emergencies or crises than others, and thus an analysis of who the most vulnerable are and how they are responding to a shock or crises identified by the early warning community is essential before a response can be prepared. Remote sensing data can provide detailed information on where and by how much cereal production have declined due to drought, but it can say nothing about the likelihood that the people living in that area are suffering food insecurity as a result. The most direct measure of peoples' ability to feed themselves is food consumption. Unfortunately, there are significant methodological difficulties in directly measuring food intake for large numbers of people in diverse communities, even during periods of calm, therefore it is impractical during emergencies. Consequently, less direct measures or indicators must be used (Young et al., 2001). These include:

- significant shifts in the major sources of food which cannot be compensated for adequately by other sources;
- the impact on nutritional status.

Shifts in the nutritional status of children are particularly significant indicators of a widespread underlying food security problem that has resulted from these changes. Data on acute malnutrition has therefore become the basis for many decisions of the intensity, characteristics and timing of humanitarian aid during a crisis. Figure 9.1 shows that in many regions that may experience food insecurity, assessments are carried out every year. In places where FEWS NET has identified early warning indicators that suggest an impending problem, assessments are the first things that are done to determine the scope and severity of the problem. Here we will briefly describe these assessments, how they are carried out, and the information they provide to FEWS NET for their analysis.

The term assessment refers generally to surveys conducted in a statistically valid way for determining the severity of a crisis, advocacy or triggering a response and program planning, monitoring and evaluation. The exact design of a survey depends on the intended use of the information, which in turn is determined by the organization that carries them out. Some organizations gather information to determine the composition of the food aid provided, others try to determine the underlying causes of malnutrition with a survey. A recent World Food Program survey in Darfur, Sudan, covered the entire crisis-affected population to estimate the impact of the overall humanitarian response on food security and nutrition (Young and Jaspars, 2006). Early warning systems, whose main focus is to provide information of an impending crisis, use information from assessments in order to monitor nutritional status in the most vulnerable populations, and to detect a deteriorating situation, in order to prevent a serious famine in the future. Once a relief operation has started, survey information is used to target operations and to monitor impact of the activity.

Assessments of malnutrition typically measures malnutrition of children under the age of five. The rationale for estimating under five years mortality rather than all-age mortality is that the under five years population is an early warning population, in that mortality is likely to rise in this population before it rises in the general population. Surveys typically measure acute malnutrition with two measurements: mid upper arm circumference and weight-for-height. A child's weight varies with their height and sex, so any raw measurements must be standardized. For each child, their measurements are compared with a reference value drawn from international growth standards. These are assumed to reflect normal child growth under optimal conditions, regardless of ethnicity, socio-economic status or feeding. Recent studies have shown that the obvious differences in different populations in weight and height become evident at the onset of adolescence, before then the growth of children in different populations can be very similar (WHO, 2006).

Mid upper arm circumference changes little between one and five years of age, which means that the actual measurement can be used instead of standardizing using a growth reference. Used with weight-for-height, this measurement is very useful as it requires little training and equipment and does not depend on prior knowledge of the child, such as when he or she was born. Weight for height is a widely used index that measures either acute recent weight loss or failure to gain weight within a relatively short period of time due to lack of nutrients. Recovery can be rapid once optimal feeding, health and care are restored. There are many malnutrition-induced diseases, which are described, along with ways to measure them, in Young and Jaspars (2006).

In order to assess malnutrition effectively, nutrition surveys must be conducted in a way that they are comparable, robust, and statistically valid so that the results can be generalized across the entire population. Over the past two decades, methods for collecting information on nutritional status have been standardized probably more than any other crisis indicator, with the result that acute malnutrition has become one of the most reliable indicators used in emergencies (Young and Jaspars, 2006). The standardized sampling framework used in most surveys is called a two stage cluster 30 by 30 method, resulting in sampling of 900 children. The design calls for selecting 30 cluster, or groupings of neighboring households, and then measuring 30 children within each cluster. The first household within a cluster is selected randomly, and the children within that household are measured, and the remaining households in the cluster are selected by visiting the closest households until the sample of 30 children is complete.

This survey methodology is well suited for emergency situations because it does not require a list of households in the target population, and because the households do not need to be organized in a regular pattern. It is also quicker than a traditional random sample because travel is required only between 30 different locations. The clusters are selected by listing the population according to smaller administrative or geographic units, typically villages, with their population size and cumulative population total. Thirty clusters are randomly selected from the list using a technique known as 'probability proportionate to size', which ensures that all children have an equal chance of being selected, regardless of the size of their community. This sampling design precludes the need for calculating sample size, as 30 clusters of 30 children always produces results of acceptable precision (Young and Jaspars, 2006). Drawbacks and other details of this methodology are detailed in the excellent Young and Jaspars 2006 report.

Early warning organizations such as FEWS NET use nutrition survey information in their analysis and reporting, but do not conduct or implement the surveys themselves. FEWS NET is fully integrated into the humanitarian community and thus is usually very involved in organizing and coordinating any response that may be required into a food security crisis. Because nutrition surveys are the most reliable and systematic way to determine the severity, distribution, and required response to food insecurity, they remain a key part of the work that FEWS NET does in providing timely and actionable information to decision makers.

9.3 FEWS NET Products

To serve the diverse community of stakeholders involved in ensuring food security in a region, FEWS NET produces a wide array of reports and briefings focused on different audiences. FEWS NET personnel make numerous trips to troubled regions to ensure that the information is accurate, comprehensive and is distributed to as many people as possible. Table 9.2 shows the variety of reports focused on various levels that FEWS NET produces. In order to ensure that the information of impending food insecurity is taken seriously by decision makers, it needs to be easy to interpret and consistent across all products. This is far harder than it seems, as these reports are written by different people at different geographic locations. By having a clear, understandable, and actionable message, FEWS NET can promote consensus by all the various stakeholders in a region. There are a wide variety of distribution mechanisms, including the web site, email, hard copy, briefings, networking, and dialogue opportunities.

Using assessment information and information gathered in the field, FEWS NET works to expose the causes of food insecurity so that efforts can be focused towards monitoring the dynamics of the most important underlying causes. If food

Table 9.2 FEWS NET reports and products

Country	Regional	Home office (DC)
National Alert	Regional Alert	Executive Overview Brief (EOB)
Monthly Report	Regional Report	Food Security Information Brief (FSIB)
Targeted messages	Regional FSIB	Ad-hoc Briefings
Periodic Thematic reports	Periodic Thematic reports	Periodic Thematic reports
Ad-hoc Information requests	Ad-hoc Focused reports	Ad-hoc Focused reports
Trip reports	Trip Reports	Trip Reports
Assessment Reports	Assessment Reports	Climate analysis from NOAA and USGS
Data and Maps	Data and Maps	Comparative Data and Maps

production variability from year to year causes food insecurity among rural population, then monitoring of rainfall, cropped area and yields is a key focus. Variations in prices that result from interannual differences in production may also cause acute problems in communities of wage laborers in towns and cities, thus local, national and international food prices must be monitored to determine the vulnerability of these populations. Figure 9.2 shows another way to relate production, consumption

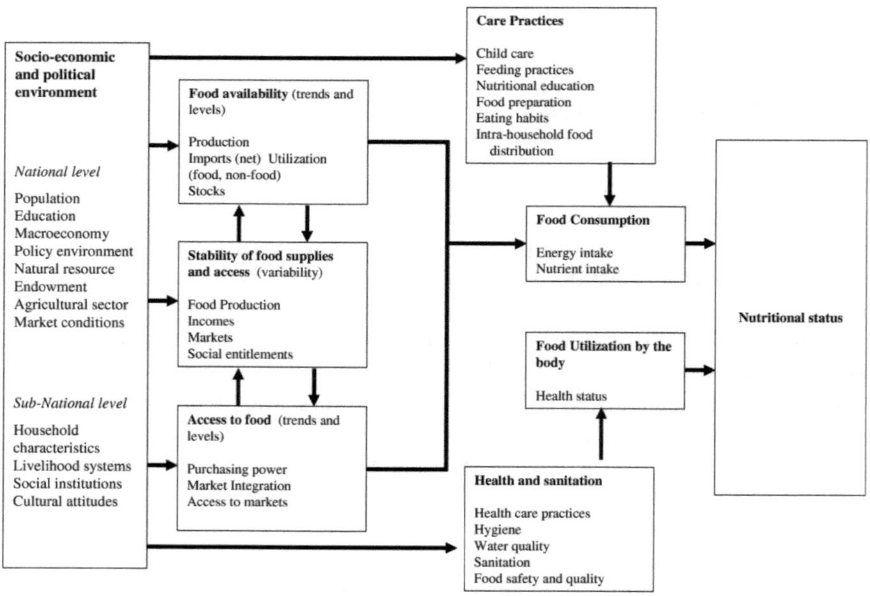

Fig. 9.2 Connecting socio-economic and political environment to nutritional status. From (IAWG on FIVIMS, 1998)

and utilization of food to overall health. The diversity of information needed to accurately portray the environment in which food availability and food access by different parts of the community, a rather long list of characteristics are important. These are all mentioned at different times in the monthly reports.

9.4 Elements in Monthly Country Reports

The primary information flow in FEWS NET about food security is from the country representatives to the home office in Washington D.C. Each country representative has the primary responsibility to monitor the situation in the country, to gather primary and secondary data, analyze the data and provide a minimum number of reporting elements each month to enable the home office to conduct comparative analysis across all countries. The reporting elements include:

- Food insecurity severity
- Geographic extent of food insecurity
- Immediate causes
- Underlying causes
- Food security outcomes – malnutrition, morbidity, mortality, destitution – from nutrition surveys
- Timeline of agricultural season
- Depending on the season, an agriculture update, possibly including rainfall conditions
- climatic hazards (floods, sand storms, cyclones, etc)status of crops (phase of development, impact on food security)
- pests – birds, insects, etc may impact production
- Reported food aid need and suggested amounts
- Food security scenario and outlook

In this rather long list, agricultural information is a fairly minor component of the overall picture. Remote sensing can give important information about food production, but says very little about food access or underlying vulnerability (Dilley, 2000). Researchers associated with FEWS NET have done work attempting to link more strongly variation in production to overall food insecurity (Boudreau, 1998; Verdin et al., 2005) (of which it can be a strong predictor), and of the variation in food prices (Brown et al., 2006). FEWS NET representatives, however, focus on providing information on the food security situation, leaving complex production and comparative analysis to analysts in Washington D.C.

Recent developments at FEWS NET have focused on improving the connection between the elements of the report and the different reporting mechanisms. By providing a minimum set of deliverables for each country, FEWS NET can better integrate the analysis across countries and compare the situation between regions. This comparative analysis is very helpful for decision makers at the international level who may need to allocate scarce resources to the locations where it is most needed.

9.4.1 Alert Levels

As part of its routine reporting, FEWS NET assigns every country an alert level. This alert level allows decision makers to know immediately the status of the most vulnerable, and ensure that attention is focused on the places that are most in need. FEWS NET works hard at both having an appropriate alert level for each country, and ensuring that when the situation improves the country's alert is downgraded. As many food insecure countries are in perpetual crisis of one kind or another, this is more difficult than it seems. The Alert levels that FEWS NET uses are the following:

- **Emergency**: A significant food security crisis is occurring, where portions of the population are now, or will soon become, extremely food insecure and face imminent famine. Decision makers should give the highest priority to responding to the situations highlighted by this Emergency alert.
- **Warning**: A food crisis is developing, where groups are now, or about to become, highly food insecure and take increasingly irreversible actions that undermine their future food security. Decision makers should urgently address the situations highlighted by this Warning.
- **Watch**: There are indications of a possible food security crisis. Decision makers should pay increasing attention to the situations highlighted in this Watch, and update preparedness and contingency planning measures to address the situation.
- **No Alert**: There are no indications of Food Security problems.

If a country remains on emergency for too long, attention cannot be drawn to its current condition. Great efforts are made to downgrade a country's status as soon as possible so that when things change for the worse, additional resources can be transferred as appropriate.

9.4.2 Executive Briefings

The Executive Briefing is a summary based on FEWS NET regular monitoring and reporting. The document provides executive decision-makers with an overview of the food security situation in Africa, focusing on countries covered by FEWS NET as well as non-FEWS NET countries (where possible). Countries currently under 'Alert' are highlighted and prioritized for urgent action. Issued approximately every month, the briefing is accompanied by a two hour presentation near the USAID offices in Washington D.C. These presentations are given by senior FEWS NET staff, gathering input and presenting analysis from all of FEWS NET partners and representatives from all regions.

The front page of the Briefing has a regular format, shown in Fig. 9.3. Summaries of the key points for each FEWS NET country are given, listed by alert level. A map of the alerts is given and a timeline of significant events for the three primary regions, Southern Africa, Greater Horn of Africa (GHA), and the Sahel (West Africa). The map also includes markers which show where FEWS NET has a presence and

USAID FROM THE AMERICAN PEOPLE

Executive Overview of Food Security in Sub-Saharan Africa
July 25, 2007

FEWS NET

FEWS NET Alert Status

Emergency

Highest Priority—Urgent Action Required

Chad: Low availability of, and access to, land at camps for internally displaced persons (IDPs), combined with government promises to improve security in some villages, are encouraging many IDPs to move from camps to areas with more arable land but fewer humanitarian services.

Ethiopia: Restricted trade and movement of people and livestock, combined with a high risk of flooding, will likely result in a humanitarian crisis for some of Somali Region's extremely vulnerable population.

Somalia: Cereal prices are increasing unseasonably in the south with expectations of a below-average *gu* harvest in August. Demand for food, resources and services is already high in rural areas, as many households remain displaced from Mogadishu. The closure of the main market in Mogadishu has affected both the import capacity of traders and trade flows to other regions of Somalia.

Zimbabwe: The 2006/07 harvest will only cover 55 percent of cereal requirements. Poor policies, hyperinflation and foreign currency shortages will further limit food access, especially in the southwest and urban areas.

Warning

Urgent Action Required

Djibouti: Poor March to May rains have stressed animal conditions in inland pastoral areas. Income-earning opportunities for the urban poor are decreasing as the hot season progresses. Increasing food costs are significantly limiting food access for the urban poor and in most rural areas.

Kenya: Pastoral recovery is at risk as recommended interventions have not been implemented by the government and donors. Some pastoral and marginal agricultural households will likely experience significant stress before the start of the short rains in October.

Watch

Preparedness and Monitoring Required

Mozambique: Food prices are above average in the south as a result of maize production deficits, and food access will be difficult for households in the region until early 2008. The second season is progressing reasonably well in many southern and central regions, and initial harvests in some areas are mitigating current levels of food insecurity.

Sudan (southern): Tensions over inter-tribal conflict and cattle raiding persist in numerous areas, and could affect local agricultural activities. Disarmament is needed to enhance peace-building activities. One million people are currently food insecure as the hunger season progresses.

Uganda: Improved civil security in the north has increased household access to land and markets. However, a highly contagious livestock disease affecting sheep and goats in northeastern Karamoja Region is jeopardizing a major local source of food and income.

Food Security Alerts in Africa

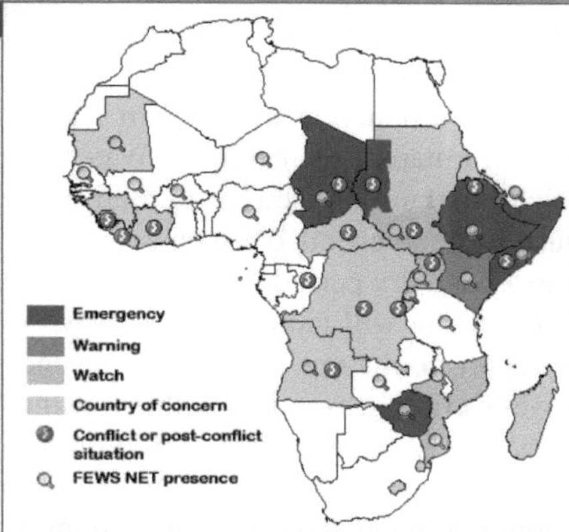

Significant Events Timeline

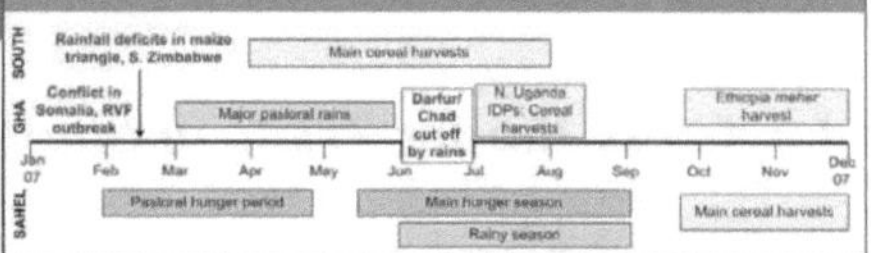

Food Aid Needs and Beneficiaries

Country	Population at Risk	Food Aid Beneficiaries
Chad	732,370 (incl.150,000 IDPs)	549,370 (incl. 176,850 purchasing subsidized food)
Djibouti	150,000	121,250
Ethiopia	9 million (estimated)	8.6 million (PSNP + emerg.)
Kenya	2.4 million	924,000
Mozambique	520,000	215,000
Somalia	1.5 million (incl. IDPs)	150,000
South Sudan	1.7 million	1 million
Uganda	2.1 million (incl. IDPs)	1.7 million
Zimbabwe	Not available	560,000 (WFP and C-SAFE)

FEWS NET is a USAID-funded activity. For more information, please visit www.fews.net or email: info@fews.net.

Disclaimer: The views expressed in this publication do not necessarily reflect the views of the United States Agency for International Development or the United States Government.

Fig. 9.3 Example of the Executive Overview for July 25, 2007

locations that are experiencing problems due to conflict. The significant events are primarily agricultural, with planting, harvest and hungry periods noted for each region. Particular items of interest are also included on the timeline, such as regions where food prices are abnormally high, problems with the agricultural season such as a late start of the rains, and other events. Finally, a summary of the food aid need statistics are given for all countries requiring aid during the period, as a guide for planners who must acquire and ship food aid globally.

The back side (or second page) of the Briefing is an approximately 600 word summary focusing on a particular issue of interest. The issues covered are usually regional in nature and timely given the date when the Briefing is issued. For example, in July 2007 a summary of the possible impact of the ongoing El Nino event on cereal production in southern Africa is discussed. The summary is accompanied by a map of the extended Water Requirements Satisfaction Index (WRSI) for maize croplands, based on rainfall estimates and end-of-season climatology. Because El Nino events often characterized by an early start of the rains followed by an extended dry period from January to March, there is concern that Zimbabwe and Mozambique will have poor harvests, resulting in widespread food insecurity. As the report points out, the El Nino event is weakening and therefore an average to above-average season is possible if rainfall continues. The implications of possible below-average production on food security are outlined explicitly, as the regions where rainfall is likely to be below normal are already food insecure due to the ongoing political and economic instability.

9.4.3 Cross Border Trade in Southern Africa

In southern Africa, FEWS NET and the United Nation's World Food Program (WFP) have funded a comprehensive series of monthly reports funded by FEWS NET that describes the quantities and direction of cross-border trade of maize, rice and beans in the region. Critical to understanding the variation in availability of these grains, trade between the countries has long been unmonitored. In 2004, the Technical Steering Committee of the Cross Border Food Trade Monitoring Initiative, with funding from USAID and the World Food program, began to collect data through a network of border monitors based at selected border points. Borders throughout the region were surveyed and the most active and important borders selected for monitoring. The border monitors record data on a daily basis, and transmit it to a central location every week for collation and analysis. Currently, the informal cross border trade monitoring system includes twenty nine borders, with new borders being added as necessary. Data from borders surrounding Malawi are collected and managed by FEWS NET and WFP Malawi, while the rest of the borders are managed by the Technical Steering Committee. A strong link between the quality of the growing season and cross border trade is a positive sign of functional markets as economies in the region respond appropriately to price variations.

9.4.4 Other Reports

FEWS NET produces a wide variety of other reports outside of the formal monthly bulletins and executive briefings. These include current food security analysis, projected hazard identification and analysis, analysis of effects on livelihood systems, food security outlooks (projections and scenarios) and a variety of reports on the specific impact of particular hazards. Seasonal assessments and projections as to the outcome of an agricultural season in regions that are particularly vulnerable are also often solicited from FEWS NET partners. Focused studies, baseline and contextual information on new regions, systematic agricultural monitoring when the results are unusual or important, and emergency and other assessments are all part of FEWS NET's suite of reporting. The objective of these reports is to improve the information source for decision makers who are trying to determine whether or not to provide assistance, what form the assistance should take, what kind of assessments should be requested, etc. A wide variety of information sources is needed to understand complex problems such as those that FEWS NET is trying to address.

9.5 Decision Support Through Networking

In the past, FEWS has been generally successful in reaching a narrow set of decision-makers, perhaps many more in Washington DC or in NGO and donor communities than within African governments. During the past five years, FEWS NET has decided that to reach a broader set of African decision makers, it had to produce more 'market' driven information, capable of meeting the changing needs of a new generation of African decision makers. FEWS NET must develop a better understanding of those needs while demonstrating how more analytical information, such as food economy knowledge, can help inform a broader agenda of policy challenges that African policy makers face. For instance, a better understanding of the changing nature of household livelihoods can help policy makers identify practical and lower cost ways of enlarging income and productivity opportunities of the poor.

Decision-making also takes place outside of governments. FEWS NET information through the Internet or through networks can be useful to the private sector, commercial producers, traders, and transporters of food products. Greater optimism and confidence in the ability of African farmers and commercial networks to maintain a supply of food to rapidly growing urban markets, where, in light of recent policy reforms, improved urban/rural terms of trade should ensure a much larger rural production response. In the Sahel for instance, it is generally agreed that FEWS information helps to open markets, assisting the private sector to increase food availability, as urban population increased one hundred fold in less than fifty years. The increased food supply response has meant reduced price volatility when poor rainfall reduces grain harvests, and savings to US taxpayers in the form of reduced requirements for food aid.

FEWS NET has provided information to progressively help decision makers make more informed decisions – for instance, food security policy, poverty reduction strategies, the lifting of trade barriers, and information on where to focus rural development projects. The work plan process must build on FEWS NET's reputation for reliable and relevant information. The methods of information dissemination to policy makers will depend on the targeted audience. In addition to the continually updated FEWS NET website, FEWS NET will continue to bring together governments with all other relevant actors in more policy-oriented networks in joint missions, briefings and workshops.

In the past, 'networking' has always been a part of FEWS NET's work. Then, as now, networks were considered as a means for information sharing and joint validation of recommended actions. FEWS NET's recent renewed focus on network building has also tried to bring about consensus between government and non-governmental parties on food security assessments and coordinated responses. The networks should be built around government institutions, or alternatively enabled to provide capacity enhancing support to weakened government early warning and response systems. In this regard, several different kinds of networks and country situations have developed:

- Networks that are little more than donor/NGO working groups with limited local ownership through local civil society organizations.
- Formal food security entities, mostly involving government agriculture ministries under the direction of a government office or a regional inter-governmental organization.
- Country-level networks that come together periodically by a convergence of interests during periods of crisis that stimulate temporary, but not necessarily longer term, network for coordination.
- Country situations where there are not yet consensus-serving networks and where government exhibits sensitivities to some aspects of building networks for information sharing and policy influencing purposes.
- A variant in which the 'network' is actually a series of special interest networks, each working in different but related food security domains and with separate (if overlapping) memberships.

It is clear that there are a wide variety of 'networking' constraints and opportunities in each FEWS NET country and region. Selecting which network to work with and how, are important early decisions that affect how FEWS NET carries out its work.

An example of how networking can facilitate FEWS NET's work can be seen in Tanzania, when FEWS NET brought together individuals to form the Food Security Information Team (FSIT). The network was responsible for vulnerability assessments, monitoring and evaluation of food insecurity in the country and for providing recommendations for decisions and response planning. It was an 18-member group composed of government departments, local and international NGOs, including FEWS NET/Tanzania. FSIT has developed its own terms of reference and developed regular action plans. In preparation of the establishment of the FSIT, FEWS NET organized a study visit to Kenya in March 2000. During the visit the FSIT members reviewed the collaborative systems at national and district levels in Kenya that brought together Government, UN Agencies, Donors and non-governmental organizations (NGOs) to generate early warning information and how this information is translated into response plans and implementations. In 2006, the FSIT was still functioning and Tanzania was still very food secure, despite strongly variable weather for crop production and an increasing population.

9.6 Challenges for FEWS NET in Decision Support

All of FEWS NET's activities are focused on decision support. The program is structured to provide actionable, well-supported and clear information that motivates appropriate action in response to the food security problems it sees in the field. The food security experts that FEWS NET employs have come to realize, however, that being so closely tied to USAID's Food for Peace program, with its overwhelming focus on bulk grain shipments, is a disadvantage. In its analyses of food security problems, FEWS NET focuses on both the causes of a food security crisis and on identifying appropriate responses. With its primary audience being decision makers at USAID's Food for Peace, who has almost exclusively grain at its disposal, the complexity of its message gets lost in discussions about the quantity of grain that should be shipped. Analysts at FEWS NET would like to be able to advise a much larger constituency who can provide livelihood support before the crisis occurs, who can implement urgently needed monetary programs in exchange for work, along with other responses that have been shown to be effective. Without this broadening of audience, FEWS NET is often frustrated that its nuanced analysis is boiled down to the number of food insecure people which is then used to calculate how much grain will be needed to feed them.

Another significant challenge for FEWS NET is chronic malnutrition. In some countries malnutrition levels among children is routinely above World Health Organization emergency guidelines despite good harvests, low prices and a calm political situation. In these situations, food aid will not improve the nutritional status of the population in question. Chronically high malnutrition levels are becoming more common, particularly in urban populations. The conceptual frameworks that FEWS NET uses are not well suited to identifying appropriate responses to these problems, particularly because its focus is to guide decision making for emergency action. It has few ways of contributing to the larger development debate. FEWS NET's experts can discourage action, but when the global media starts reporting a problem and it becomes politically expedient to respond, USAID may do so despite FEWS NET's advice. FEWS NET would like to broaden its audience in order to improve the underlying livelihood situation in which it conducts its work, instead of continually responding to short term crises.

9.7 Summary

This chapter focused on the reporting mechanisms that FEWS NET uses to support decision makers. Information flows from FEWS NET in country representatives through reports to Washington DC, to tailored products to a myriad of decision makers at the local, national, regional and international levels. The timing of information and the requirements of decision makers was presented, along with a summary of reports that attempt to meet these requirements. A summary of how the humanitarian community determines the severity of the crisis was presented. FEWS NET's suite of reports and reporting mechanisms was described along with a summary of their content. Networking and its critical role in FEWS NET's work was also described.

References

Boudreau, T.E., 1998. The Food Economy Approach: A Framework for Understanding Rural Livelihoods. RRN Network Paper 26, Relief and Rehabilitation Network/Overseas Development Institute, London.

Brown, M.E., Pinzon, J.E. and Prince, S.D., 2006. The sensitivity of millet prices to vegetation dynamics in the informal markets of mali, Burkina Faso and Niger. Climatic Change, 78: 181–202.

Buchanan-Smith, M., 2000. Role of early warning systems in decision making processes. In: D.A. Wilhite, M.V.K. Sivakumar and D.A. Wood (Editors), Monitoring Drought: Early Warning Systems for Drought Preparedness and Drought Management. World Meteorological Organization, London, pp. 11.

Buchanan-Smith, M., Davies, S. and Petty, C., 1994. Food Security: Let Them Eat Information. Institute for Development Studies, London, England.

Choularton, R., 2005. Improving Decision Making and Response Planning: A Framework for Contingency and Response Planning. FEWS NET, Washington DC.

Dilley, M., 2000. Warning and Intervention: What Kind of Information does the Response Community Need from the Early Warning Community, USAID, Office of US Foreign Disaster Assistance, Washington DC.

IAWGonFIVIMS, 1998. Guidelines for national food insecurity and vulnerability information and mapping systems (FIVIMS): background and principles, UN Food and Agriculture Organization, Rome.

Verdin, J., Funk, C., Senay, G. and Choularton, R., 2005. Climate science and famine early warning. Philosophical Transactions of the Royal Society B: Biological Sciences, 360: 2155–2168.

WHO, 2006. World health organization child growth standards. Acta Paediatrica, Special Supplement, 450pp.

Young, H. and Jaspars, S., 2006. The Meaning and Measurement of Acute Malnutrition in Emergencies: A Primer for Decision Makers. Overseas Development Institute, London, England.

Young, H., Jaspars, S., Brown, R., Frize, J. and Khogali, H., 2001. Food-Security Assessments in Emergencies: A Livelihood Approach. 36, Humanitarian Practice Network (HPN), London, England.

Chapter 10
Use of Remote Sensing in FEWS NET Country and Regional Offices

Most of FEWS NET's primary data gathering, analysis and communication activities occur at the offices in each FEWS NET country. Its national representatives are primarily responsible for both knowing what is happening in the country and providing evidence which can be used to document the situation. Country representatives are primarily social scientists, trained in agricultural economics, nutrition, development studies, even anthropology. This training provides each FEWS NET representative adequate expertise in food security analysis to run operations in the field. They rarely also have the technical expertise to use remote sensing data as well.

FEWS NET has therefore extended its field personnel by employing four Regional Scientists through the USGS to provide technical assistance in the use of operational remote sensing products for food security analysis. As of 2007, FEWS NET has twenty five representatives in country offices, and four regional offices with regional representatives. In each regional office, there is also who is responsible for assisting with integration of biophysical information with food security analyses. The four Regional Scientists work in partnership with institutions in the Sahel, Greater Horn of Africa, Southern Africa, and Central America to provide technical assistance in the use of operational remote sensing products for food security analysis by FEWS NET, regional and national early warning systems, and international organizations. Formally, they:

- ensure timely and uninterrupted access to NDVI, RFE, and derivative products (e.g., WRSI, SOS, Moisture Index) by food security analysts in the region;
- participate in the interpretation of satellite-derived products to identify drought and flood anomalies of potential significance to regional food security;
- support the development of capacity at regional institutions to produce remote sensing and derivative information products that add value to FEWS NET continental products through higher spatial resolution, greater specificity of information content, or other enhancement; and
- assist in the production of regular bulletins and special reports in association with overseas and/or US-based FEWS NET partners.

Because FEWS NET seeks to create a strong network of informed food security professionals in each country, working with regional partners and networks, the presence of these field scientists improves the integration of remote sensing into FEWS NET's work.

There are many challenges to using remote sensing information in a country office. The need for accurate, accessible information regarding the local agricultural conditions or biophysical hazards such as floods or drought is great, but the data available often is poorly suited to the demands. A recent professional review of FEWS NET data users showed that they want high resolution data with a long time series at daily intervals. Most of all, they want data that reflects the conditions they see on the ground and are very unhappy when the global datasets that are available do not do so. It is very frustrating for FEWS NET representatives in the field who invests considerable effort in explaining what an indicator like the Water Requirement Satisfaction Index is to decision makers, including how to read the plots, providing accessible imagery in a timely manner, and when they succeed at having decision makers look at it regularly, it does not reflect conditions on the ground as they know them. The reasons for imagery to depart from actual conditions are numerous, including sensor error or malfunction, incorrect model parameters, changing crop planting dates or crop composition that are not reflected in the model, systematic bias in the product, or simply inadequate spatial resolution. Despite these challenges, remote sensing data is in demand in most country FEWS NET offices, and the products can provide an important perspective on agricultural conditions in the region.

The review of data users showed that there is a great diversity of analysis demands and logistical constraints. Participants of the review wrote of the differing demands required by analysis that can track both slow onset hazards such as drought and extreme events such as cyclones and flooding. They described varying climate regimes which may be poorly captured by the available datasets. The digital infrastructure available to them and their end-users places limits on the size of data sets that are available. The following comment is representative reviewers' observations.

> I work at multiple spatial and temporal scales. As mentioned, I think many of the answers to these questions [regarding remote sensing data requirements] are highly driven by applications and product type. Applications of data for flood monitoring are obviously dependent on high temporal frequency/short latency data such as rainfall, rainfall forecasts, stream flows, run off anomalies. Vegetation/crop monitoring and modeling are not as time dependent and can utilize longer periods of both latency and frequency. The spatial requirements are also variable by region. We are increasingly experiencing the need for having the ability to monitor at much finer scales . . . whether it be vegetation and rainfall monitoring over small land masses or cropped area delineations for small localized fields . . . finer spatial resolutions are becoming more of a necessity than a luxury.

This chapter summarizes the responsibilities and activities of the regional scientists in FEWS NET and describes how remote sensing data is transferred into the region, distributed by the regional representatives and FEWS NET representatives, and ultimately used by decision makers. It will describe the approaches and tools that they have derived in order to improve the ability of these local users to derive meaningful information from the remote sensing products in the field to improve decisions.

FEWS NET's information gathering, distribution, networking and action to ameliorate food insecurity are, by definition, local events, thus our focus for this chapter will be local.

10.1 Building Capacity

The capacity strengthening and institution building elements of FEWS NET are perhaps the project's most challenging objectives. FEWS NET has to be comprehensive and creative in determining where and how networks can act to enhance country capacity because the demand is large and resources are constrained. Network processes such as conducting a food security assessment can lead to consensus and transparency in analysis and estimates. FEWS NET is already taking some key steps to demonstrate that reliable and locally available capacities exist for decision makers. Any work activity of FEWS NET and networking partners is seen as an opportunity for capacity strengthening. FEWS NET information, whether accessed through the Internet or rural radio, can be employed by users to expand capacities at all levels from villages to capitals. When networks are enabled with information to arrive at consensus contingency plans, then capacities for coordination have been strengthened, and arguably, the ability to disseminate information to decision makers also strengthened through the consensus.

FEWS NET provides training to senior and technical local government staff on a variety of technical subjects, mostly focused around conducting analysis and providing support to technical teams. For example, they provided the following training in the year 2000 in West Africa:

- Thematic mapping training for two senior database officers in Southern Sudan to enable enhanced visual quality of the graphics in food security reports;
- Trained Mali's National Early Warning System technical staff to download, use and interpret NDVI data and on the use of free software for opening, mapping and analyzing data;
- Training of key network partners in Mauritania was done during the year, including training Food Security Commission agents on how to use free price analysis and data management software; and
- Extension Agents in Mauritania on how to use ArcGIS (GIS/thematic software) and free WinDISP software in order to conduct satellite imagery analysis;

These training sessions are just a few of the many given each year by FEWS NET personnel, particularly by USGS field scientists who have expertise in remote sensing and GIS. They report that the turnover of the government people whom they train is such that they really need to provide training on an annual or more frequent basis. As training is provided to key partners on vegetation index and other analytical remote sensing datasets, the partner's acceptance of these datasets as a proxy for food production grows as does their effectiveness as an integrated part of the food security information system.

With public sector budgets in most FEWS NET countries severely constrained, following years of economic adjustment and low growth, erosion of capacity has occurred throughout the civil services in many governments. It is probably not realistic to think about measuring progress in capacity strengthening in terms of increased budgeting outlay from FEWS NET countries for offices devoted to early warning, disaster response, or contingency planning. In order to be sustainable in the long term, centers or organizations that can conduct geospatial analyses for early warning organizations need to be funded by local governments for a wide range of applications, not just food security. FEWS NET works to support such centers in the effort to increase the effectiveness of local networks. A good example of such a center is the Regional Center for Mapping of Resources for Development (RCMRD) in Nairobi, Kenya. This center has grown by over 200% in the past five years and has become a vigorous partner in much of what FEWS NET does in the region. By demonstrating the utility of geospatial data for wider economic development and for the provision of critical information to the agriculture sector of the economy, such centers can become viable organizations that improve the understanding of and drive demand for remote sensing data and analysis in the region.

10.2 Remote Sensing Data and Local Food Security Specialists

The remote sensing data products that FEWS NET uses are created in the United States and posted at the USGS Africa Data Dissemination Service (ADDS) server (see earlywarning.usgs.gov). Every ten days, updated jpg images of the products (such as the rainfall anomalies and WRSI updates) are also emailed to FEWS NET representatives for download. The binary data as well as the jpg imagery are available for download from the ADDS server via ftp. Some more technically capable FEWS NET representatives download the binary data to analyze current conditions. In the 1990s, FEWS NET funded the development of WinDISP, a free custom software package that is simple and easy to use. The software enables its users in the field to manipulate binary images, make differences from long-term means, from previous and other analyses simply using a graphical user interface. Although it has become rather dated, WinDISP continues to work well with the standardized image formats produced by the USGS.

Most FEWS NET representatives, however, do not usually manipulate the binary data, relying instead upon the jpg images created from the data for their report. The limitation of this approach is that the data are regional, at a coarse spatial resolution which may not show the appropriate level of detail needed. In addition, specialized maps cannot be created that appropriately focus attention where it is needed for decision making in the region. A good example can be seen for Mozambique during February of 2007. The Zambezi River basin received over 200% of its normal rainfall since the beginning of the 2007 rainy season, resulting in a flooding situation for low-lying counties next to the river. It is hard to see this with the standard FEWS NET flooding product, however, as it accumulates

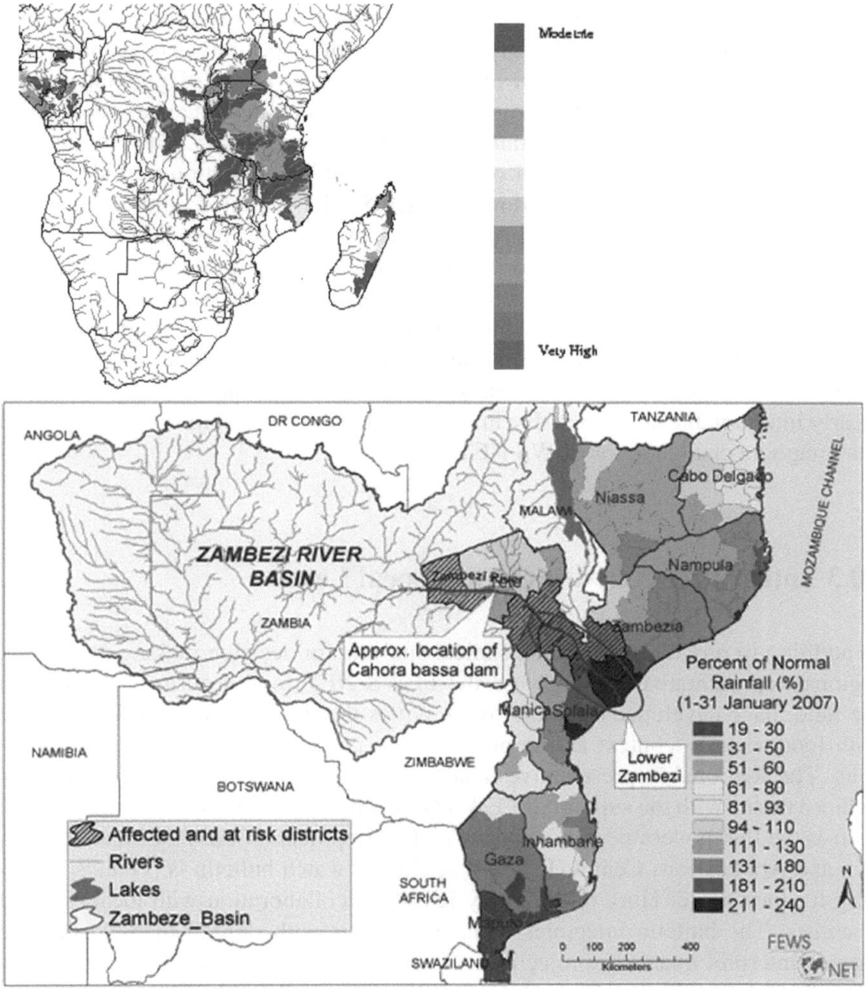

Fig. 10.1 Flooding from the BERM model for southern Africa, Feb 1–10, 2007 showing only moderate floods with a much larger spatial extent (**top**) compared to analytical figure showing areas of flooding from the Mozambique Country Report during the previous month, February 9, 2007 (**bottom**)

rainfall only over the past ten days, not the entire period. Figure 10.1 shows the standard USGS Basin Excess Model product for the first ten days of February, 2007 compared to the specialized analysis that was in the February Mozambique monthly report.

Without access to the original data and the ability to analyze and produce figures, the real situation on the ground cannot be demonstrated effectively for decision makers. Even the state-of-the-art, best available continental products developed to show particular problems like flooding and drought have fixed temporal periods and means, and thus have difficulty clearly showing specific regional problems. Mozambique has a complex growing season, with a variety of growing season lengths, flooding and cyclone hazards, and food production calendars, depending if you are in the northern or central/southern part of the country. Thus to make an effective figure, specific selection of means and of accumulation periods for the analysis was required. This was done by FEWS NET-affiliated rainfall experts in the United States for inclusion in the report, not by users in Mozambique.

A significant challenge, then, for FEWS NET's effectiveness in working through its local networks is the ability to train its local representatives and key partners in the effective use of remote sensing information. Significant communication can occur using effective maps based on remote sensing, but knowledge, skills and data access are all required in order to transform data produced at the global level into clearly interpretable and actionable maps. Transferring these skills to the local level is an ongoing challenge for FEWS NET.

10.3 Integrative Activities at the Local Level

In addition to providing specific technical analysis for decision making, the USGS regional representatives, along with the FEWS NET food security specialists in the field, have developed a number of activities that integrate climate information with food security context and implications, to produce truly interdisciplinary results. The Regional Representatives analyze climate trends and local vulnerability to those trends with the support of FEWS NET representatives and regional partners such as the Intergovernmental Authority on Development (IGAD)'s Climate Prediction and Applications Centre (ICPAC). A climate watch bulletin is produced quarterly for the Greater Horn of Africa by ICPAC in collaboration with local regional scientists. The bulletin integrates climate forecasts with rainfall maps to provide easy-to-interpret quarterly projections of agricultural conditions which are developed through the Climate Outlook Forum. Work by the USGS regional representatives has improved the integration of such climate information with food security analysis.

10.3.1 Climate Outlook Forum

The regional Climate Outlook Forums are an important mechanism for the formulation and dissemination of seasonal climate forecasts. The meetings bring together climate scientists, operational forecasters, and climate information users to formulate a consensus forecast and to discuss the implications of probable climate outcomes for climate-sensitive sectors such as agriculture and ultimately food security. NOAA's Climate Programs Office and the United Nation's World Meteorological Organization have financed these in Africa and other regions for nearly ten years. During this time the seasonal forecasts have significantly improved, particularly in years with a strong signal from the El Nino Southern Oscillation (Brown et al., 2007).

FEWS NET participates in the meetings and through the work of its scientists transforms the probabilistic forecasts that are the official output of the meetings into rainfall forecasts, which are then used to drive crop models to show the possible impact of variations in rainfall on the performance of a particular crop. This information is then used to estimate the food security situation for the next six months for planning purposes. Figure 10.2 shows the results of using the FEWS NET developed Forecast Interpretation Tool (FIT) to translate the Climate Outlook Forum Climate Outlook Forum forecast into estimates of expected rainfall and expected rainfall anomalies. The FIT was developed in collaboration between the USGS regional representative in the Greater Horn and FEWS NET scientists at the University of California. Sector specific interpretations were analyzed to provide more specific analysis on its potential impact on the pastoral, agro-pastoral and agricultural zones.

Building on the awareness of the utility of forecasting, recent research has enabled the projection of vegetation index dynamics (Funk and Brown, 2006) and dynamical and statistical projections of rainfall (Funk et al., 2007) based on observations up to four months into the future. These datasets can be immediately incorporated into operational analyses of food security because unlike traditional probabilistic forecasts, they are in exactly the same format and units with the same meaning as commonly used observations.

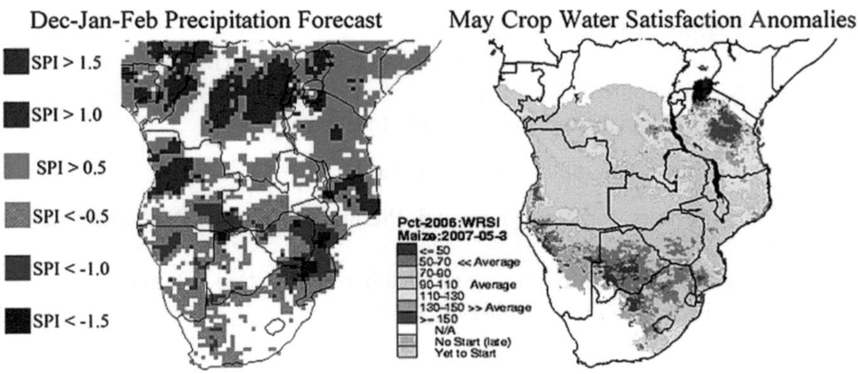

Fig. 10.2 Standardized Precipitation Index (SPI) rainfall forecasts (Funk et al., 2007), based on November observations of sea surface temperatures, precipitation and wind fields. The right panel shows a common index of crop performance, end-of-season maize Water Requirement Satisfaction Index (WRSI) anomalies

The climate outlook forum, like the one conducted by ICPAC, is a good example of the networks that are forming in highly climate sensitive regions such as the Greater Horn of Africa (GHA). The quarterly forum was initially organized by ICPAC within the framework of the USAID funded project 'Sustainable generation and application of climate information, products and services for disaster preparedness and sustainable development in the Greater Horn of Africa'. The forum is an international framework initiated by National Oceanic and Atmospheric Administration, Office of Global Programs (NOAA-OGP) and a range of partners such the United State Agency for International Development's Office for Foreign Disaster Assistance (USAID-OFDA) and the United Nation's International Strategy for Disaster Reduction (UN/ISDR). ICPAC's basic operations and the development of financial and administration systems are supported by USAID/East Africa.

Contributors to the consensus climate outlook included:

- Representatives of the Meteorological Services from nine GHA countries
  - Insititut Geographique du Burundi
  - Meteorologie Nationale de Djibouti
  - Eritrea Meteorological Services
  - National Meteorological Services Agency of Ethiopia
  - Kenya Meteorological Department
  - Rwanda Meteorological Service
  - Sudan Meteorological Authority
  - Tanzania Meteorological Agency
  - Uganda Department of Meteorology
- Climate scientists and other experts from national, regional and international institutions and organizations, including
  - ICPAC
  - Drought Monitoring Centre, Harare
  - Columbia University's International Research Institute (IRI)
  - World Meteorological Organization
  - University of Nairobi and Beijing Climate Centre
- Additional climate data inputs were provided by
  - National Centers for Environmental Prediction/Climate Prediction Center (NCEP/CPC)
  - European Center for Medium range Weather Forecasting (ECMWF)
  - the United Kingdom Meteorological Office
  - USGS
  - FEWSNET

Because all these local, regional and national actors are involved, the product is widely used in the region and trusted.

The Food Security Outlook Forum (FSOF) is an attempt to use the seasonal rainfall forecasts provided by the climate forum to understand potential food security outcomes. Precipitation forecasts are put into a food security context and then used to derive possible food insecurity scenarios. The food security context is provided by an in-depth analysis of livelihoods in current food security hotspots and of underlying causes of food insecurity. These underlying causes may be climate or non-climate factors (such as civil wars, pest and diseases outbreaks, high prices levels, poor integration to regional markets, and land degradation). The FSOF has relied heavily on two important inputs: (1) the reports of current food insecurity and its causes, prepared by food security networks in each country; and (2) the rainfall forecast issued by the forum following intense joint working sessions with some world meteorologists and climate forecasters from each country were able to capture the observed variability, and that these were integrated into the outlook effectively.

At the beginning of each Forum, partners begin by evaluating the success of the previous outlook in order to improve the reliability and utility of the process. The outlook that was issued in March 2005 (forecasting conditions to July 2005) was compared to the situation prevailing in the second half of August 2005. This outlook was found to have been accurate in most areas. It proved, however, to have been inaccurate in some areas, either because it was too pessimistic (e.g., northwestern Kenya; eastern Ethiopia; and northern, central and eastern Somalia) or too optimistic (Tanzania). The FSOF and its products, still being improved, intend to provide useful scenarios to inform decision-making and planning.

FEWS NET is at the center of these network building exercises, maximizing synergy between the individuals and institutions all working on the same issues in the same region. The complexity of these overlapping organizations and institutions can inhibit collaboration, but much can be done to assist in collaborative and constructive engagement of the individuals who represent the institutions. Regional representatives are focused on providing linkages between the climate data, conclusions which they support, and the food security impacts.

10.3.2 Improvement of Local Rainfall Data

An important goal for remote sensing scientists at the local level is to improve the rainfall estimate. The Regional Representatives support RFE improvement activities by collecting local observations of rainfall at as many locations as possible. By collaborating with meteorological services, historical rainfall time series data can be obtained and used to improve the satellite observations in the area. Improved observation networks can be established and maintained through time. These improved data are needed to understand climate's contribution to food insecurity of vulnerable pastoralists and marginal agricultural communities in the region, and to determine if there are long term trends in the rainfall amounts which may undermine food security in the long term.

The Regional Scientists provide agro-climatic inputs to regular bulletins/reports which address food insecurity at a regional level, as well as improve the data locally. They provide feedback for the Africa Weather Hazards and Benefits Assessment conducted in the United States, as well as local monthly reports, thematic reports, ad-hoc decadal rain watch bulletins and flood advisories. Rainfall data can be used to inform water resources development and agricultural production planning efforts. Through the integration of considerations of climate risk in food security analysis and reporting at national and regional level, the region can improve its ability to factor climate risk into food insecurity estimates.

10.3.3 Crop and Rangeland Monitoring with Local WRSI

Having a regional implementation of a continent-wide or global remote sensing based product is one way to improve the accuracy and reliability of monitoring datasets. Regional Scientist Tamuka Magadazire, in collaboration with partners at the University of California Santa Barbara, have implemented a version of the WRSI that can incorporate knowledge about the local climate. The GeoWRSI is a geo-spatial, stand-alone implementation of the Water Requirements Satisfaction Index (GeoWRSI), as it is implemented by the USGS. The program runs a crop-specific water balance model for a selected region in the world, using raster data inputs. The program produces a range of outputs which can either be used qualitatively to help assess and monitor crop conditions during the crop growing season, or can be regressed with yields to produce yield estimation models and yield estimates. Other tools are available to post-process the GeoWRSI outputs so that they can be used in yield estimation models (Magadzire and Funk, 2006). The key inputs to the WRSI are crop maps, start and end of season parameters, and rainfall maps can all be customized using the best available information. Running this model locally and then comparing with the operational, global USGS WRSI model is a good way to improve both the usage and understanding of the model locally. The output can be validated with actual crop status at various crop stages (field crop assessments), as well as with production estimates at the end of the season.

A companion tool to the GeoWRSI is called the Water Balance Post Processor (WBPP) is a program that facilitates yield modeling and estimation from some water balance models. The WBPP uses outputs from a water balance modeling program and reformats them into summaries that are compatible with historical yield records. It can basically be used to facilitate yield modeling and yield estimation from water balance models. The program has four primary uses:

1. Combined analysis of historical yield records and historical water balance analyses to produce data tables that are ready for regression modeling. This regression modeling would result in yield estimation models.
2. Extracting summaries from up-to-date water balance analyses so that these summaries can be used as inputs into water balance models in order to derive up-to-date yield estimates and forecasts.
3. Combining outputs from the point-based FAO AgrometShell and the raster-based GeoWRSI into a set of improved water balance outputs.
4. Calculating summaries from water-balance analyses for specific regions of interest such as districts, provinces or countries.

Water Balance modeling programs such as the Agrometshell (AMS) and the GeoWRSI produce outputs that are useful in food security analysis, in particular for yield estimation. The outputs from these programs are geospatial datasets that are in point vector format (in the case of the AMS) and raster grid format (in the case of the GeoWRSI). These outputs are generally not compatible with historical yield data, which tends to be produced and archived by administrative units such as provinces and districts. Because of this difference, translating the outputs from these water balance models into a format that can be used directly for yield estimation is a laborious and time-consuming process. The WBPP considerably simplifies the process of comparison by enabling the user to select outputs from the water balance models, and automatically reformats these outputs into a format compatible with historical yield records. It also combines the water balance outputs with the historical yield records, thereby simplifying the regression process. Finally, it combines the point-based AMS with the raster-based GeoWRSI to produce enhanced water balance outputs.

By coupling the GeoWRSI-based yield statistics with on-the-ground measurements and estimates from other remote sensing products such as NDVI, FEWS NET can have more diverse approaches in yield estimation, promoting the 'convergence of evidence' approach. In order to have good local yield figures to compare the satellite-based estimates to, FEWS NET is working with partners such as AGRHYMET, the Niger Ministry of Agriculture and Ministry of Animal Resources to transcribe the primary datasets metadata and create a database of comparative figures.

10.3.4 Validation of WRSI Product

To ensure that the WRSI is as useful as possible, validation efforts are undertaken in the regions where the product is used most frequently. A USGS/USDA/FEWS NET/Ministry of Agriculture from Kenya mission took place in June-July 2005. The mission's aim was to validate the Water Requirement Satisfaction Index (WRSI) outputs. Through field verification, the performance of these products in the maize growing zones of Kenya and Tanzania can be improved (FEWSNET, 2005).

The main results of the recent mission were GPS-linked digital photos for selected maize fields (Fig. 10.3) and other land cover types along the route followed during the trip. Meetings and interviews were also held with the Ministry of Agriculture extension officers and farmers to obtain information on start of the season, types of seeds used, fertilizer use and harvesting dates. By and large, it was confirmed that the conditions simulated by the WRSI matched well with the actual conditions on the ground.

Adjustments to the WRSI monitoring tool were made at the end of the mission to take into account the real field conditions. For the Kenyan highlands, for example, the tool was adjusted after finding that the growing season was over 6 months long, and that most planting was done in February in southwestern Kenya. This knowledge was critical in making the tool more realistic. It was also recommended

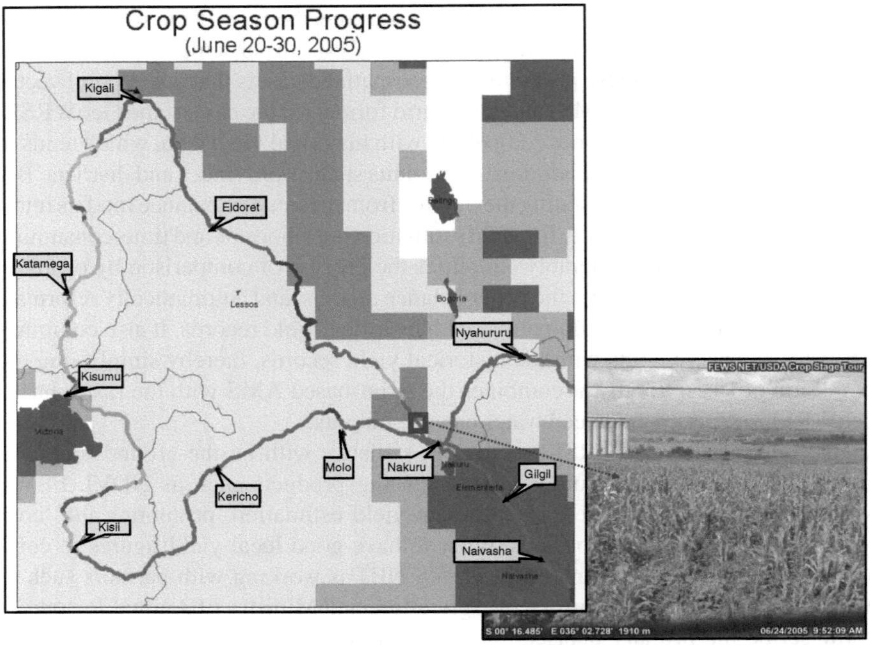

Fig. 10.3 The joint crop tour mission map in Kenya and Tanzania, with the Africover herbaceous crop overlay on WRSI. Images courtesy of Curt Reynolds, USDA

that new maize varieties and farming practices currently being introduced in the region be regularly monitored in order re-adjust the WRSI tool and to complement it with other mechanisms if required.

10.3.5 Improved Accessibility of Geospatial Biophysical and Socio-Economic Datasets

In order to effectively use remote sensing data, local datasets need to be developed that collect information, capacities and standards that will enable the integration of nutrition, economic, and biophysical information to better support to relevant field operations in Africa (Rusu, 2006). Improved geospatial data exchange and food security status mapping with spatial analysis are a critical focus of FEWS NET. Maps that juxtapose the food security status and its context with biophysical information is an important first step to better understanding of food security reports. Regional representatives work on data collection and preparation of country census maps, administration, topography, land use, land cover, livelihood and other relevant geospatial data. Development of country contextual map posters and web-based information with integrated food security information is a key way to disseminate to stakeholders critical food security messages. A wide variety of geospatial information, including accurate administrative borders, coastlines, population densities, urban extent, farmland, markets and other information must be available locally for use in such illustrative maps. Support from NOAA and USGS to ensure that remote sensing information is easily imported into GIS software is a critical first step to populate these databases, and Chemonics International is moving to a web structure that can store and display spatially explicit information. Eventually, FEWS NET hopes to have an integrated system where raster, vector and point data are available via the internet to all, regardless of their location.

10.4 Flow of Information from Satellite to Analysis to Decision Maker

The satellite-derived information that FEWS NET uses originates from US and European sensors. Scientists in the US have been funded by FEWS NET to develop products that can be used to estimate crop production variations from year to year. These products have been produced to be continental in scope and static in their methods of production. Thus the NDVI, for example, is calculated in the same way every ten days for everywhere on the planet, regardless of its efficacy in identifying the local problems. The data is processed automatically using algorithms and production software at NASA Goddard Space Flight Center in Greenbelt, MD, and is delivered to the USGS EROS data center in Sioux Falls South Dakota. The Rainfall Estimate (RFE) is produced at NOAA in Camp Springs, MD, and derived products are produced at EROS after NOAA delivers the RFE. Anomalies and jpg images of the products are created and emailed to the Regional Representatives. The satellite binary data is made available on the Africa Data Dissemination Service (ADDS) server for download by the public and by FEWS NET representatives.

Subsets of the standard products are used by FEWS NET representatives in their monthly reports to illustrate the current state of crop production, but often the standard product cannot tell the whole story. Local information is included in the analysis, such as local rain gauge information that records excess precipitation, such as in the example given above. Analysis of the satellite imagery is conducted in the field to illustrate the current conditions, and these images are put with explanatory text into the FEWS NET product and distributed to the decision makers in the country. Decision makers often include the president of the country, ministers of agriculture, food security, and health and regional and local civil servants who are critical for a functioning, responsive government.

FEWS NET uses a variety of distribution mechanisms. The primary mechanism is the web page, http://www.fews.net and email. The web page has a number of reports and data links where users can obtain information on all of the wide variety of activities that FEWS NET is involved in. The web site has evolved through time, changing and growing with the focus of the organization. Recently, Chemonics has begun to change the underlying infrastructure of the web site to improve the rapidity and ease of information transfer between the field offices and from the field offices to the home office in Washington D.C. By utilizing networking software that enables instantaneous file sharing, the web site can become a tool for improved information flow.

In many countries where FEWS NET works, internet access is slow, unreliable, and frequently unavailable and thus email, although important, is not always the most convenient form of communication. Field representatives prepare color copies of each report per month, per country and deliver them to key partners such as government ministry offices, local government officials and non-profit food security organizations. These reports are used as critical input to local and national decision making, particularly when there is a problem, and decisions need to be made regarding where resources should be spent to estimate vulnerability.

10.5 Challenges for FEWS NET

The humanitarian field is always changing. Not only are the ideas that food security experts use to identify and respond to crises always evolving, but the context in which humanitarian relief is funded, planned for and delivered also changing. The political landscape, in particular, has changed significantly in the past five years which impacts how FEWS NET does its work. This is because its main focus is supporting the humanitarian decision making of USAID personnel, who must respond to the larger political and economic pressures on the US government. In order to respond to these larger forces, FEWS NET adjusts its activities and its reporting, particularly when funds are tight.

The challenge for FEWS NET is to ensure that these changing realities are communicated to its personnel in the field. This is critical not only for data gathering and analysis that is done in the field to be relevant and useful, but also for the local FEWS NET offices to understand decisions taken in the central offices that affect their work. Communication from the center to the periphery is difficult, but there is considerable diversity in the connectedness of the different FEWS NET offices. Some are considerably more isolated than others, depending on the relationships between the FEWS NET national representatives, where the office is located and the length of time of employment of the personnel involved. In general, however, much of the broader context is not communicated, leaving frustration and isolation in the field. Improved communication could be achieved with more travel and meetings between the various partners. This takes resources that are difficult to find with the tight budget constraints of the past few years.

10.6 Summary

In this chapter, remote sensing information is used to identify and explain food production deficits in the local FEWS NET offices, and how networks of organizations and individuals work together to determine the impact of these deficits on food security were described. A key part of FEWS NET's activities is to strengthen these networks. A wide variety of meetings, reports, information products and web site information are used to improve the flow of information into and out of localities. Remote sensing data and products can be greatly improved by incorporating the information and understanding of users who know their environments very well. Decision makers profit from using remote sensing information by having fact-based evidence to illustrate their situation graphically, how it relates to their neighbors, and to previous years. By providing the actual data itself, food security specialists can request and get access to detailed analysis that can improve their ability to communicate and get an appropriate response to impending food security crises.

References

Brown, M.E., Funk, C., Galu, G. and Choularton, R., 2007. Earlier famine warning possible using remote sensing and models. EOS Transactions of the American Geophysical Union, 88(39): 381–382.

FEWSNET, 2005. GHA Food Security Bulletin: Crop forecasting and monitoring, Chemonics International Washington DC.

Funk, C., Verdin, J. and Husak, G., 2007. Integrating Observation and Statistical Forecasts over sub-Saharan Africa to Support Famine Early Warning, American Meterological Society, San Antonio TX.

Funk, C.C. and Brown, M.E., 2006. Intra-seasonal NDVI change projections in semi-arid Africa. Remote Sensing of Environment, 101(2): 249–256.

Magadzire, T. and Funk, C., 2006. The Water Balance Post Processor. University of California, Santa Barbara.

Rusu, S., 2006. Humanitarian Information Network Africa Workshop. Office for the Coordination of Humanitarian Affairs and ReliefWeb Project, Nairobi, Kenya.

Chapter 11
Population Datasets

Population estimates are a critical part of estimating the impact of a biophysical hazard. If a flood, for example, submerges a three square mile area along a river, FEWS NET needs to be able to estimate how many people live there or have fields and crops that have been destroyed. Population data is usually collected in districts or administrative regions, which usually are unrelated to the biophysical characteristics of the region. Food security analysts need gridded population data which disaggregates national census totals to maps that can give the number of people living in each region. It may be that no one lives in the submerged region, and thus the hazard has no victims and no response is required. Because FEWS NET's scale of analysis is at the country and regional level, these gridded population estimates need to be spatially extensive (global) and at a fairly fine resolution (at least one kilometer resolution). Population density maps are thus a critical tool, used at every level of analysis, to estimate the number of people at risk to a particular hazard.

Although FEWS NET primarily uses the local, official population figures by administrative region, it has entered into an agreement with the producers of LandScan data so that it may use the most up to date population density maps. The LandScan Global Population Project has developed a worldwide population database at 30 by 30 minute resolution based on the best available census counts, distributed across the landscape using probability coefficients of the population living in some areas over others, with road proximity, slope, land cover, nighttime lights and inhabited structures as the input. This chapter will describe the population data in the regions where FEWS NET works, the LandScan data as an improvement on census polygons, and the difficulties of using population data in a complex, consensus environment.

11.1 Using Population Data to Estimate Food Aid

Population figures have always been a critical part of famine early warning systems. Knowing the number of people who live in a particular area is a critical part of knowing how many people may be at risk during a crisis because of a biophysical hazard. Assessment processes estimate the number of vulnerable people, spot demographic anomalies such as elevated mortality among children, improving sampling frames for food security assessments, nutritional surveys and rapid field assessments. Figure 11.1, for example, shows how FEWS NET assigns vulnerability by district instead of by grid point. In order to figure out how many people are in each region, FEWS NET requires accurate population information. Once a food crisis has been identified, demographic information continues to be important in determining the appropriate food aid response by using population data by age and sex and in intervention evaluation. Food security specialists are expected to come up with best-guesses of populations either at risk of famine or in need of assistance in non-contiguous areas that do not fit administrative boundaries and therefore cannot be associated with official population estimates.

FEWS NET has a core set of demographic needs, which include up to date population figures, at a sufficient resolution, by age and sex cohort. A number of subgroups are of interest to FEWS NET about whom there is little information (Watkins et al., 2006). These include pregnant and lactating women, children under five years of age, orphans, those infected by HIV/AIDS or other illnesses that reduce the ability of the individuals to absorb nutrients. These are considered to be vulnerable groups because of increased nutritional needs, but estimating the proportion of the total population that fall into these groups is difficult. Fertility rates, mortality and migration rates are also of interest to FEWS NET. Mortality can be arrived at from vital statistics, death registries and surveys, but because mortality can vary significantly between communities, particularly with children, errors can be significant. All these statistics vary spatially, both in value and quality (Watkins et al., 2006).

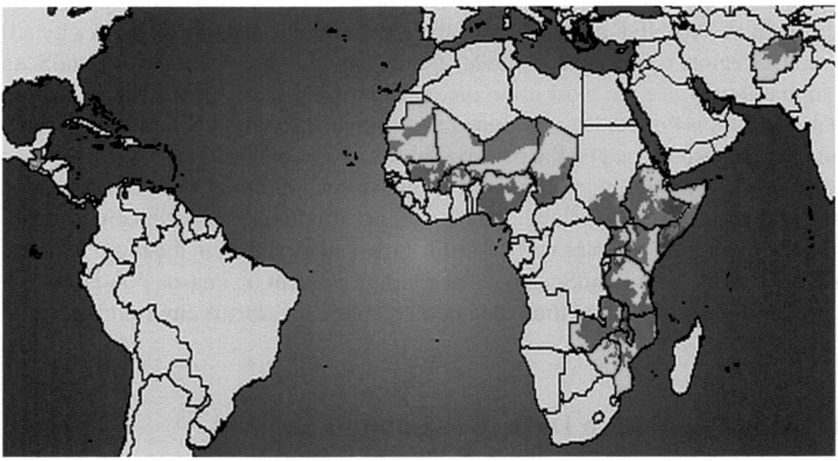

Fig. 11.1 Global food security status, March 2008. FEWS NET typically identifies sub-national districts as being food secure or insecure in order to more easily determine the number of people needing assistance. In the next five years, FEWS NET will move towards an even finer resolution, sub-district level of analysis as improved population data becomes available

FEWS NET has therefore been working to incorporate into its processes a gridded population dataset that can be used with other geographic data currently used at the country level. Knowing the population density is a critical step in providing accurate and reliable estimates of people in need. Without these estimates, variation in the number of people at risk can be extremely large. Typical problems include assuming that an administrative region's population is evenly distributed across its area. When a corner of a particular district is affected by a drought, the entire region's population is assumed to be affected when only a small portion is affected in reality. Gridded population data will help ameliorate this problem, reducing over and underestimation of food aid needs.

Problems with the underlying census data itself are also common. Regions that experience frequent and/or severe food crises often have many problems of governance, including the inability to perform an accurate and timely census. Census figures of these countries are often verified by international organizations, resulting in competing population figures.

The local government may, as a matter of pride or politics, be extremely interested in maintaining the original figures no matter how inaccurate they may be. To complicate this picture, population movements within the country itself may also be high due to chronic instability and environmental hazards, resulting in large actual differences between the known population figures and actual. FEWS NET works in a country by invitation, thus negotiating with these various actors to come up with reasonable and agreed-upon population figures is required and can become a complex process. LandScan data offers a way for FEWS NET to improve the spatial accuracy of these datasets, and to provide a scientific justification for their analysis which can be accepted by all parties. FEWS NET is also working to adjust the LandScan dataset to known differences between the census figures used in the base dataset and the figures that FEWS NET uses, maintaining the utility of gridding and distribution while allowing for variations from the standard datasets.

11.2 Global Population of the World and LandScan Data

LandScan gridded population data was first released in 1998 by the Oak Ridge National Laboratory (Fig. 11.2). LandScan is based on data developed by the Global Rural-Urban Mapping Project (GRUMP), which was funded by a consortium of partners, including the Center for International Earth Science Information Network (CIESIN), Columbia University; International Food Policy Research Institute (IPFRI); the World Bank; and Centro Internacional de Agricultura Tropical (CIAT). The Global Population of the World (GPW) Project, part of a larger global database effort, collected the best available census counts (usually at province level) for each country and distributed the population across the district, producing a spatially distributed population at a 30×30 minute geographic grid (approximately 1 km) globally. The population database provides global and continental subsets of population distribution currently containing estimates for three time periods: 1990, 1995, and 2000 (Dooley, 2005).

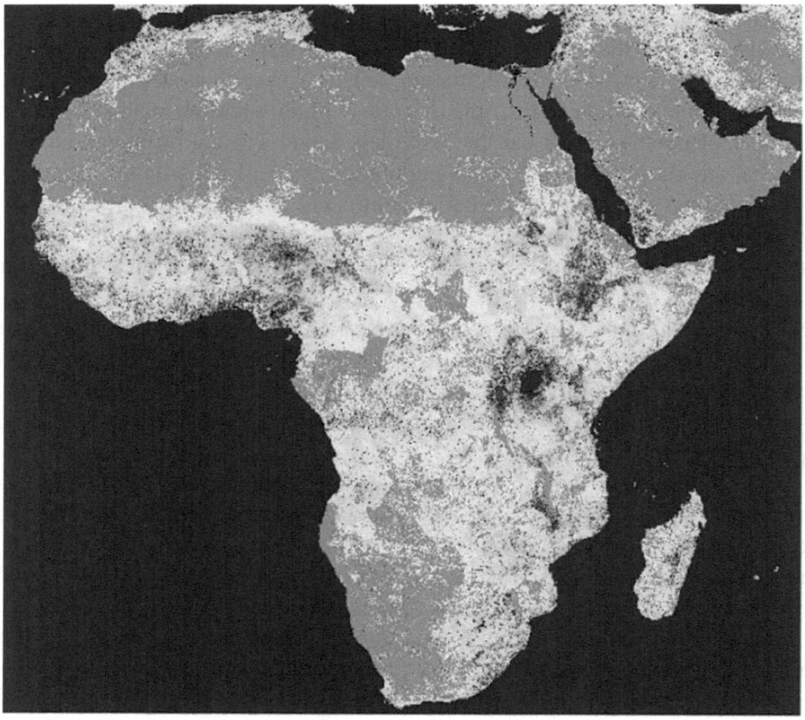

Fig. 11.2 LandScan data from 2005 for Africa. Darker areas have denser populations

Unlike the various GPW modeling efforts which have essentially distributed human population onto grids based on the relative area of a grid cell in – or across – particular administrative units, Oak Ridge National Laboratory's LandScan effort employs an extensive asymmetric spatial model to distribute global population. The LandScan model produces population grids by utilizing weighting criteria based on known population centers, the radiance calibrated night-time lights, distance to transportation infrastructure such as roads, elevation, slope suitability categories, land cover and other socio-environmental parameters. In addition, LandScan utilizes very high resolution 1–5 m global satellite imagery databases in both the modeling, e.g. for delineating the extents of urban areas, identifying dwellings and assigning population to that region instead of other areas of the map, and for a verification and validation process. Remote sensing is an essential source of two input variables, land cover, nighttime lights, for high-resolution panchromatic imagery. The following sections describe the basic input to the product.

11.3 Census Counts

All population counts, even the most sophisticated, high-resolution official censuses of like that of the United States, are stochastic estimates. Accuracy and precision are limited by the census takers' access to homes and even to whole neighborhoods; by the census takers' understandings of personal work and travel habits; and by the frequency with which censuses can be undertaken. These limits are exacerbated in many nations due to lack of resources and, all too often, outright manipulation of census figures to meet political objectives (Watkins et al., 2006). In addition, many nations are reluctant to release detailed census counts, and some release only a national total. For most of the world, the best available official census data are at province level (i.e., one administrative division below national) and of varying age, sometimes decades old. A few nations release high-quality census counts for sub-provinces, but only a few release the geometry of sub-province boundaries in digital form, such as can be found in the U.S. Census TIGER files (Bright, 1998).

The variable quality of census figures from country to country presents a major challenge to global population distribution efforts such as LandScan. Official census counts must be acquired from published sources and evaluated skeptically. Fortunately, for most countries the demographic literature is surprisingly rich, deficiencies are recognized by scholars, and adjustments have been proposed in literature (Watkins et al., 2006). In addition to published articles and reports, the World Wide Web has become an invaluable resource in locating and acquiring population data and understanding consensus and disagreement among demographers. Ultimately, ORNL analysts must choose a single number for each nation or province based on their own professional judgments of arguments and evidence offered by demographers. LandScan's purpose is not to count people and certainly not to count them in their nighttime residences, but to distribute populations based on their likely ambient locations integrated over a 24-hour period for typical days, weeks, and seasons. This is a very different objective than a typical census that is concerned primarily with residences (Bright, 1998).

11.4 Other Input Variables

Calculation of the probability coefficient for each cell depends on publicly available databases offering worldwide coverage of roads, slope, land cover, and nighttime lights at scales of 1:1,000,000 or larger and resolutions of 1 km. or finer. The sources and characteristics of current databases used by LandScan are discussed in this section, and are described further in the 1998 LandScan documentation, which can be obtained from the ORNL website. Although other inputs may be used by the current LandScan dataset (circa 2007), the inputs described below remain at its core.

11.4.1 Census Figures by Administrative Region

Accurate administrative boundaries are an integral part of the LandScan population distribution modeling process. Every year, LandScan incorporates administrative boundary changes for countries that have amended their boundaries. For these countries, international second order administrative boundaries are changed, and in many cases their digital accuracy improved, for population distribution. These boundaries coincide with new census counts provided by the International Programs Center (IPC) of the United States Bureau of the Census and represent the most recent census for all countries.

11.4.2 Roads

Transportation networks (i.e., roads, railroads, airports, and navigable waterways) are primary indicators of inhabitation. As a single indicator, roads are preferred because of their vital role in human settlements with or without other forms of transport. It would be helpful to know the location of all roads and to calculate road densities as suggestive of population densities, but this is not possible for most of the world. The United States is an exception due to the availability of Census TIGER files which include the geometry of local roads and even some private driveways and farm roads. The best universal coverage of road networks comes from the National Imagery and Mapping Agency's (NIMA) Vector Smart Map (VMAP) series. VMAP-Level 0 (formerly Digital Chart of the World) is publicly available and covers the entire world at 1:1,000,000 scale. The current version of LandScan includes VMAP-Level 1 data (1:250,000 scale) in regions where tiles become available, typically in the developed world (Bright, 1998).

11.4.3 Slope

Slope is an important variable in calculating the LandScan population probability coefficient because most human settlements occur on flat to gently sloping terrain. Even in regions noted for hillside settlement, relative measures of slope may correspond (inversely) with population density. The ideal measure of slope would be the area (at resolutions approaching the typical size of individual home sites) in each slope category, expressed as a percentage of LandScan cell area. LandScan's slope resolution is limited, however, by data availability and by the processing burden that would be required for global coverage. Hence, LandScan employed the Digital Terrain Elevation Data (DTED) Level 0 Terrain Data at 30 arc second resolution for the first version, and used it to calculate a single gradient for each 1 km LandScan cell (Bright, 1998). Subsequent versions have used Shuttle Radar Topography Mission

(SRTM) elevation data which is available at 90 m globally, thus allowing enhanced suitability analysis for settlement prediction.

11.4.4 Land Cover

Perhaps the best single indicator of population density is land cover type. With local knowledge and well-structured in situ sampling one conceivably might determine average densities per unit of area for each land cover type which then could be multiplied times the total area occupied by that type. In most regions population would range from extremely low density in desert, water, wetlands, ice, or tundra land cover to high density in developed land cover associated with urban land use. Arid grasslands, forests, and cultivated lands would range in between. Globally, of course, such rigorous in situ sampling is infeasible, especially in politically sensitive areas. LandScan analysts assign relative weights to each land cover type and employ these weights in calculating the probability coefficient for each cell (Bright, 1998).

Even at 1 km resolution, land cover can be a good indicator of relative population density, and its efficacy improves as resolution approaches the typical size of individual home sites. For example, the National Oceanic and Atmospheric Administration's (NOAA) Coastal Change Analysis Program (C-CAP) has demonstrated that high intensity developed and low intensity developed land cover can be distinguished reliably for coastal regions of the United States with Landsat Thematic Mapper (TM) imagery at 30 m resolution (Dobson et al., 1995).

The best global land cover database available to LandScan during its development was the U.S. Geological Survey's (USGS) Global Land Cover Characteristics (GLCC) database derived from Advanced Very High Resolution Radiometry (AVHRR) satellite imagery at 1 km. resolution (Loveland et al., 1991). The LandScan Land Cover Database is derived from the U.S. Geological Survey's (USGS) Global Land Cover Characteristics (GLCC) database with the following substantial modifications:

1. The LandScan Land Cover Database has been georegistered at 30 arc second resolution in a common grid for the entire globe. The original GLCC database was in Goode's Homolosine projection.
2. Considerable effort has been devoted to reconciling the positional accuracy of diverse global databases. Mismatches among databases were most conspicuous on coastlines. In the final LandScan Land Cover Database, the NIMA's World Vector Shoreline (WVS) at 1:250,000 scale took precedence over other databases.
3. The LandScan Land Cover Database contains a much improved urban class. The USGS urban class was replaced with two new classes: developed and partly developed. The developed class is composed of GLCC's urban cells plus all cells included in the Census Bureau's P-95 urban circles. The partly developed class is derived from Nighttime Lights of the World and contains all cells with a

Table 11.1 LandScan Land Cover classes

LandScan land cover classes	
1 Developed	15 Mixed Forest
2 Dry Cropland & Pasture	16 Water
3 Irrigated Cropland	17 Herbaceous Wetland
5 Cropland/Grassland	18 Wooded Wetland
6 Cropland/Woodland	19 Barren
7 Grassland	20 Herbaceous Tundra
8 Shrubland	21 Wooded Tundra
9 Shrubland/Grassland	22 Mixed Tundra
10 Savanna	23 Bare Tundra
11 Deciduous Broadleaf Forest	24 Snow or Ice
12 Deciduous Needleleaf Forest	25 Partly Developed
13 Evegreen Broadleaf Forest	28 Unclassified
14 Evergreen Needleleaf Forest	

frequency value of 90% or greater. The partly developed class typically includes suburban areas, small towns, and scattered industries, airports, etc (Table 11.1).

For many areas, the current version of LandScan has used high-resolution land cover data, National Geospatial-Intelligence Agency's (NGA) Vector Map of the World version 1, CIB, and/or scanned maps. These extremely high quality land cover maps are not available publically as they are classified. As the NGA actively improves its land cover database, LandScan incorporates these products annually into its database to reflect these changes.

11.4.5 Populated Places and Nighttime Lights

VMAP-Level 0 contains three categories of human settlement features. Two of them are point features distinguished only as 'named' or 'unnamed' populated places; the other consists of polygon boundaries for larger urban areas. Attributes for named populated places and populated polygons provide the name but not the population count for each place. Populated polygons originally were digitized from small-scale maps, sometimes aeronautical charts dating from the 1970s, and they are now notoriously imprecise and out of date. LandScan matched the populated polygons with nighttime lights (discussed in the following section) and assign a greater probability weighting for LandScan cells containing both features than that for cells containing only nighttime lights.

To overcome several deficiencies of the previously gridded databases, LandScan used satellite data produced by the Defense Meteorological Satellite Program (DMSP) which measures nighttime light emanating from the earth's surface at 1 km resolution. LandScan uses the DMSP Operational Line

Scanner (OLS) data which measures light intensity (Elvidge et al., 1999); (Sutton, 1997), processed and provided by NOAA's National Geophysical Data Center

(NGDC). Frequency data cover the Northern Hemisphere and South America, but most areas south of the equator are limited to a binary value indicating lights present versus no lights present.

Investigating the efficacy of nighttime lights for estimating population in the United States, Sutton (1997) found in the United States, about 17% of the population, occupying about 90% of the land area, is dispersed too sparsely for detection (i.e., adjusted pixel value of 1) by this particular sensor. Sutton (1997) further investigated the correlation of nighttime lights with population density and his model accounted for 25% of the variation in population density. Thus at the high end of the population/light spectrum, no further distinction of population densities is possible once light saturation occurs. At the low end of the spectrum, no further distinction is possible in pixels with undetected lights (Watkins et al., 2006).

In Africa and Central America, the regions of interest to FEWS NET, the majority of the population lack access to electricity at night. Thus LandScan replaces the nighttime lights in these regions with a feature-space extraction technique used on extremely high resolution imagery (1 m or less) that can detect rooftops and other constructed buildings and then uses these structures to distribute the population across the land surface. The LandScan project has acquired high resolution imagery of the entire land surface, and the project is in the process of replacing the nighttime lights function with high resolution data globally. Because FEWS NET will be purchasing a license for the LandScan dataset, they will have access to this new dataset as it becomes available.

11.4.6 Urban Density Factor

Urban agglomerations of 25,000 people or more are covered by one or more circles including at least 95% of the population, as defined by the US Census bureau. These circles are known as P-95 circles and each contains at least 5000 people, cannot be less than 0.3 and no more than 1.0 nautical miles in diameter, and the amount of unpopulated ground within each circle must not exceed 20 percent of the entire circle. The circles are used to match the point locations and diameters of P-95 circles with nighttime lights. This increases the probability weighting for LandScan cells containing both lights and high density circles over cells containing only nighttime lights. The associated P-95 population values proportionally increase the probability weighting, but absolute P-95 values are not employed in the final calculation of LandScan cell populations.

11.4.7 Coastlines

Considerable effort is required to reconcile the positional accuracy of diverse global databases, and mismatches among databases are most conspicuous on coastlines. Globally, LandScan coastlines are based on NIMA's World Vector Shoreline (WVS) at 1:250,000 scale. Typically, this coastline differs somewhat from the related line representing the seaward boundary of administrative units, and both of these differ from the land/water boundary indicated on the Global Land Cover gridded database. In the final LandScan Global Population Database, Version 1.1, the WVS takes precedence, and no population is apportioned to cells extending more than one-half cell beyond the WVS coastline.

11.5 LandScan Population Model

Best available census counts (usually at province level) are allocated to 30 by 30 minute cells through a 'smart' interpolation based on the relative likelihood of population occurrence in cells due to road proximity, slope, land cover, and nighttime lights. Probability coefficients are assigned to each value of each input variable, and a composite probability coefficient is calculated for each LandScan cell. Coefficients for all regions are based on the following factors:

- Roads, weighted by distance from major roads,
- Elevation, weighted by steepness of slope categories,
- Land Cover, weighted by type with exclusions for certain types,
- Nighttime lights of the World, weighted by frequency, and
- Dwelling and structure identification from high resolution imagery.

The resulting coefficients are weighted values, independent of census data, which can then be used to apportion shares of actual population counts within any particular area of interest. Coefficients vary considerably from country to country even within a particular region. Control totals can be based on any administrative unit (nation, province, district, minor civil division) or arbitrary polygon for which census data are available. The resulting population distribution is normalized and compared with appropriate control totals to ensure that aggregate distributions are consistent with census control totals. Successful operation of the model has been demonstrated for various control totals, control areas, and weighting values (Bright, 1998) (Fig. 11.3).

11.6 FEWS NET Evaluation of LandScan

FEWS NET funded an evaluation of LandScan data to determine if it is sufficiently accurate for its work. To test the data, a comparison with official census figures from Burundi was carried out by Watkins et al. (2006). Official census figures from 129 Communes, the third administrative level, were compared to LandScan 1km data to determine whether LandScan data can be used to predict small-area urban populations as well as rural. Watkins et al concludes:

- LandScan population estimates are consistently lower than official population estimate, by a factor of 20% in this case;

Fig. 11.3 Close-up of West African LandScan data from 2005. Regions with one person per square mile are minimized to highlight higher population areas. Notice the effect of using road and river networks in producing the data

- LandScan is a highly effective small area estimator of official rural populations, once the overall difference between the official and LandScan figures has been adjusted for;
- LandScan population estimates are somewhat weaker in urban areas than rural areas, requiring a slightly different algorithm to adjust for differences (Watkins et al., 2006).

Although the official population figures in Burundi are likely to be overestimated, FEWS NET must work with them through agreement with the national government. Thus in order to work with LandScan FEWS NET will need to adjust the base population figures. An additional objective in working with LandScan data is to generate population estimates for small units for which official population data are not available. To do this effectively, FEWS NET has funded the production of interactive software that enables users to fit the LandScan data to the local or regional datasets currently in use by local experts who generate them through field work.

In addition to matching already in used population datasets, the LandScan database at 1 km resolution is too coarse in some situations to accurately locate small-scale settlement typical of many agricultural areas (Tatum et al., 2004). As information on settlements is gathered and become available digitally through GIS and other databases, LandScan can be amended. In other situations, LandScan should be amended to show the movement of large numbers of people from one region into another during a crisis. At the moment, there is no way for FEWS NET to show known population movements against a realistic population surface. By developing and training its personnel in the use of this new software, FEWS NET will improve its ability to estimate accurately populations at risk.

11.7 Challenges for FEWS NET and Future Plans

Knowing accurately the population in a particular region of a country accurately is extremely important to FEWS NET's work. Only recently has it begun to use gridded population datasets such as LandScan, however. One of the reasons for this is the need to develop a consensus and agreement with the local government regarding how many people are at risk. Official census figures from a particular national government have always been the basis and starting point for any population estimate. This has restricted FEWS NET to using official, national and district-level statistics instead of the state-of-the-art gridded datasets that have recently become available. FEWS NET must be able to use locally-derived population data in order to continue to work in a region. The case cited in the previous section is a case in point. Burundi's national census figures are known to be rather higher then is likely to be true. This has a number of implications for calculating food aid deliveries, which are, on average, to Burundi's advantage.

In addition to official government figures, FEWS NET often also uses informal population surveys that are conducted in between formal government censuses to know what the population is in a particularly at-risk region. FEWS NET often works in regions that are politically unstable, where a government may not have conducted a regular national census in several decades. There is a clear need for more international resources to provide knowledge, information technology and resources so that the national statistical offices can be integrated into each country's emergency preparedness and response organizations.

Not only is it important to know who usually lives where, but FEWS NET is also extremely interested in knowing about displaced populations. They need to know both where the displaced people came from (to reduce the population density in those areas) as well as where they are now (to ensure those areas get more assistance and attention). Worldwide, millions of people are displaced annually because of natural disasters or social upheaval. Reliable data on the numbers, characteristics, and locations of these populations can bolster humanitarian relief efforts and recovery programs. Using sound methods for estimating population numbers and characteristics, and ensuring that the data are geographically referenced for projection onto maps is also important. FEWS NET works to ensure that surveys and other data gathering activities are properly integrated into existing population databases for improved decision support (Cutter et al., 2007).

Given the problems mentioned above, FEWS NET has contracted with LandScan to provide software into which the official census figures and ad hoc population surveys of displaced people can be imported to form a new population baseline. The new population data is then combined with previously known data and the model run on the fly, providing disaggregated population figures in a gridded one kilometer map. The software is run on a web server which enables vector calculations with user-provided vectors, queries and map generation without the actual population data ever being downloaded locally. This will both enhance the usability of the data for FEWS NET's widely spread personnel as well as reduce its cost, as the license for use is far less then for that required to download the entire dataset.

11.8 Summary

This chapter summarizes the use of population data in early warning and humanitarian communities in their efforts to estimate the numbers of people at risk during emergencies. FEWS NET typically uses census data from the local communities in which they work, but they are also exploring the use of gridded population datasets such as LandScan. This data allows for much more accurate identification of populations at risk and a more accurate idea of the locations and density of urban centers. Although FEWS NET uses official census figures developed through consensus with local and regional decision makers, they plan to amend the LandScan dataset to use both official statistics and scientifically produced population models that apportion population across the land surface. Improved population datasets will help FEWS NET more accurate estimate populations at risk for food insecurity in the regions in which it works.

References

Bright, E.A., 1998. LandScan Global Population 1998 Database, Oak Ridge National Laboratory, Oak Ridge, TN.

Cutter, S.L., Arnold, M., Balk, D., Hovy, B., Kwan, M.-P., Mayer, J.D., Rain, D.R., Rodriguez, H., Torrey, B.B., Turner, II, B.L., Weeks, J.R. and Zuberi, T., 2007. Tools and Methods for Estimating Populations at Risk from Natural Disasters and Complex Humanitarian Crises. The National Academies Press, Washington D.C., 175pp.

Dobson, J.E., Bright, E.A., Ferguson, R.L., Field, D.W., Wood, L.L., Haddad, K.D., Iredale, H., Klemas, V.V., Orth, R.J. and Thomas, J.P., 1995. NOAA Coastal Change Analysis Program; Guidance for Regional Implementation Version 1.0, National Oceanic and Atmospheric Administration.

Dooley, J.F., 2005. An inventory and comparison of globally consistent geospatial databases and libraries, UN Food and Agriculture Organization, Rome.

Elvidge, C.D., Baugh, K.E., Kihn, E.A. and Davis, E.R., 1999. Mapping city lights with nighttime data from the DMSP operational linescan system. Photogrammetric Engineering and Remote Sensing, 63: 727–734.

Loveland, T., Merchant, J., Ohlen, D. and Brown, J., 1991. Development of a land-cover characteristics database for the conterminous U. S. Photogrammetric Engineering & Remote Sensing, 57(11): 1453–1463.

Sutton, P., 1997. Modeling population density with night-time satellite imagery and GIS. Computers, Environment, and Urban Systems, 21(3/4): 227–244.

Tatum, A.J., Noor, A.M. and Hay, S.I., 2004. Defining approaches to settlement mapping for public health management in Kenya using medium spatial resolution satellite imagery. Remote Sensing of Environment, 93: 42–52.

Watkins, B., Jordan, L., Mwangi, M. and Rose, R., 2006. Practical approaches to population estimation for Famine Early Warning and Food Security Analyses. Kimetric Ltd., Nairobi, Kenya.

Chapter 12
Food Markets and Prices

In many regions where FEWS NET works, the majority of the population grows food in subsistence rain-fed agricultural systems as all or part of their income generating activities (Galvin et al., 1997). Access to food for these farmers as well as people who do not farm involves small, informal markets where grain is bought and sold. Food prices are therefore influenced in these markets both by local production and by the price and availability of food produced elsewhere. FEWS NET is working towards monitoring the coupled response of food prices to food production through predictive economic modeling. This chapter will describe how FEWS NET uses price data and the predictive model and how it may be used in FEWS NET's analysis.

The most rural agriculturalists in semi-arid Africa have a flexible response to food supply and demand. Farmers typically sell a portion of their crop on the market after harvest, save a portion for consumption, and purchase food from the market as their own supplies diminish later in the year. This interaction with the market tends to amplify the response of market prices to the production of low-cost, locally grown grains such as millet. The farmer's flexibility in timing the sale of grain provides a linkage between grain prices in the spring and summer and the vegetation conditions during the previous summer's growing season. When cereal prices rise generally as they have done in recent years, however, farmers with cereal stocks to sell can significantly benefit. Rainfall variability from year to year changes how much grain is available for sale in the market, which has implications for future food security of the region.

Other people in the countries where FEWS NET works have no agricultural land and thus are completely dependent on the price of food in the market to determine access. When these individuals also work in the agricultural sector as laborers, vulnerability to periodic crashes in the agricultural labor market due to weather-related crop failure, coupled with simultaneous sharp increases in food prices can cause extreme food insecurity. Variations in the global commodity prices also have an important role in determining local food prices, particularly in Southern Africa, a region well integrated into the global food economy. When the South African maize harvest goes poorly and commodity prices are high generally, regions with weak economies and poor terms of trade can have difficulty in importing sufficient grain, resulting in localized price increases.

Higher food prices can cause food insecurity among the most vulnerable in a population even in times with adequate or even abundant food supplies (Sen, 1981). Early warning of these price increases can enable organizations to increase food or income assistance in order to reduce the loss of lives and livelihoods as well as the cost of providing these services (FEWS, 2000). FEWS NET has a variety of strategies to monitor food price variations, although these vary from region to region, and to integrate remote sensing information into these observations for improved price prediction and monitoring. This chapter will focus on prices, price monitoring and how FEWS NET will begin to use remote sensing in the monitoring process.

12.1 Monitoring for Early Warning

FEWS NET has developed food market systems to regularly monitor changes in the price of food in order to understand variations in food access in the context of food security. The structure and performance of food markets are central to the normal flow of food from field to table. In contrast to FEWS NET's normal focus on household-level dynamics as the unit of its food security monitoring and assessment, market price analysis examines the performance of key components in the larger food marketing system (e.g. retailers, wholesalers, transporters, the market policy environment, infrastructure, financial markets, etc.), and their roles in assuring food security through providing access. The system identifies factors that affect the market's retail functions at the national or regional, or macro-economic levels, and examines factors such as import/export policies, internal trade and movements of food, infrastructure issues, financial environment, etc. It can identify indicators that could be monitored to identify earlier than has previously been possible potential causes of the system's failure. It can also assist in identifying market-specific remedial interventions that might be implemented when food insecurity threatens an area, and the market system's failure appears imminent.

In the past five years planning and preparation for crises, particularly in regions that experience problems frequently, has increased in importance (Buchanan-Smith et al., 1995). Clearly anticipation of future problems makes for better integration with humanitarian aid sources that require lengthy negotiation, early purchasing, mobilization, and shipping of food aid (Choularton, 2005). The necessary precision of the information needed for planning and forecasts of possible future food insecurity is far lower than that required to estimate current food aid needs. Recent research has enabled the projection of vegetation index dynamics up to four months into the future (Funk et al., 2006) that could be incorporated into operational analyses of food security. FEWS NET is working to develop methods that apply the projections to planning for future food aid to enable the estimation of spatial locations and severity of future problems in regions at risk.

12.1.1 RATIN Analysis for Cross-Border Trade

One of the most successful examples of a market-based monitoring system that has been implemented by the USAID Regional Economic Development Services Office (REDSO) for East and Southern Africa office is the Regional Agricultural Trade Intelligence Network (RATIN) activity (see www.ratin.net) in the Horn of Africa and the USAID Regional Center for Southern Africa-funded cross-border food trade monitoring activity in Southern Africa. Both provide insights on better understanding food markets (retail/wholesale volume, trade flows, pricing impulses) for early warning purposes, and for informing development programs oriented at agricultural development and food security.

RATIN is a collaborative effort comprising two USAID projects: FEWS NET, which focuses on bringing in the crop production and trade information for food security analyses, and Regional Agricultural Trade Enhancement Support Program (RATES), which focuses on changing trade policy to enhance regional trade in maize. It was developed to help reduce regional food insecurity by strengthening the ability of markets to provide access to affordable food to poor households and improve food availability through providing adequate incentives to producers. It came out of the realization that although there are households among the food insecure that are structurally poor and are heavily dependent on food donations, there are also market-dependent households who are able to purchase food if it is available at the right time, price and quantity in the local markets through enhanced effective competition.

Competition in the supply of food commodities (maize, beans, rice and cooking bananas) is enhanced by regional trade in these commodities. However, trading opportunities for most food and livestock cross-border traders are limited by the need to plan for or justify incurring certain cross-border marketing costs in the face of uncertain receipts due to lack of timely market information. Consequently, the major task of RATIN is to supply traders with improved early warning marketing and trade information that would lead to more efficient and competitive transactions in food trade between surplus and deficit regions in East Africa.

Small and medium scale cross-border traders account for over 80 percent of regional trade in maize, beans and rice in East Africa. Consequently, they are the main target group of the RATIN information. To easily distribute RATIN information to hundreds of small and medium cross border traders of cereals and pulses in East Africa, the traders have been organized into 30 loose associations each of which receive RATIN's monthly bulletin and disseminate it to all members. RATIN also disseminates trade related information through TV and radio (Uganda and Kenya through partners), cellular telephone SMS services (East Africa), newspaper (source of information for reporters), and the www.ratin.net website. Figure 12.1 shows how these organizations interact within RATIN, and how they distribute the information it collects.

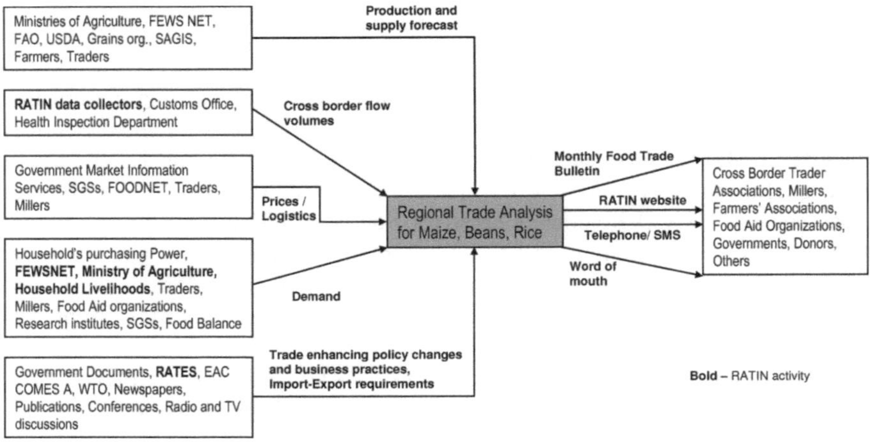

Fig. 12.1 Activities of the Regional Agricultural Trade Intelligence Network (RATIN). Boxes on the *left* show the information source, *arrows* from the source to the analysis show the type of information. *Arrows* from the shaded trade analysis box in the *middle* to the target groups on the right shows the dissemination pathways

12.1.2 RATES Program

The Regional Agriculture Trade Expansion Support (RATES) program is a 5-year program funded by USAID's East Africa regional office based in Nairobi. The RATES program is designed to increase value/volume of agricultural trade within the East and Southern Africa region and between the region and the rest of the world. RATES focuses on developing commodity-specific regional trade initiatives through innovative private sector/public sector alliances and partnerships and works primarily through regional trade flow leaders such as regional trade associations, national-level trade organizations, private companies and individual entrepreneurs. RATES is currently supporting activities in specialty coffee, maize and pulses, cotton and textiles, livestock and dairy sectors.

The program is designed to increase value/volume of agricultural trade within the East and Southern Africa region and between the region and the rest of the world. RATES focuses on developing commodity-specific regional trade initiatives through innovative private sector/public sector alliances and partnerships and works primarily through regional trade flow leaders such as regional trade associations, national-level trade organizations, private companies and individual entrepreneurs. Through policy advocacy, lobbying, public relations and marketing, RATES is developing a regional dialogue regarding trade issues in East and Southern Africa.

A functional regional market information and intelligence system can provide commodity traders a network to take advantage of trading opportunities and mobilize them for better trade policy advocacy. RATES will expand and sustain regional agricultural trade flows through strengthened human and institutional capacities by:

- Expanding agricultural trade in selected sub-sectors
- Increasing institutional capacity to sustain agricultural trade
- Expanding private sector contribution to regional trade initiatives

FEWS NET focuses on developing these networks because they provide critical intelligence during a food security crisis. These networks also enable an efficient way of distributing assistance as traders have a large network of transportation methods that can be utilized to move food aid from the capital city to small regional markets where it can do the most good.

12.2 Price Modeling for Earlier Early Warning

FEWS NET is now moving towards having access to predictions of market price dynamics through incorporating remotely sensed vegetation into quantitative price models. Remotely sensed vegetation data is used qualitatively by the food security community to link vegetation indices to agricultural yields (Fuller, 1998; Funk et al., 2005, 2006). The justification for using satellite data in famine early warning activities came from research on net primary production (NPP) (Prince et al., 1990; Diallo et al., 1991; Justice et al., 1991; Prince, 1991). At first primary production estimates from satellite data were used to determine if an area was experiencing a decline in food availability. Later, early warning organizations moved towards additional indicators based on a wide variety of locally gathered social and economic data. Unfortunately no direct connection between remotely sensed indices and food access has been established.

This section summarizes a paper that describes a price-vegetation model that can be used to create maps of price predictions, with data from West Africa as an example (Brown et al., 2008). The intention is to provide a methodology which will enable the use of satellite observations of vegetation as a proxy for food production together with food prices to quantitatively capture the impact of variations in production on variations in prices. The model presented here does not attempt to capture trade flows, exchange rate influences, cross border trade or other macro-economic influences on the market. It only attempts to explicitly model the production portion of the price variation, letting the rest of the complexity of the market being predicted by economic methods of autocorrelation (Markov).

Integrating economic modeling into FEWS NET's activities will allow regions such as West Africa to improve the functioning of their market information systems, and increase the effectiveness of government organizations that are involved in food security. FEWS NET currently uses no economic modeling at all, relying entirely on observations of prices. This is certainly extremely accurate, but in the context of the Food Security Outlook activity, having some insight about future price movements will be very helpful. Because most economic models provide little skill more than one month into the future, coupling biophysical projections with economic information can increase these models ability to estimate future changes.

The model is presented first, with its data and outputs. An analysis of the relative impacts on local prices of local conditions and millet of export regions is then presented. Maps of prices across the region, which are made possible by the spatially continuous nature of remotely sensed vegetation data, are then compared with specific market price observations. The impacts of mapping forecasted prices on the ability of regional, national and international governments and organizations to respond to market movements due to environmental conditions are discussed in the final section.

12.2.1 Model Price Inputs

The price data used as the basis of the model are monthly millet prices from 445 markets in Niger, Mali and Burkina Faso. The data were obtained from local market price monitoring organizations through the USAID's Famine Early Warning System (FEWS) (May, 1991; Chopak, 1999). The data have been kept in the local currency (CFA), which is fixed at the same exchange rate with the French franc in all three countries. The datasets used were monthly millet prices from 445 markets in Niger, Mali and Burkina Faso. The data were obtained from local market price monitoring organizations through the Famine Early Warning System (FEWS). The data have been kept in the local currency (CFA), which is fixed to the French Franc at the same exchange rate for all three countries. The data series vary in length but have similar means and standard deviations (Fig. 12.2). The 223 markets with fewer than 50 months of data out of the total possible of 300 months were excluded. 56% of the data in the three countries are either missing or could not be used due to a lack of sufficient consecutive months of data. There were 29,731 monthly millet price data points available for analysis.

Although the prices in the three countries are similar as is expected, there are significant differences in the number of observations per country in different time periods, as seen in the lower panel of Fig. 12.2. Particularly after 2000, there are many more Niger stations than in Mali, and Burkina Faso has only one station (Ouagadougou). These differences will have an impact on the model's errors, and greatly reduces the ability of FEWS NET to analyze changes in agricultural and livestock prices in the region. Although it is likely that there is more information regarding market prices in the region, these are not reported to the centralized database and are thus unavailable for historical analysis.

Although the FEWS database also includes monthly data for maize, rice, sorghum, wheat, peanuts, and a variety of other local and imported products, this study considered only millet prices. Millet is the most frequently purchased in West African rural areas when own-grain production is low (Jayne et al., 1996). It is also the most widely grown and most readily available grain. Market locations are shown on a map of NPP in Fig. 12.3. The millet price data were deflated with a national annual consumer price index (CPI) (IMF, 1999, 1999, 1999), interpolated across months. The CPI was used to relate changes of consumer's purchasing power over time in the three countries where data were used (Jayne et al., 1996). The deflation removes inflation trends from the data and corrects for the impact of 50% devaluation of the CFA in 1994.

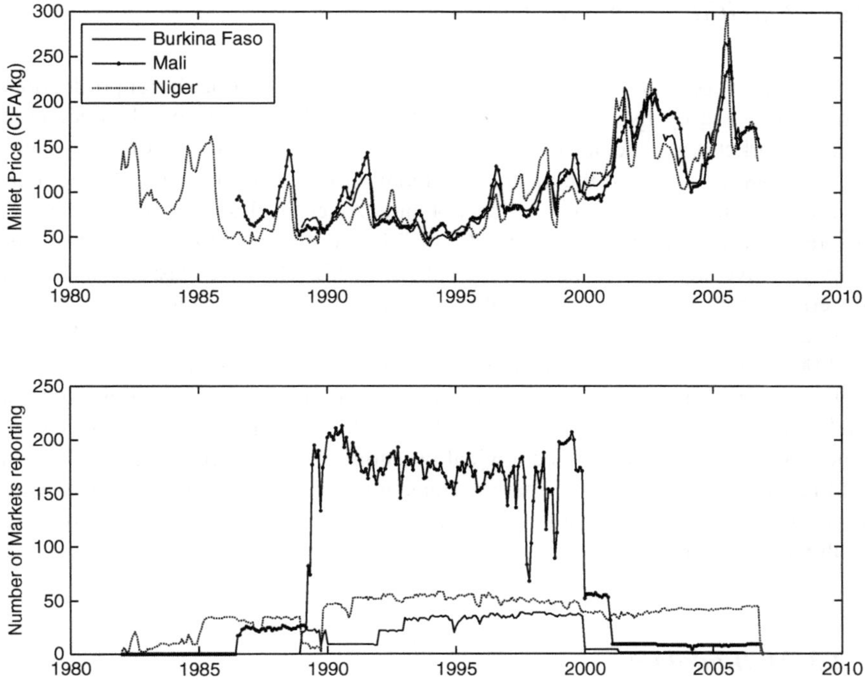

Fig. 12.2 Millet prices in West Africa. *Top panel* shows the averaged price time series from the three countries, and the *lower panel* shows the number of observations that go into each mean

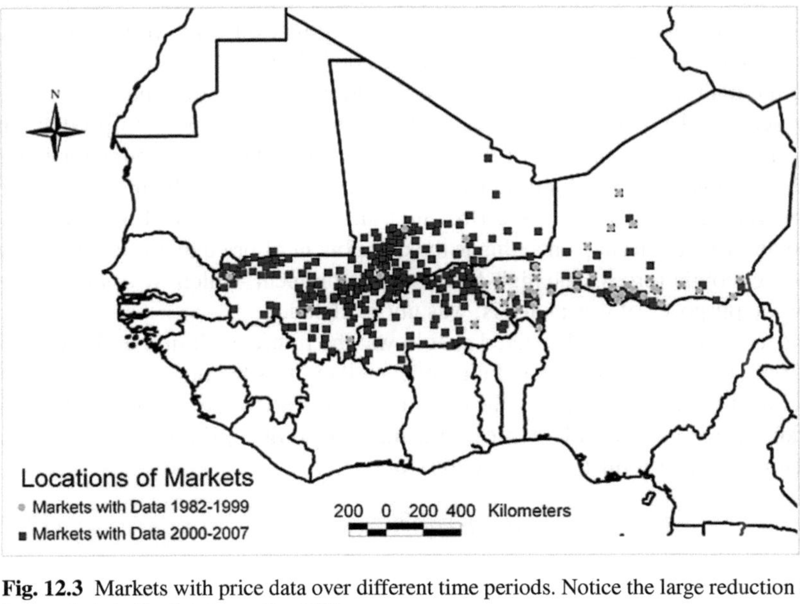

Fig. 12.3 Markets with price data over different time periods. Notice the large reduction in reporting markets in the database after 1999

12.2.2 Satellite Vegetation Data

In the model, vegetation index data was used as a proxy for agricultural production. A mean of a five by five-pixel box (40×40 km) of AVHRR NDVI data centered on each market was calculated from monthly maximum value NDVI composites (Holben, 1986). A map of the means of annual NPP was used to classify the markets into those in similar agroecological potential zones, in classes of $100\,\mathrm{g\,m^{-2}yr^{-1}}$. The NPP data were obtained from the GloPEM model from the Department of Geography, University of Maryland (Prince et al., 1995). The nine classes range from a desert environment with an NPP of $0–100\,\mathrm{g\,m^{-2}yr^{-1}}$ (class 1) to the sub-humid Sudanian zone in the south of Burkina Faso and Mali that has an annual NPP of 800–900 grams of carbon per meter per year ($\mathrm{g\,m^{-2}yr^{-1}}$) (class 9). The NPP classes parallel latitude bands and are continuous across the landscape from East to West (shaded map behind the points in Fig. 12.3). The NPP data provided a classification of markets having similar agroecological potential. An alternative to the NPP data could be the livelihood zones, where regions with similar ways of making a living are grouped together. Operational implementation of the model will probably use livelihood zones instead of NPP.

12.2.3 Methodology

Previous work has shown that there was a negative linear relationship between vegetation productivity and millet prices in Mali, Burkina Faso and Niger during the 1980s and 1990s (Brown et al., 2006). Empirical Mode Decomposition was used here to isolate components with different cycles of fluctuations (seasonal, interannual, errors) in order to model the environmental effect on price from stochastic economic effects (Huang et al., 1998; Pinzon, 2002; Pinzon et al., 2005). By partitioning the data into components that had variations at the same timescale as the growing season, the seasonal component that is most related to crop production could be isolated. By using satellite-derived vegetation data to account for these seasonal variations instead of a dummy variable, the estimate of the modeled price was improved (Deaton et al., 1992; Brown et al., 2006). The model uses Markov theory to describe the trend, since this is the component of price that is most clearly driven by economic forces. The high-frequency component, which we assume to be noise for the purposes of this analysis, was not modeled.

The Markov property can be stated as follows: the future evolution of the system depends only on its current state and it is independent of its history. This means that to predict next month's prices only the current month's prices are needed. Although the Markov property is usually expressed using probabilities, Eq. (12.1) shows the relationship as a deterministic linear regression.

$$price_t = \alpha + \beta\,(price_{t-1}) + error \qquad (12.1)$$

A linear model was used instead of the probabilistic form of the Markov property to estimate more precisely the price dynamics. An exploration of the use of probabilistic Markov theory for this application can be found in (Brown et al., 2006). The linear estimation is very similar to an autoregressive moving average model (ARMA), but the coefficients here are estimated for each market.

Huang et al. (1998) introduced the Empirical Mode Decomposition (EMD) as an alternative to standard decomposition technique for representation of nonlinear and nonstationary data that show clear physical scales or frequency content. The Empirical Mode Decomposition technique is empirical, intuitive, direct, a posteriori, and adaptive, with the decomposition functions based on and derived from the data (Huang et al., 1998, 1999). Unlike Fourier decomposition (Wilks, 1995; Trefethen et al., 1997), the EMD takes the basis for the signal from the data themselves. The goal of the EMD is to decompose the signal into individual intrinsic modes of oscillation (Huang et al., 1998, 1999; Pinzon et al., 2001). At any given time, the data may be represented by many different, coexisting modes of oscillation, each one superimposed on the others.

The components were isolated using a sifting process repeated as many times as is required to reduce the extracted signal to an intrinsic mode function (Huang et al., 1998). The sifting allowed the expansion of the time series into modes that would reveal the principal frequencies or scales that dominate the signal. The components of an EMD are usually physically meaningful, since each mode is defined by the physical data themselves, and are additive, so summing all components recreates the original time series. The EMD technique was applied to both the NDVI and the price time series and three components were extracted from each: the trend, the seasonal component characterized by a 12-month period, and a component that consisted of high frequency oscillations with a nearly Gaussian distribution, treated here as error and discarded. By cleaning the data of the high frequency information (noise), a much improved estimation of the price can be obtained from the Markov analysis.

It was assumed that the seasonal price component isolated using EMD was most related to variations in the environment measured by the NDVI. The NDVI seasonal profile was used to model the price seasonal component using two linear regressions. The regressions relate the green-up of vegetation during the spring months to the increase in prices leading to the harvest, and the senescence of vegetation to the decline in cereal prices during the harvest. This was done to associate the price with the NDVI in order to take advantage of the NDVI's spatially complete information about the variations in the environment. The coefficients from the regressions between the NDVI and price profiles were applied to the monthly time series of NDVI data to estimate the monthly price seasonal profile.

Using the EMD decomposition technique, the following procedure was used in order to model the prices.

1. Decomposition of both the mean NDVI of a 5×5 pixel box around each market and the accompanying millet price time series each into two components: a trend and a seasonal component. See Fig. 12.4 for an example of the seasonal and trend components of the millet price.
2. A mean seasonal profile was created for NDVI and price for centered on the location of each market by averaging all available months of data from all years

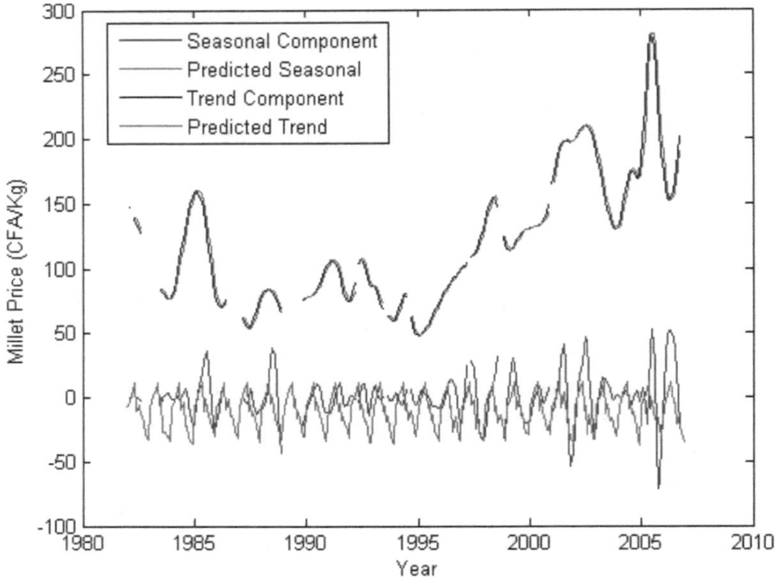

Fig. 12.4 EMD model illustration, showing one month prediction using Markov and NDVI-based seasonal component model. Notice that the seasonal component varies around zero, whereas the trend does not

for each market. Linear regression was performed from May to August and from August to December between the mean price seasonal profile and the mean NDVI seasonal profile. The coefficients from these regressions were then applied to the NDVI data for all years to construct a continuous vegetation-based price seasonal time series from May to December for all years. The slope of the mean price profile from January to May was used to inform the period when the NDVI has little information due to the onset of the dry season, and the price is increasing as supply in the markets diminish.

3. The reconstructed seasonal price component was subtracted from original price data and the EMD decomposition re-run with the new price dataset, providing a new trend component and enabling the identification and removal of the noise component.
4. The trend component from step 3 represented the economic portion of the price that could not be explained by NDVI changes. The trend was predicted using the relationship between the current and next month price based on a strong Markov relationship (Equation 12.1).

The trend and seasonal components were then summed to reconstruct the historical price time series. By basing the portion of the price that has seasonal characteristics on the NDVI, the price reconstruction was improved over traditional economic models that either remove this portion or use an invariant climatology (see (Deaton et al., 1992)). In addition, the model provided a spatially continuous map of prices derived from the satellite remote sensing data.

12.2.4 Results

The EMD decomposition of the price data allowed the separation of the overall trend from the yearly seasonal increases in price. The trend component was able to isolate non-environmentally related movements in price very effectively, reducing the errors of the linear model of price. Figure 12.5 shows the original price time series, the decomposed price and the actual compared with the predicted price from the market in Matameye, Niger. Notice the low millet prices in 1993 and 1994, influenced by a 50% devaluation of the currency and resulting high inflation rate for the next few years in all three countries (Kelly et al., 1995; Coulibaly et al., 1998; Yade et al., 1999), and a very large millet crop as a result of one unusually wet year (Tucker et al., 1999; Nicholson, 2000).

The results of the EMD model are summarized in Table 1 using RMSE. The EMD model is compared to results using price only on raw data, as presented in

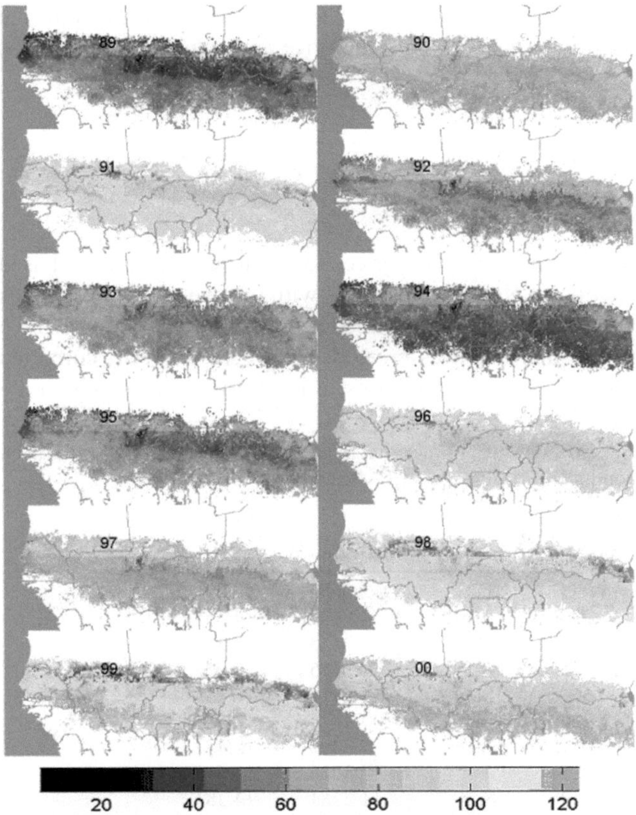

Fig. 12.5 Maps of millet prices, in CFA/kilogram, for August from 1988 to 1999, the period with the most robust and complete price time series data (Brown et al., 2008). Used with permission

the methodology, where the current month's price was used to predict the next month's price with the raw data. Note the errors for the simple, price-only model were seasonal, increasing during the harvest and post-harvest months of September – December, and were larger in the Sahel and smaller in the more humid zones. In addition to the linear model, the standard deviation of the noise component is presented, since the errors in the model should be less than the noise level present in the data (Table 12.1).

The EMD model output is compared to results using price only on raw data, as presented in the methodology, where the current month's price was used to predict the next month's price with the raw data. By modeling the seasonal cycle of prices using the seasonal information from remotely sensed vegetation data, it was possible to explain more fully the impact of differences in the environmental conditions on millet prices. When the local growing season was poor, lower millet production resulted in less price seasonality. The prices after harvest did not fall as much as normal, but instead continued to grow throughout the following year, depending on the extent of the shortfall. The EMD decomposition allocated this overall price increase to the trend portion of the price instead of in the seasonal portion, thus the small seasonality during drought years in both the price and vegetation signal was appropriate.

12.2.5 Maps of Prices

To produce spatially varying maps of millet price, seasonal price estimates based on the by-pixel transformation of NDVI were scaled with the averaged price trends from each NPP class. Figure 12.5 shows the model output for the August price for 1989–1999, and because the map was created using data that was corrected for inflation, it shows real prices. Large increases in food prices in the region are

Table 12.1 Root mean square errors (RMSE) of the point model vs the actual price of millet at all markets by NPP class and month. RMSE is given in CFA/Kilogram

NPP class $gm^{-2}yr^{-1}$	EMD model	Linear model	Month	EMD Model	Linear Model
0–100	9.0414	15.0431	Jan	5.0684	9.8988
101–200	8.7617	17.3906	Feb	4.7132	14.9558
201–300	7.8712	13.9025	Mar	4.3145	10.3543
301–400	8.2227	14.0754	Apr	4.3715	8.6072
401–500	9.1502	13.6635	May	4.6919	10.114
501–600	8.494	11.2062	Jun	7.442	10.383
601–700	8.7809	9.9719	Jul	7.8496	11.4727
701–800	9.8487	10.4214	Aug	10.8616	11.691
801–900	7.0015	10.2596	Sep	11.2781	14.3812
			Oct	12.3036	18.093
			Nov	11.9864	19.1073
			Dec	10.3871	19.2006

evident from the large variation among the years due to the averaged price trends. The variation within the year was much smaller, although the price in December was nearly always below 90 CFA/kg and the price in August was frequently above 90. For wage laborers, landless rural and urban poor who are dependent on the market to obtain food, these large seasonal and interannual price changes significantly reduce their access to food.

12.3 The Meaning of Price Maps

Famine early warning integrates assessments of household vulnerability to shocks (Boudreau, 1998), indicators of the economic and political situation, and measures of unfavorable meteorological conditions (Verdin et al., 2005). The model presented above demonstrates that by linking food prices and remotely sensed vegetation index data in quantitative relationships, the importance of rainfall-driven variations on the evolution of prices can be assessed. This relationship can be applied by providing an improved price model for areas of rainfed agriculture, which is likely to be valuable for planning purposes. Because the results of the model show that rainfall dynamics, as captured by vegetation index data, can improve the economic model in some regions, the model should improve the interpretation of the input variables taken individually and provide an improved leading indicator of changes in food prices. Maps of food prices can be integrated with other data to create an enhanced leading indicator of changes to food prices for planning purposes.

Food price variations are critical to food security, both directly and as an indicator of current trends. The dynamic relationship between supply, which depends in part on rainfall, and prices observed in specific markets allows the calculation of spatially continuous maps of prices. By coupling two indicators into one map, the intensity and extent of price-induced vulnerability can be identified. This extrapolation of prices in a few markets to the entire region using satellite data could be exploited by food security organizations to enable the need for humanitarian assistance and its magnitude to be estimated for the immediate future.

The price models that use real-time vegetation observations provide information with only moderate accuracy, but adequate to be incorporated into planning responses that benefit from anticipation of need. In fact the reduction in skill by generalizing across productivity zones was found to be only moderate, but the gain in knowledge of the spatial extent of anomalous prices exchange provided by the model made the sacrifice of accuracy more than justified. Many vulnerable communities are isolated geographically and so the extension of price data from sentinel markets to provide estimates of future prices in areas that have no markets is likely to be particularly useful.

Price maps produced from this model, however, are only effective in regions that have closely coupled production and marketing systems such as the informal millet markets in the Sahel. Regions where most of the food is imported, or in countries or times when food prices are controlled by outside events (by centrally fixing prices or during political or economic crises, for example), the model will capture a smaller portion of the changes in food prices. Thus modeled millet prices are of limited accuracy in urban areas and in areas with significant trade in commodities into and out of the region.

The analysis presented above shows that the quality of the growing season affects the price of millet at the annual and the seasonal time scales. If the growing season is characterized by erratic, sparse rainfall, higher cereal prices are the result, but with well-distributed, abundant rainfall the region experienced lower prices. Additional research using vegetation information from surplus-producing zones improves the price model, using the local information to create relationships which enable these maps is being undertaken. The required accuracy of these anticipated prices is far lower than that required for determining immediate food aid requirements. Price projections need only give planners a general idea of which direction food prices are likely to move. This information can be used to direct the on-the-ground assessments needed to respond in an accurate and timely manner and can help food program organizations begin the large task of building up grain reserves in the most threatened areas.

12.4 Role of Trade in Food Provision

In most of sub-Saharan Africa, regional trading in food commodities is the primary source of local grains in markets. When one locality is suffering from a drought-related production deficit, a neighboring region is having a good season and thus can export grain into the area. This natural response to elevated prices cannot occur if the border is closed, if the roads are impassable, if civil unrest or political upheaval make it unlikely that the traders will be able to return safely with their product, the market system otherwise does not function. What is more common, however, is that cereals are imported into the markets from other regions, but the price of the grain is so high that local residents are unable to purchase sufficient quantities to meet their needs.

12.4.1 Role of Prices in Southern African Crisis of 2001/2002

Monitoring the retail price of the main staple foods in Southern Africa has been particularly important, given the role that retail prices play in household food security, especially during the crisis that occurred in 2001/2002 in parts of the region. Conditions in Zimbabwe, Malawi and Zambia were of concern given that household purchasing power had been low since May 2002. Nominal prices continued to remain high and increase throughout 2002 until the next harvest in 2003, which worsened food access for a large number of households in Zimbabwe, Malawi and Zambia, as well as parts of Lesotho, Swaziland and Mozambique. The abnormally high maize prices in much of the region acted as a strong incentive for producers and exporters in northern Mozambique, South Africa and Tanzania, who invested in significant productive capacity for the next year.

In Zimbabwe, Malawi and Zambia, retail prices were generally higher than average in food insecure areas, and continued to rise throughout 2002 in many markets. Reasons for these abnormally high cereal prices include low maize availability in markets, a greater demand from heavier than normal purchasing by rural households who produced less that usual and higher initial prices going into the 2002/03 consumption season. Other reasons for high prices were that all or parts of these countries suffered a two consecutive poor agricultural seasons, increasing market demand – in terms of amount and timing earlier in the season – for staple foods (such as maize and maize meal). Second, varying degrees of poor macroeconomic conditions these three countries, especially rising inflation (with Zimbabwe with the most acute situation with an annual inflation rate of 144% as of mid November 2002) that are supporting the upward pressure on prices. Finally, the slow rate of commercial imports has generally supported the high level of retail maize prices. For many reasons – including the unwillingness of the region to import genetically modified (GM) foods, obstacles from national governments (including inefficient institutional arrangements and poor policies) and an inadequate response from some donors – the arrival of food in Zimbabwe, Zambia and Malawi was slow and inadequate during that time.

Even though basic commodities like maize meal and cooking oil were available in the markets in Zimbabwe during this period, accessibility remained a major problem. This was further compounded by very high unemployment, low income and high inflation rates. Figure 12.6 shows the extremely high prices in urban areas in 2003–2004. These prices significantly reduce access to food, particularly when wages do not keep up with commodity inflation.

In this type of situation, elevated food prices are the primary cause of vulnerability. Although the modeling approach presented earlier won't be able to capture the

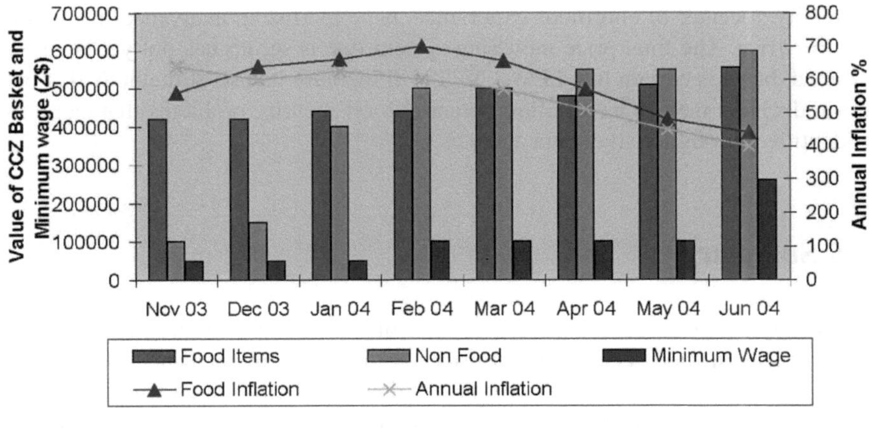

Fig. 12.6 Price of food and other necessary goods compared to minimum wage and inflation rates. Sustained inflation has significantly eroded people's ability to access food

enormous volatility of the prices in Zimbabwe, they will provide needed guidance for much more normal response to the markets to impending scarcity. In times of crisis, even a small reduction in production locally may cause an exaggerated increase in prices because demand is strong and fear is high. FEWS NET is interested in developing new tools that will provide quantitative guidance that can inform decision makers of market-driven food insecurity.

12.5 Challenges for FEWS NET

Markets are a key way that FEWS NET directly monitors variations in food access. FEWS NET must show evidence for its conclusions regarding food security, and market prices, although imperfect, are a key way that FEWS NET has in documenting variations in food access. In most countries in Africa, however, much of the economic activity that occurs is outside of formal control by the government (Pederesen et al., 1999). FEWS NET samples the grain prices through their own networks, and thus the prices that they are reporting on are from the open-air, informal markets typical of the region. Prices obtained from such market surveys are highly related to overall economic activity and can be used to infer as to the overall cost of food for the most vulnerable. FEWS NET has found that is very challenging to evaluate market price movements in the most isolated areas due to the high cost of visiting these areas. A much higher allocation of resources must be made if FEWS NET is to obtain sufficient sampling of the smaller markets which are critical to understanding food access in the more remote regions.

FEWS NET has only recently begun to require market data analysis in its reports for all countries every month. With the implementation of standard reporting requirements and an integrated information system that will permit immediate delivery of all data obtained in each country, the use of economic models which require the most up-to-date market data is now possible. FEWS NET has begun to collaborate with personnel at the Department of Agricultural Economics at Michigan State University, a center of enormous experience and expertise in analysis of economic data in Africa. The final price modeling system that is set up has not yet been determined, but it is certain that FEWS NET will increase its use of methods that can inform decision makers as to future potential food security problems that can only be identified through market data analysis.

12.6 Summary

In this chapter, we discussed the critical role of markets in providing food and the role of prices in food security. Food prices are a key component to food security analysis, because access to food for rural residents is often through small, informal markets where grain is bought and sold. Food prices are influenced in these markets by local availability of cereals and by the price and availability of food produced elsewhere. Cross-border trade and transportation costs are other factors that affect food prices in local markets. Until recently, food price analysis has been at the individual market level and conducted in the context of the monthly reports. Time series of local prices were plotted and the implications of significant increases or decreases on food access noted in the context of an analysis of food security. FEWS NET has begun focusing on providing food security outlooks for the next six months. This future orientation creates a need for price projections that can both fill in information that is missing from real-time market reporting, and provide information regarding future price directions. This chapter presents methods and approaches for estimating prices and coupling them with NDVI observations for improved modeling.

References

Boudreau, T.E., 1998. The food economy approach: A framework for understanding rural livelihoods. London, Relief and Rehabilitation Network/Overseas Development Institute: 32.

Brown, M.E., Pinzon, J.E. and Prince S.D., 2006. The sensitivity of millet prices to vegetation dynamics in the informal markets of Mali, Burkina Faso and Niger. Climatic Change, 78: 181–202.

Brown, M.E., Pinzon, J.E. and Prince, S.D., 2008. Using Satellite Remote Sensing Data in a Spatially Explicit Price Model. Land Economics, 84.

Buchanan-Smith, M. and Davies S.M., 1995. Famine Early Warning and Response, IT Press.

Chopak, C., 1999. Price Analysis for Early Warning Monitoring and Reporting. Harare, Zimbabwe, FEWS: 78.

Choularton, R., 2005. Improving Decision Making and Response Planning: A Framework for Contingency and Response Planning. Washington DC, FEWS NET: 8.

Coulibaly, O., Vitale, J.D., and Sanders, J.H., 1998. Expected effects of devaluation on cereal production in the sudanian region of Mali. Agricultural Systems, 57(4): 489–503.

Deaton, A. and Laroque, G., 1992. On the behavior of commodity prices. Review of Economic Studies, 59: 1–23.

Diallo, O., Diouf, A., Hanan N.P. and Ndiaye, 1991. AVHRR monitoring of savanna primary production in Senegal, West Africa: 1987–1988. International Journal of Remote Sensing, 12(6): 1259–1279.

FEWS, 2000. Framework for food crisis contingency planning and reponse. Arlington, VA, FEWS-ARD: 29.

Fuller, D.O., 1998. Trends in NDVI time series and their relation to rangeland and crop production in Senegal. International Journal of Remote Sensing, 19(10): 2013–2018.

Funk, C., Sanay, G., Asfaw, A., Korecha, D., Choularton, R., Verdin, J., Eilerts, G. and Michaelsen, J., 2005. Recent Drought Tendencies in Ethiopia and Equatorial-Subtropical Eastern Africa. Washington DC, Famine Early Warning System Network, USAID: 11.

Funk, C.C. and Brown, M.E., 2006. Intra-seasonal NDVI change projections in semi-arid Africa. Remote Sensing of Environment, 101(2): 249–256.

Galvin, K. and Ellis, J., 1997. Climate patterns and human socio-ecological strategies in the rangelands of Sub-Saharan Africa. Global Change and Subsistence Rangelands in Southern Africa, Gaborone, Botswana, IGBP Committee.

Holben, B., 1986. Characteristics of Maximum-Value Composite Images from Temporal AVHRR Data. International Journal of Remote Sensing, 7(11): 1417–1434.

Huang, N.E., Shen ,Z. and Long, S.R., 1999. A new view of nonlinear water waves: the Hilbert spectrum. Annual Review of Fluid Mechanics, 31: 417–457.

Huang, N.E., Shen, Z., Long, S.R., Wu, M.C., Shih, H.H., Zheng, Q., Yen, N.-C., Tung, C.C. and Liu, H.H., 1998. The empirical mode decomposition and the Hilbert spectrum for nonlinear and non-stationary time series analysis. Proceedings of the Royal Society of London, 545: 903–995.

IMF, 1999. Burkina Faso: Statistical Annex. Washington DC, International Monetary Fund: 47.

IMF, 1999. Chad: Recent Economic Developments. Washington DC, International Monetary Fund: 126.

IMF, 1999. Mali: Selected Issues and Statistical Index. Washington DC, International Monetary Fund: 82.

Jayne, T., Mukumbu, M., Duncan, J., Staatz, J., Howard, J., Lundberg, M., Aldridge, K., Nakaponda, B., Ferris, J., Keita, F. and Sanankoua, A., 1996. Trends in Real Food Prices in Six Sub-Saharan African Countries. East Lansing, MI, Michigan State University.

Justice, C.O., Dugdale, G., Townshend, J.R.G., Narracott, A.S. and Kumar, M. 1991. Synergism Between NOAA-AVHRR and Meteosat Data for Studying Vegetation Development in Semi-arid West Africa. International Journal of Remote Sensing, 12(6): 1349–1368.

Kelly, V., Reardon, T., Diagana, B. and Fall, A.A., 1995. Impacts of Devaluation on Senegalese Households: Policy Implications. Food Policy 20(4): 299–313.

May, C.A., 1991. Update Report to USAID/N'Djamena on the Market Information System (SIM) in Chad. Washington, FEWS: 86.

Nicholson, S.E., 2000. Land surface processes and Sahel Climate. Reviews of Geophysics, 38(1): 117–139.

Pederesen, P.O. and McCormick, D., 1999. African business systems in a globalising world. The Journal of Modern African Studies, 37(1): 109–135.

Pinzon, J., 2002. Using HHT to successfully uncouple seasonal and interannual components in remotely sensed data. SCI 2002 Conference Proceedings Jul 14–18, Orlando, Florida, SCI International.

Pinzon, J., Brown, M.E. and Tucker, C.J., 2005. Satellite time series correction of orbital drift artifacts using empirical mode decomposition. Hilbert-Huang Transform: Introduction and Applications. Huang, N.: 167–186.

Pinzon, J., Pierce, J.F. and Tucker, C.J., 2001. Analysis of Remote Sensing Data Using Hilbert-Huang Transform. SCI 2001 Conference Proceedings.

Prince, S.D., 1991. Satellite Remote Sensing of Primary Production: Comparison of Results for Sahelian Grasslands. International Journal of Remote Sensing, 12(6): 1301–1311.

Prince, S.D. and Goward, S.N., 1995. Global Primary Production: A Remote Sensing Approach. Journal of Biogeography, 22: 815–835.

Prince, S.D., Justice, C.O. and Los, S.O., 1990. Remote Sensing of the Sahelian Environment. Brussels, Belgium, Technical Center for Agriculture and Rural Cooperation.

Sen, A.K., 1981. Poverty and Famines: An Essay on Entitlements and Deprivation. Oxford, Clarendon Press.

Trefethen, L.N. and Bau, D., 1997. Numerical Linear Algebra. Philadelphia, Society for Industrial and Applied Mathematics.

Tucker, C.J. and Nicholson, S.E., 1999. Variations in the Size of the Sahara Desert from 1980 to 1997. Ambio, 28(7): 587–591.

Verdin, J., Funk, C., Senay, G. and Choularton, R., 2005. Climate science and famine early warning. Philosophical Transactions of the Royal Society B: Biological Sciences, 360(1463): 2155–2168.

Wilks, D.S., 1995. Statistical Methods in the Atmospheric Sciences, an Introduction. San Diego, Academic Press.

Yade, M., Chohin-Kuper, A., Kelly, V., Staatz, J. and Tefft, J., 1999. The role of regional trade in agricultural transformation: The case of West Africa following the devaluation of the CFA Franc. Nairobi, Kenya, Michigan State University: 34.

Chapter 13
Scenario Development and Contingency Planning

In their 1995 book on the gap between early warning and response, Buchanan-Smith and Davies detail the difficulties of providing information that the complex political and economic actors that will ensure prompt and effective response to an impending crises. In the past five years, FEWS NET has put much greater emphasis on planning for crises by national governments and regional organizations. There has been a significant increase in the number of countries that are implementing routine planning processes that incorporate a variety of actors who are involved in responding to a crisis. Geospatial data is a source of information for these planning processes that currently are not used to their fullest potential, particularly in regions that are susceptible to biophysical hazards. This chapter will discuss how FEWS NET works to improve the humanitarian community's response to its warnings.

Contingency planning is defined as the process of establishing program objectives, approaches and procedures to respond to situations or events that are likely to occur, including identifying those events and developing likely scenarios and appropriate plans to prepare and respond to them in an effective manner (UN IASC Taskforce on Contingency Planning and Preparedness). In the past few years, every non-profit, national government and program has done some sort of contingency planning for emergencies, even if the plans are not very sophisticated or in depth. Effective contingency planning must be integrated into on-going regular operational planning processes, allowing organizations more time and flexibility in the face of an emergency. By incorporating projections or forecasts of the environmental monitoring products that planners already use, improved understanding of both the probability of particular events and the planning process can be achieved (Table 13.1).

An important part of the contingency planning process is scenario development. Scenarios are used in the planning process to identify possible response requirements for hazards of varying severity as an integrated part of food security early warning. Integrated with concepts of food security, analysis of vulnerability is primarily focused on identifying hazards that may reduce a particular household's ability to remain food secure. Remote sensing analysis and environmental climatology can also play a role in identifying trends and establish the normal variation in an area of interest. A wide variety of rainfall and vegetation products are currently

Table 13.1 Contingencies considered in the planning process

Common contingencies derived from the planning process	
• Hurricanes/Cyclones	• Internal Displaced Persons
• Floods	• Border Closure
• Earthquakes	• Epidemics
• Droughts	• Volcanic Eruptions
• Famine	• Tsunami
• Internal Conflict	• Landslides
• War	• Crop Failure
• Refugees	• Food aid pipeline breaks
• Economic Collapse	• Peace
• Logistical Bottlenecks	• Prepositioning

used in hazard detection through monitoring of food production and weather dynamics. These tools can provide the basis for improved scenario development and contingency planning.

The planning process itself can be an invaluable tool for building consensus between government, donors, and humanitarian organizations and for developing the relationships and understanding needed for effective emergency response. By giving more time to develop consensus, contingency planning can expedite resource mobilization and ultimately response, especially for major emergencies like the one that Ethiopia faced in 2002/03. An active contingency planning process enables individuals, teams, organisations and communities to establish working relationships that can make a critical difference during a crisis. By coming together in a contingency planning process, people develop common understandings of the different problems facing each team member, of each other's capacities, of varying objectives and organizational requirements, and many other issues that facilitate effective collaboration in a crisis (Choularton, 2007).

13.1 Background on Contingency Planning and Scenarios

Contingency planning is a tool for preparing for potential crises where organizations develop scenarios and then plan to respond to those scenarios. The scenarios developed in contingency planning outline the potential humanitarian consequences of a particular event or series of events, including how many people may be affected, for how long, where they located, to name a few. Based on this analysis, planners can determine what type of assistance will be needed (food aid, water and sanitation, health care, etc), how many people will need it, where the assistance will need to be brought, etc. These plans are then used to help actively prepare for potential crises, including resource mobilization (Choularton, 2007).

When organizations respond to a humanitarian crisis, an assessment is among the first things that needs to be done. Humanitarian organizations need to know what has happened, where it happened, why it has happened, how people have been affected, for how long they will be affected, etc. The answers of these questions form the basis for developing an emergency response strategy and plan. When the planners develop a contingency plan before the emergency, none of the answers of these questions are known because the crisis has not yet occurred. Thus, in lieu of an assessment, contingency planners use scenarios as a basis for planning.

The major objective of the scenarios is to provide enough information for decision makers to prepare for and implement early and effective measures to respond to humanitarian needs. It is not necessary to be precise or to provide all possible permutations in each scenario. Scenarios are descriptions of situations that could occur, and are informed by a set of assumptions about a situation that may require humanitarian action.

Until recently, scenarios have rarely been used in early warning estimates because until recently FEWS NET has focused on providing information about current conditions. Much effort has gone into estimating current aid needs, including numbers of vulnerable people, where they are located, amount of aid they require, the type of aid that would be most effective, etc. These numbers have always been forwarded on to decision makers so that they may make good decisions as to how their assistance dollars will be spent. The problem with this approach is that it takes three to six months between the decision to provide assistance and the assistance arriving. U.S. food aid, which represents fifty seven percent of world food aid deliveries, is even slower as it prioritizes shipping food from the U.S. at high expense rather than providing cash for purchase of food in the affected countries or their neighbors (Murphy, 2002). Appropriation, purchasing, shipping and distribution of food must all occur before the food arrives locally. The US requires a long-lead time to get the aid in the right place, and thus having appropriate information early is very important to effective response. Scenarios can provide much better information much earlier, because it takes the knowledge of the socio-economic status of each locality and estimates what the effect of different crop production outcomes, food price regimes, trade scenarios, political changes and other events that may affect food security in the area.

There are a number of different approaches to scenario building. In the humanitarian community the most common is the best, most likely and worst case scenario approach. This approach builds scenarios which normally describe differing levels of severity of food security crises although this is not always the case. Table 13.2 shows the best, most likely and worst case scenario summary for a drought. Scenarios of different intensities allow planners to examine and plan for different scales of the same potential crisis or emergency. The approach is easy to understand, and provides a starting point for analysis of the impact of emergencies that can become overwhelming complex.

13.1.1 Information Requirements for Planners

Decision makers have very different information requirements at different points in preparing for action. The closer you get to beginning to act, the higher the detail and precision of information is required. Because humanitarian action often takes months to prepare for, the level of precision required as planning is taking place is

Table 13.2 Examples of the best, most likely and worst case scenarios

	Drought Scenario	Climate Monitoring product	Scenario Product for each case
Best Case	No Drought	SPI between -1 and 1	Historical probability of normal rainfall amounts
		No NDVI anomaly	Statistical projections of next season based on the current situation
Most likely or Middle Case	Moderate drought affecting one part of the country	SPI between -1 and -2	Climate conditions and typical persistence
			Analog years and probability of their occurring in historical record
Worst Case	Severe drought affecting large areas of the country	SPI between -2 and -3	Probable extent of spatial anomaly
		Drought affects most of country and neighbors	Statistical probabilities of worst case occurring given current climato-logical conditions

far lower than right before action occurs. Thus, the question that is asked routinely before any hazard appears is: Do I need to prepare for a potential food security crisis? Because the consequences of answering incorrectly to this question are fairly mild, the precision of information required nine or more months before a crisis occurs is quite low. As decision makers move towards the point of action, the questions change to: What responses are needed? How many people could be affected?

These questions require information about the spatial extent of the hazard, the severity compared to previous events, and a quantitative assessment of the number of people who may live in the affected area. Finally, as planning for all the actions required to respond to a crisis moves forward, the question becomes: How many people in each district need each type of assistance? To answer these questions for immediate action or food delivery, specific assessments on the ground need to be conducted in order to gather information on the numbers of vulnerable people who will require intervention. The assessments must be planned well before the crisis is at its peak, however. Early warning is central to the system working correctly.

13.2 Scenarios and Early Warning

Currently, three different levels of analysis have been identified in the FEWS NET household livelihoods-based analytical framework, resulting in a conceptual model of the social and economic landscape that can later be used for planning purposes. They include:

- *Zoning*: In this initial stage, the country is sub-divided into generally homogenous areas/populations of similar food security livelihood strategies, opportunities, and constraints. This stage is usually accomplished in a few weeks, and is carried out through a survey of key informants by an expert (or experts), and partners who are receiving hands-on training in the method. This first step serves to delineate an initial geographic framework of areas and specific livelihoods, which then serves to organize and receive the information collected in the second, profiling phase. Livelihood maps are produced at the conclusion of this phase. In many cases, this first stage of work is immediately followed, without a clear break, by work on the second stage, described immediately below.
- *Profiling*: In this stage of development of the framework, data collection, structured surveys built from knowledge gained in the zoning phase, and interviews are applied in the zones, to validate and refine, to fill-in with statistical data, and perhaps to substantially modify, the initial zoning decisions. Profiling provides a more concrete definition, using objective and subjective inputs, of the characteristics of normal and exceptional livelihood food-related strategies and behaviors. The profiling information provides an initial basis for informed and insightful monitoring and assessment of food security and vulnerability, as well as for identifying possible priority indicators that can be used as efficient monitoring tools.
- *Scenario Modeling*: In this phase, a quantitative baseline of information on key livelihood and food security conditions is developed, so that a range of crisis scenarios (e.g. what if there were a severe drought, steeply rising food prices, or a closure of the border?) can be modeled and tested to assess the potential impacts of specific shocks and hazards on food security and vulnerability. This stage of development of the framework requires a substantial input of time and effort, but is intended to produce very specific insights and conclusions about local coping abilities, livelihood dynamics, the need for external food or non-food aid, the behaviors that can be expected following a shock, etc (USAID, 2005) (Fig. 13.1).

Although remote sensing data are used in the development of livelihood profiles to identify regions with similar agro-climatic zones, biophysical data remains marginal to the development of scenarios during contingency planning processes. This is a lost opportunity and results in less grounded and less realistic scenarios because variations in seasonal food production can be one of the most important variants that determine the outcome of a food security crisis. The application of biophysical data can provide important insights into the variability of crop production. A good example of the use of remote sensing data to provide early information about future production deficits is the routine use of the Belg rainfall performance in Ethiopia as a proxy calculation for key Meher production a few months later. This remote sensing-based analysis can provide significant information for planners.

Many planners only have a very limited understanding of and rarely use remote sensing data themselves. Without sound biophysical basis for different seasonal crop outcomes, the resulting scenarios are completely hypothetical instead of being informed by the probability of rainfall dropping below crop requirement thresholds. In some regions of Africa, dry conditions lead to reduced crop yields two out of

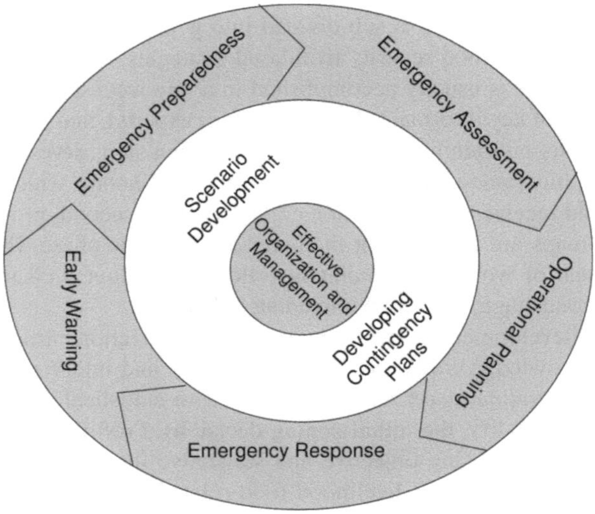

Fig. 13.1 Contingency planning, emergency response and scenario development continuum

every three years, with extreme dry conditions occurring every seven years. Work by FEWS NET partners on long term variations in climate in the Greater Horn of Africa shows that catastrophic droughts over a 10–20 year period are likely (Funk et al., 2005). As the succession of bad years lengthens to a decade, the lack of rain saps the resilience from livelihoods otherwise designed to be drought tolerant, pushing many populations over the edge into famine. Humanitarian organizations are focused on intervening in these regions rapidly and effectively to reduce economic disruption, and the loss of lives and livelihoods in the affected areas.

13.3 Ethiopia Contingency Planning Scenarios for 2006

In 2006, Ethiopia became a test bed for contingency planning for FEWS NET. The country had experienced drought, had negative export conditions, high food prices, internal conflict during the previous decade and was facing another year of high food aid needs in 2006. To anticipate the needs of the region, a contingency planning effort was launched to estimate the impact of various outcomes of the long rains on food security in the country. As a result it is essential for government, donors and humanitarian organizations to work together to ensure they are adequately prepared to respond to the human suffering encountered during these times. The planning was conducted under the leadership of the Ethiopian Disaster Prevention and Preparedness Commission as the first of an annual multi-agency and multi sector contingency planning process to plan for each year's aid needs to improve response and coordination among the various agencies. This process brought together all emergency task force groups (sectors: food, water and environmental sanitation, health and nutrition and agriculture and livestock) in a planning exercise where potential crises were analyzed, scenarios developed and response plans made. The objective of the national contingency planning process was first to provide decision makers with a reliable estimate of potential humanitarian needs in 2006, and second to foster improved collaboration and preparedness within the sectoral taskforces with the aim of improved emergency response when needed.

Three main scenarios were developed by each sectoral emergency taskforce: best, mid-case and worst case scenarios. In each sector a detailed analysis of the hazards affecting people's lives and livelihoods was conducted. While a great deal of analysis and assessment work went into the development of the scenarios, it is important to note that the results are indicative and meant to be used as planning tools, rather than in a predictive manner. This is especially true for the numerical results which represent approximations of need based on the best available analysis, rather than precise estimates. The main *meher* season and pastoral areas assessments as well as other assessments in November and December will provide the more definitive estimates of needs for 2006.

13.3.1 Best Case Scenario

The best case scenario assumed the 2005 *meher* season would result in good agricultural production in most of the country. In addition, it assumed that the October – December (*deyr/hagaya/dadda*), March – May (*belg/gu*) and 2006 Meher rains would be normal. Under this optimistic scenario, humanitarian needs were expected to be much lower than in recent years. However, pockets of crisis would remain in highly vulnerable areas suffering the lingering effects of the multiple shocks they sustained in previous years. These areas include Bale, East and West Hararghe of Oromyia Region, most of Afar Region and parts of Somali Region. Chronic food insecurity would persist in much of the country in addition to these areas including in Tigray, Amhara, Oromyia and SNNP regions. Under this scenario, about 6.5 million people were anticipated to face high levels of food insecurity. In addition to food insecurity, it is envisaged that these and other households would face malnutrition, disease, water shortages, sanitation problems, and seed and input deficits which should be addressed by the humanitarian community.

13.3.2 Mid-Case Scenario

The mid-case case scenario assumes that pastoral rains (*deyr/ hagaya/ dadda*) between October and December 2005 would be poor or fail. In addition, this scenario assumed the *meher* rains would end 2 – 3 weeks early in some areas resulting in limited yield reductions for agricultural areas. Poor October – December rains would lead to a widespread crisis in pastoral areas, especially Somali Region as well as Borena Zone of Oromyia Region and South Omo Zone of SNNP Region. Recovery in Afar's pastoral areas would also be hampered. The typical three month dry season between the main and short rains, lasting from June – September would extend to eight months, lasting until at least March 2006. Given that only three of the last ten *deyr* seasons in recent years have been normal, this scenario was a real possibility.

In addition to a serious crisis in these pastoral areas, severe conditions were also anticipated in Bale, East and West Hararghe of Oromyia Regiona, most of the Afar Region if the rains fail. Chronic high levels of food insecurity would also persist throughout the country, especially in Tigray, Amhara, Oromyia and SNNP regions. Under this scenario 8.8 million people would be food insecure and unable to meet their basic food needs. In pastoral areas as well as other areas facing severe conditions significant levels of malnutrition, susceptibility to disease, water shortages, sanitation problems, livestock losses, disruption in the educational system, seed and input deficits, and increased destitution can be expected.

Under this scenario the impact of the crisis in pastoral areas would peak during the final months of 2005 and the first quarter of 2005 (November 2005 – March 2006), as was the case in 1999/2000. While the food aid pipeline for this period was likely to be able to cover needs, adequate funding for critical non-food life and livelihood saving interventions was expected to pose a challenge.

13.3.3 Worst Case Scenario

Given the good performance of the rains in 2005, a worst case scenario was highly unlikely, in terms of food security. Thus, food sector contingency planning focused its efforts on the best and mid-case scenarios and did not develop a worst case scenario. On the other hand, Agriculture and livestock, Health and nutrition and Water and sanitation sector task forces did contemplate worst case scenarios. Given the very low probability of a worst case scenario in 2006, the national overview did not include details on this scenario, though the projected needs for those sectors that included the worst case scenario in their planning are included in the overall food aid needs described below.

13.3.4 Food Aid Needs Under Each Scenario

Under all the scenarios presented above, humanitarian aid requirements would be significant, though in the most likely scenarios, the best and mid- cases, needs would be significantly less than in recent years. While some of the food needs foreseen under this plan would be covered by the productive safety net program, additional food needs were foreseen in both scenarios.

Ethiopia faced a persistent food crisis with over 5 million people chronically suffering from high levels of food insecurity during the past few years. After a challenging start-up, 2006 will mark the second year of the implementation of the Productive Safety Net Programme (PSNP). The current official figure of 5.6 million beneficiaries for 2006 is a preliminary figure based on the document entitled 'Productive Safety Net Program: Program Implementation Manual' issued by the Ethiopian Ministry of Agriculture and Rural Development in December 2004.

The scenarios developed in the food sector assume the PSNP will cover the needs of chronically food insecure populations, while additional needs, especially in areas not covered by the PSNP, would be covered through emergency programs. The emergency caseloads indicated under the scenarios detailed here would be resourced through the humanitarian appeal. The DPPC will be primarily responsible for implementing emergency food aid interventions in *woredas* without the productive safety net program with local governments, WFP and other emergency partners.

Excluding the cost of the productive safety net program, humanitarian needs in food, are likely to be over 51 million US dollars in the best-case scenario. Figure 13.2 shows the percentage of the population in each *woreda* who will need food aid under the emergency during the year, in the best-case as well as the mid case scenarios. The scenarios estimate food aid requirements fairly well compared to actual aid delivered for three years. Food aid calculations for each scenario include full rations (cereal, pulses, and vegetable oil) and provision for blanket supplementary feeding for up to 35 percent of the overall population. The duration of assistance is calculated using historical average figures for each month as a percentage of the

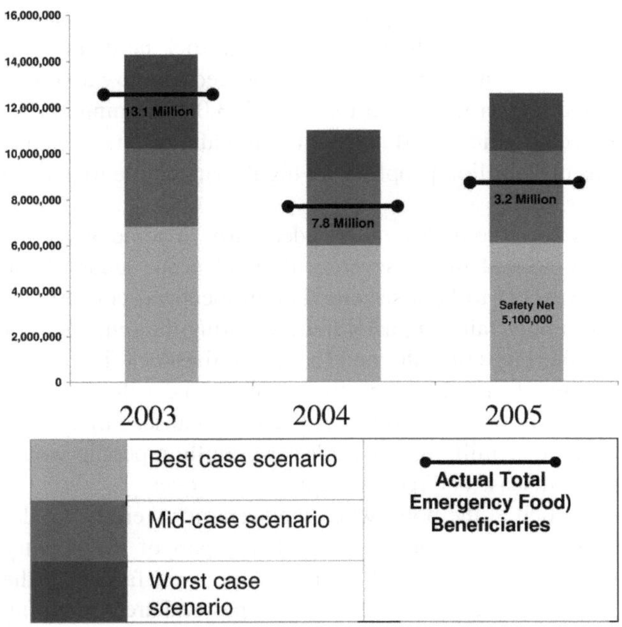

Fig. 13.2 Food aid scenarios and the actual food aid delivered to Ethiopia in 2003–2005

figures above. August is the month where 100% of beneficiaries are expected to be assisted (which are the numbers noted in Fig. 13.2 and above). For the other months, a percentage of the total beneficiary figure is used based on what percentage of the total beneficiary population was assisted in the past in that month and converting that beneficiary population to metric tons using the standard rations for each commodity.

This methodology results in a good early estimate of the numbers of people who will receive food aid. Unfortunately, this may or may not relate to how many people are food insecure and should be given assistance. This is because the methodology used to develop the scenarios are exactly the same as the way humanitarian food aid needs are estimated, and the process is conducted by the same people. Thus the scenarios can be seen as self-fulfilling in some ways. In order to meaningfully evaluate the success of the scenario process, it must be compared to the number of people who actually needed assistance or who were malnourished. Unfortunately, all data on malnutrition that are available in the country already are being used in the food aid need estimation process in order to reach as many people as possible with these programs.

13.3.5 Agriculture and Livestock

Seed security and animal health are key factors in the livelihoods of most rural Ethiopians, and thus the needs of these sectors are detailed in each scenario by experts who are familiar with requirements. Even in the best case scenario for 2006, where good *meher* rains led to a good 2005 agricultural season and good pastoral rains in the remainder of 2005 supported the recovery of livestock herds in areas hard hit by drought in recent years, emergency assistance in seed security and livestock health were still required. Under this scenario, about 213,000 farming households, or 1,065,000 people faced critical seed insecurity. In addition, the livestock of about 400,000 households or 2 million people were highly vulnerable to disease and require support to recover.

In the mid-case scenario where the *meher* rains ended early in some areas and *meher* production was below normal in the severely drought prone areas of the country, seed security was expected to be a serious though localized concern. At the same time, in this scenario poor rains in pastoral areas during the remainder of 2005 would result in serious shortages of water and fodder for livestock, increasing their vulnerability to disease and decreasing their productivity. Under this scenario about 295,000 farming households, or 1,475,000 would face critical seed insecurity. In addition the livestock of about 1 million households or 5 million people would be remain highly vulnerable to disease and require support to recover.

In a worst case scenario, which at the point when the scenarios were being developed was not realistic, poor *meher* rains in a significant part of the country would result in very significant impact on seed security. Moreover, failure of the pastoral rains would lead to a widespread livestock crisis in pastoral areas resulting in widespread livestock deaths and crippling reductions in livestock productivity. Under this scenario about 416,700 farming households, or 2,083,500 would face critical seed insecurity. Such a rainfall deficit would cause the livestock of about 1.5 million households or 7.5 million people to be in need of emergency curative and preventative veterinary support to minimize significant livestock losses.

The provision of quality seeds, creation of access to locally available seeds, and launching of variety- specific operations would be key interventions areas to respond to seed insecurity under all three scenarios. In addition, in limited and specific areas affected by especially severe shortage of seeds, direct distribution of seeds would be considered as well.

In comparison with the past three year average, there was a clear tendency for substantial reductions in the emergency seed requirements in 2006. This decline was due to a much better performance of the 2005 Belg and Meher seasons. In addition to seed security programs, curative and preventative veterinary support would be essential under all three scenarios to prevent excess mortality and morbidity and mitigate against associated production losses. Since the implementation of most of the veterinary interventions are appropriate before the onset of drought and immediately following the rains (especially drought breaking rains), the minimum estimated budget is needed before the month of January for timely preparation and executing of veterinary services in 166 identified disaster prone woredas.

13.4 Climate Data and Contingency Planning

The integration of remote sensing into planning is currently quite limited. In Eastern Africa, downscaled seasonal forecasts have begun to be used in the context of scenario development to identify regions likely to experience drought and the severity of that drought according to the numerical climatological forecasts. These interpreted forecasts enable the planners to assess the severity and temporal development of a drought, depending on the point in the season that the planning is taking place. Figure 13.3 shows the results of using the FEWS NET developed Forecast Interpretation Tool (FIT) to translate the March 2007 Climate Outlook Forum forecast into estimates of expected rainfall anomalies (Husak et al., 2007). Sector specific interpretations were analyzed to provide information on its potential impact on the pastoral, agro-pastoral and agricultural zones.

Planning is carried out primarily by social scientists who are the principle responders to these crises. During the planning process, remote sensing data is used to examine the current situation in depth before the scenarios are developed. Thus the impact of the previous few years' growing season on the current food security status by sector are examined using remote sensing data. Typically planners use the traditional monitoring products of NDVI, WRSI, RFE and others. In addition to looking back, planners are now beginning to look forward with seasonal forecasts which can inform which of their scenarios are more likely. In the example given above, some

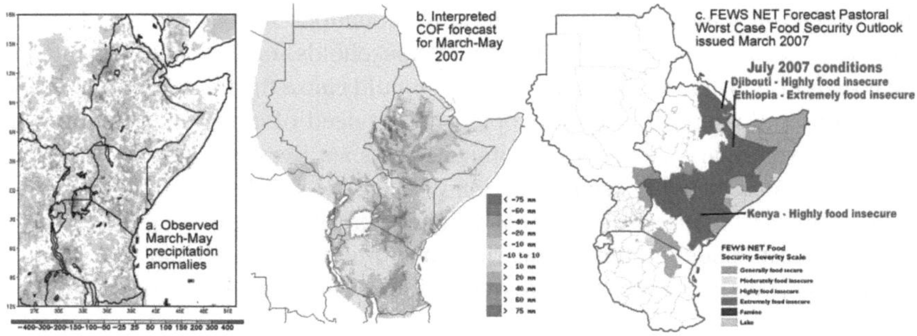

Fig. 13.3 Observed rainfall (**a**), Climate Outlook Forum forecast precipitation anomalies (**b**), and worst-case rangeland food security outlook with colors (**c**). Text on c. shows July 2007 FEWS NET watch and emergency locations for eastern Africa (Brown et al., 2007). Figure courtesy of C. Funk, UCSB and used with permission

of the sector analyses examined the worst case, but others didn't. The worst case was very unlikely when the planning was done because the rainy season was more than half over and it was unlikely that it would stop raining altogether resulting in a severely reduced production. Probabilistic forecasts are very hard to use, and thus FEWS NET has developed the Forecast Interpretation Tool (FIT) which transforms probabilities into rainfall amounts that can be easily interpreted by analysts. As data showing the statistical projection of vegetation and rainfall becomes available, it will be easier to incorporate future changes in food production as estimated by the remote sensing data products into the process.

Many of the recent advances in remote sensing that the FEWS NET team has been working on involves providing better tools for identifying hazards earlier and improving the ability of analysts to respond to those hazards. Projections of NDVI and of rainfall using statistical methods will improve our ability to provide early warning. New remote sensing products should improve our ability to identify hazards and provide early warning of impending problems. Integrated remote sensing and scenario development should improve the ability of planners to provide the emphasis on the plan that is statistically most likely to occur. These will improve response to emergencies.

Remote sensing data can provide information on the response side of the cycle as well as the warning side. The standard warning products can inform the hazard severity, extent and temporal development, which have a significant impact on priorities in response. Coupled with detailed information on the livelihood strategies of the people in each area of an affected country, this information will help planners determine who should be the first to receive aid and what type of aid is required. Identifying vulnerable groups and their location is a primary task of response planners: remote sensing data should help with this problem.

During the response, there can be significant biophysical threats to the effectiveness of aid delivery. These include areas that have been damaged by floods, roads that are impassable due to sand drifts or washed out bridges, and regions that have been impacted by high winds after a cyclone. High resolution data are often used by response planners to identify problem areas and to plan around them.

In order to identify durable solutions to vulnerability, analysis of trends in agricultural areas is fundamental to determining appropriate response. If a region will continue to experience inadequate rainfall, then programs to encourage and support people at the margins of agriculture to move to more sustainable livelihoods should be at the center of response programs. By continuing to support a depressed and declining sector of the economy, the dependency and burden of providing support will continue to increase, leaving the likelihood of donor-fatigue more and more likely. The impact of changing overall rain rates due to climate change is significant in some regions, and in Ethiopia the likelihood that the long term rainfall won't improve in the near term seems quite high.

13.5 Challenges for FEWS NET

Scenarios are not a straight-forward planning tool. When FEWS NET works with the same group of people to make policy and decisions as to how much humanitarian assistance is required as they do to develop scenarios, they run the risk of influencing the decisions they make. Scenarios have long been recognized as having 'reflexivity' – the idea that developing scenarios may influence the behavior or decisions driving the scenarios, so judgments about scenarios could reflect back on themselves, becoming either 'self-fulfilling' or 'self-denying' prophecies. Although FEWS NET uses as much evidence as possible for its decisions, ensuring that the scenarios that they build are not simply implemented without full food security analysis when the outcomes of the rains are known is very difficult. It is also not clear how the use of regularly produced scenarios in East Africa will impact the region's dependence on assistance, particularly with the implementation of the safety net in Ethiopia that enable a clear recognition of the underlying need for assistance by large numbers of people in the region.

FEWS NET operates in an environment where government responsibility, transparency and engagement with multiple stakeholders are becoming increasingly important. In order to ensure that a wide variety of issues are dealt with and a maximum amount of input is given, scenarios and planning has become a tool that enables FEWS NET to manage the process of stakeholder involvement and decision-making in the countries where it works. Integrated with Food Security Outlooks, scenario development can become a powerful tool that will enable FEWS NET to increase the participation of a myriad of non-governmental organizations, local, regional and national actors in each country, while reducing the amount of time needed to plan for a crisis (Fig. 13.4). Although this strategy is likely to improve its decision support, it may also reduce the quality of its analyses and increase the likelihood that events on the ground are not being reflected in reporting and analysis. These are problems that FEWS NET is aware of and its continued investment in and support

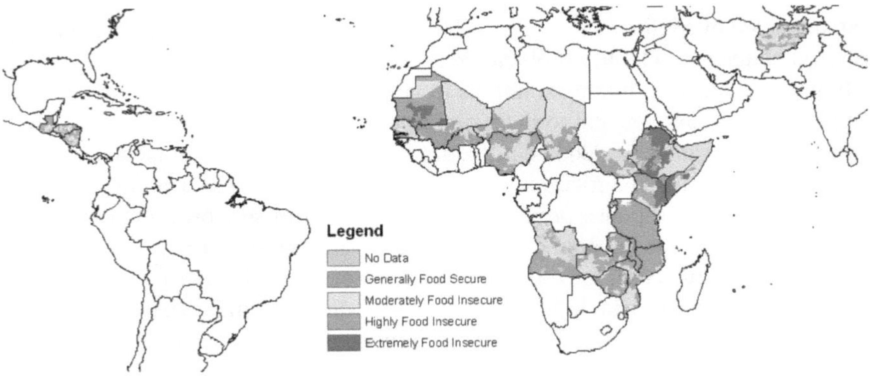

Fig. 13.4 Estimated food security conditions in February 2007, from which scenarios are developed for change during the next six months due to a variety of hazards. From Chuck Chopak, USGS science team meeting, August 2007

of its field staff will reduce the likelihood that the problem of reflexivity will reduce the value of its information in the long run.

13.6 Summary

This chapter described the contingency planning and scenario development strategy that FEWS NET uses to plan for problems in the countries they work in. The use of contingency planning and scenarios in early warning is described and an example from Ethiopia in 2006 is given. The utility of remote sensing data in these plans is identified and examples are given for different stages of the contingency planning process. Although planning is typically conducted by individuals most involved in conducting response who typically do not have experience with remote sensing data, a wide variety of different products can be useful during contingency planning. Recent developments in remote sensing and climatology analysis can support this new planning tool.

References

Brown, M.E., Funk, C., Galu, G. and Choularton, R., 2007. Earlier famine warning possible using remote sensing and models. EOS Transactions of the American Geophysical Union 88(39): 381–382.

Choularton, R., 2007. Contingency Planning and Scenario Development: A review of the practice of scenario-based contingency planning among humanitarian agencies. Overseas Development Institute Humanitarian Practice Network, London, England, 82.

Funk, C., Sanay, G., Asfaw, A., Korecha, D., Choularton, R., Verdin, J., Eilerts, G. and Michaelsen, J. 2005. Recent Drought Tendencies in Ethiopia and Equatorial-Subtropical Eastern Africa. Famine Early Warning System Network, USAID, Washington DC, 11.

Husak, G., Michaelsen, J., and Funk, C.C., 2007. Use of the Gamma distribution to represent monthly rainfall in Africa for drought monitoring applications. International Journal of Climate in press.

Murphy, S., 2002. Managing the Invisible Hand Markets, Farmers and International Trade, Canadian Foodgrains Bank and Institute for Agriculture and Trade Policy: 71.

USAID, 2005. Request for Proposals – Section C: FEWS NET Scope of Work: 2005–2010. United States Agency for International Development, Washington DC, 22.

Section IV
Case Studies

Chapter 14
Case Study: Ethiopia and the 2002–2003 Food Security Crisis

The next few chapters will highlight the use of remote sensing in case studies of specific food security crises. These case studies will illustrate how FEWS NET has used remote sensing to identify, act on and resolve a crisis. This chapter will focus on the Ethiopia food security crisis of 2002–2003. The mobilization and delivery of 1.8 million tons of food aid, the largest food aid delivery to any country to date, was achieved as a result of FEWS NET's efforts in clearly and effectively identifying the causes and consequences of the drought and a receptive humanitarian community to the message. The collection and presentation of a wide variety of biophysical and socioeconomic data during November and December of 2002 demonstrates the methods and analysis techniques that FEWS NET employs to communicate with decision makers during a crisis. The final section of the chapter summarizes an analysis of rainfall trends that show the likelihood that Ethiopia will continue to have droughts and food security crises without significant and structural economic change.

14.1 The Food Security Crisis in Ethiopia, 2002–2003

The crisis of 2002–2003 was rooted in a complex set of intertwined casual factors, many of which still persist. The drought that precipitated the crisis was the last in a series of shocks including: the lack of recovery from past droughts, subsequent environmental degradation, price collapse after the bumper harvest of 2001 caused primarily by poor market integration (inability to redistribute localized surpluses to high demand markets), high level of farmer indebtedness, the sharp deterioration in coffee prices, a livestock ban in the Gulf, and conflict and continued border closure with Eritrea. The effect of these events was multiplied by poor infrastructure, and a weak government policy environment that did not do enough to reduce their impact on farmers.

Ethiopia is the third most populous country in Africa with a population of over 60 million. Its economy is mainly dependent on crop and livestock production, and is extremely vulnerable to climatic variations. The country has two production seasons, short (Belg) rains occur from February to April and the long (Meher) rains from mid-June to mid-September. Meher season crop production contributes about 93% of the annual national production. The remaining 7% is from the Belg season (Fig. 14.1). Agriculture accounts for almost 41% of the gross domestic product (GDP), 80% of exports, and 80% of the labour force employment.

Many other economic activities in Ethiopia depend on agriculture, including marketing, processing, and export of agricultural products. Production is overwhelmingly of a subsistence nature, and a large part of commodity exports are provided by the small agricultural cash-crop sector. Principal crops include coffee, beans, oilseeds, cereals, potatoes, sugarcane, and vegetables. Exports are almost entirely agricultural commodities, and coffee is the largest foreign exchange earner. Ethiopia's livestock population was believed to be the largest in Africa, and in 1987 accounted for about 15% of the GDP. In the past two decades, livestock has been decimated by droughts, and its value to the economy shrinking due to the closure of markets to Saudi Arabia and other Middle Eastern countries in the late 1990s and early 2000s after outbreaks of Rift Valley Fever.

14.1.1 Vulnerability to Drought

Ethiopia's vulnerability to drought is a function of its low level of economic development, large inequities in access to income and assets, weak institutions and poor or non-existent service provision. The government is closely aligned with the

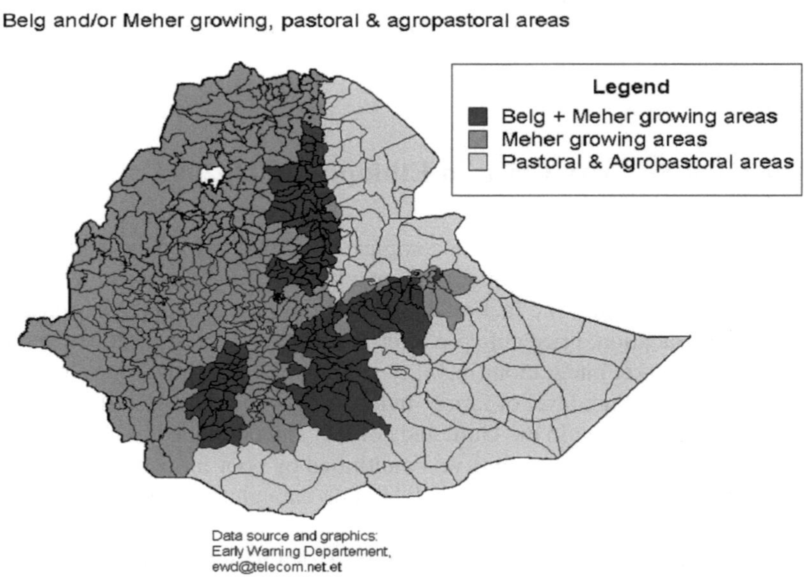

Fig. 14.1 Belg and Meher growing season regions in Ethiopia

United States so that it may rely upon it during military and food security crises, and spends significant effort and resources on these activities. These investments are to the detriment of long-term development, which is far harder to achieve.

A series of shocks before the 2002/3 season resulted in a cumulative deterioration of assets at the household level for those least able to cope. Figure 14.2 shows the relationship between March-September rainfall anomalies to food aid deliveries since 1996. This extraordinarily high sensitivity to variations in climate reflects both the large and growing portion of the population that is food insecure, but also the growing dependence on satellite remote sensing analysis to identify and quantify the number of people likely to need food. Although it is clear that monitoring the amount of food produced is a critical task, determining who needs aid and who doesn't continues to be a difficult job, since a large portion of the population is vulnerable in one way or another.

Substantial post-1997 declines in March-September rainfall have been observed in the northeast, southeast and southwestern portions of the country, a result corroborated by several independent data sources, including spatial averages of rainfall were compared with stream gauge data, reanalysis precipitation and precipitable water fields, and satellite-based vegetation and precipitation time series, outgoing long wave radiation fields and eastern African lake levels. The observed rainfall reductions have been accompanied by an increase in millions needing food aid (Fig. 14.2). More details on this analysis will be presented in Sect. 14.4 of this chapter.

During 2002, Ethiopia faced its most serious drought since 1984. Not only was this drought one of historic proportions, it came quickly after previous droughts. Whereas major droughts used to affect Ethiopia once a decade, they now occur every few years, giving little respite to the population. Recent evidence shows that Ethiopia is experiencing more frequent droughts due to a long term drying trend (Funk et al., 2005). More information on these trends will be presented later in this chapter.

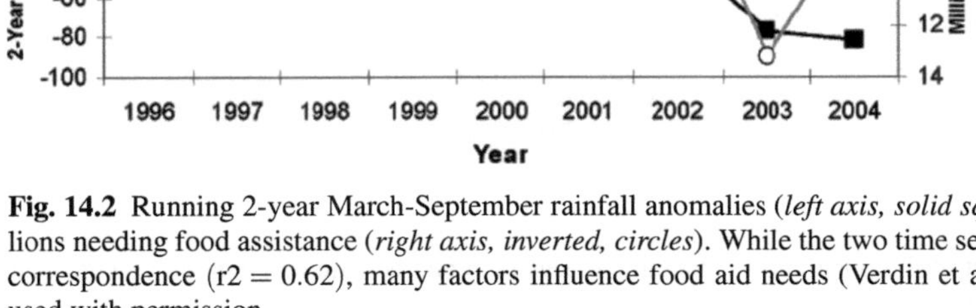

Fig. 14.2 Running 2-year March-September rainfall anomalies (*left axis, solid squares*) with millions needing food assistance (*right axis, inverted, circles*). While the two time series show a good correspondence ($r2 = 0.62$), many factors influence food aid needs (Verdin et al., 2005). Figure used with permission

14.1.2 Livelihoods Crisis

Environmental stress has been compounded by a crisis of livelihoods. Increasing destitution and impoverishment (Fig. 14.3) has been driven by a rapid succession of shocks from which households have been unable to recover. Rural families now face high and increasing levels of indebtedness exacerbated by limited labor opportunities. This crisis of livelihoods has been intensified by other shocks such as the collapse of world coffee prices, a major source of foreign earnings for the country as well as the primary source of labor and income for many poor rural households, especially in SNNPR, Hararghe and East and West Oromiya.

The scale of the 2002–2003 crisis in Ethiopia affected an unprecedented 15 million people. One factor driving the increasing scale of crises in Ethiopia is the rapid growth of the population. Over the 1990s Ethiopia's population has grown from 55 million people to almost 70 million people. Combined with other environmental factors, this has resulted in a reduction in the available crop land and ultimately declining per capita cereal production without a simultaneous increase in economic activity elsewhere. Government services have not been able to match the scale of poverty in rural Ethiopia. A weak heath care system and extremely poor access to potable water tend to compound problems of food insecurity and overwhelm the state capacity for comprehensive emergency response. While free markets offer a way to improve Ethiopia's long term food security, price trends and the dynamics of supply and demand can also cause problems. Volatile market prices, both for export crops such as coffee, as well as staple grains such as maize, are difficult for poor farmers to absorb. A severe reduction in commodity prices following the bumper harvest of 2001 resulted in large scale losses for farmers, who as a result planted much less during the 2001/02 growing season. Limited planting combined with a major drought resulted in a very poor harvest, which drove skyrocketing food

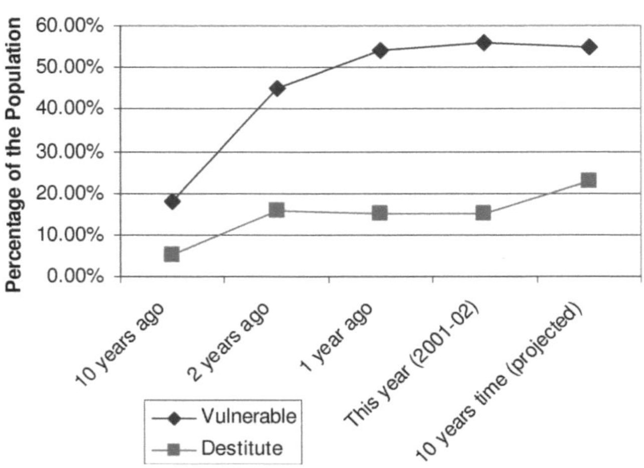

Fig. 14.3 Percent of the population vulnerable (*diamonds*) and destitute (*squares*) by percentage (source Anderson and Choularton, 2004)

prices. At the confluence of these factors in 2002/03, a massive humanitarian crisis emerged, which required enormous amounts of food aid and non-food assistance in health, water and sanitation, nutrition and agriculture.

14.2 Stages of the 2002/03 Crisis

Disaster theory, first described by Turner and Pidgeon in 1979, provides a theoretical basis for studying the origins of man-made disasters. The theory brings together relevant work based on a study of inquiries into accidents and disasters in Britain over an eleven-year period. This research shows how a disaster can have beginnings that may seem far removed from the actual events and have consequences long after the event occurs. Using the disaster model from Turner and Pidgeon, we can see the following stages:

- Stage 1 – notionally Normal starting points – the context/existing situation;
- Stage 2 – Incubation Period – accumulation of unnoticed events that will contribute to disaster;
- Stage 3 – Precipitating event – here, usually drought or flood, but also conflict or political crisis;
- Stage 4 – Onset – immediate consequences – again, difficult to determine with various impacts happening months after the 'event';
- Stage 5 – Rescue and Salvage, this is the disaster response stage;
- Stage 6 – Full cultural readjustment – this stage can be months to years later. In FEWS NET case, this stage can be long in coming as response in Stage 5 can be followed by another crisis because the underlying causes of the incubation period have not been resolved (Turner and Pidgeon, 1997).

Figure 14.4 shows the factors that caused the crisis of 2002/2003 and the issues that remain unresolved. The interplay between the functioning of the markets, poor governance and bad policies and the ever increasing numbers of people resulted in a much larger impact of poor rains in 2002–2003 then would normally be the case. It was only trailing indictors of poor nutrition, infant mortality and morbidity and lack of access to food that motivated a large scale humanitarian response. Remote sensing information was able to identify and give early warning of a poor harvest resulting from rainfall deficits in 2001–2002 and in 2002–2003, but many other complicating factors intensified the impact of the biophysical hazard. The next section will discuss the type of remote sensing data used by FEWS NET in Ethiopia to identify and forewarn of the impact of rainfall deficits and other socio-economic indicators and the accuracy of the food security warnings.

14.3 The Regional Agro-Climatic Situation

An analysis of the agro-climatic conditions of the Greater Horn of Africa (GHA) region is in marked contrast to that in Ethiopia. The USGS representative in GHA used a variety of remote sensing imagery to identify the geographic extent and severity

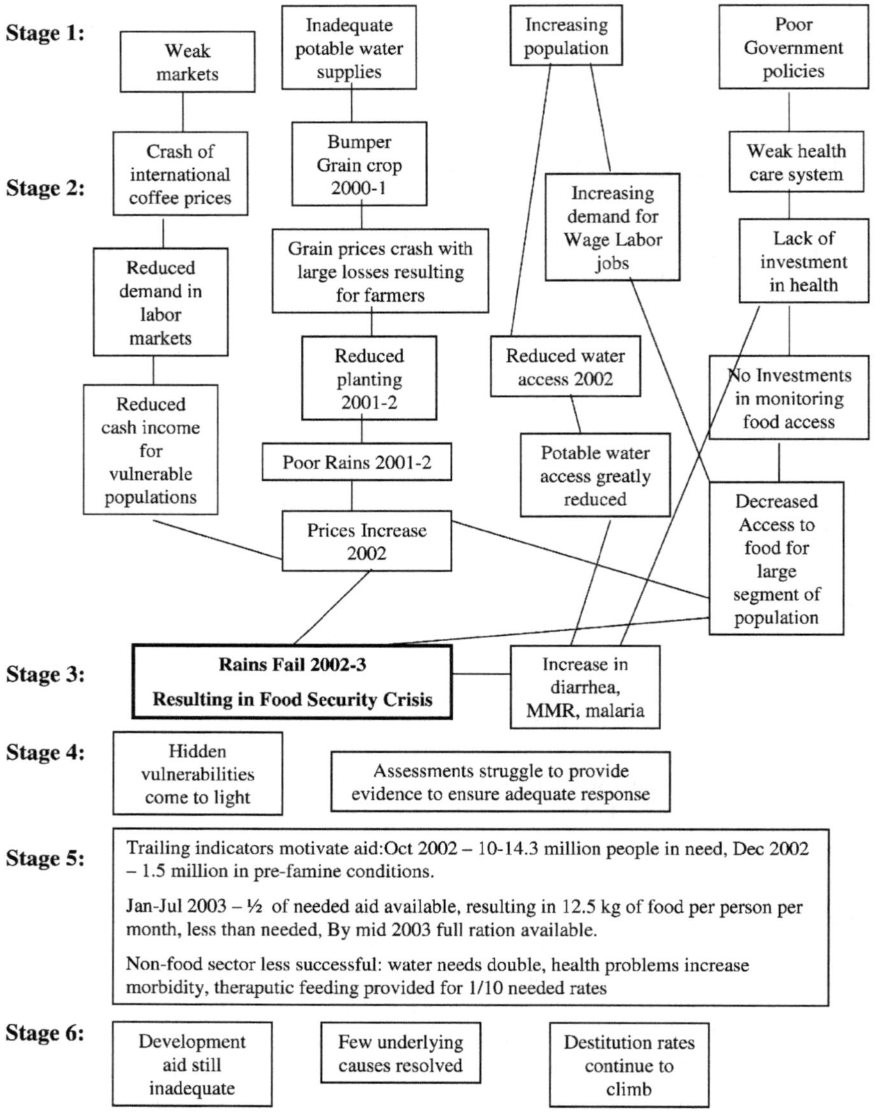

Fig. 14.4 Crisis stages after Turner and Pidgeon (1997)

of the agroclimatic risk for food insecurity. This section will describe the situation as reported by various reports in December of 2002 and what remote sensing products were used to support these conclusions. Figure 14.5 shows the areas that were deemed at risk by December 2002 according to the food security analysis that was conducted at the time.

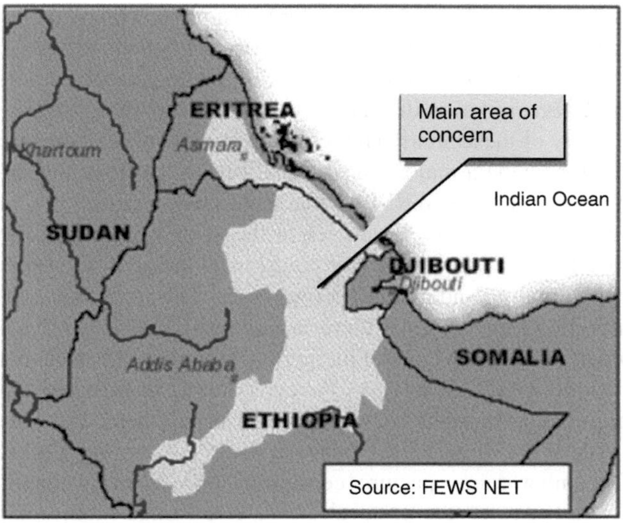

Fig. 14.5 Areas of main concern in the Greater Horn of Africa in December 2002

By the end of October, meher rainfall appeared to have withdrawn from most crop dependent areas in northern, central and eastern Ethiopia, whose rainfall usually lasts until January. While the rainfall withdrawal in these areas generally followed its normal pattern, given the late start and poor distribution of rainfall throughout the season, harvest prospects were expected to be very poor in many parts of Ethiopia (Fig. 14.6). Deyr season (late September through November)

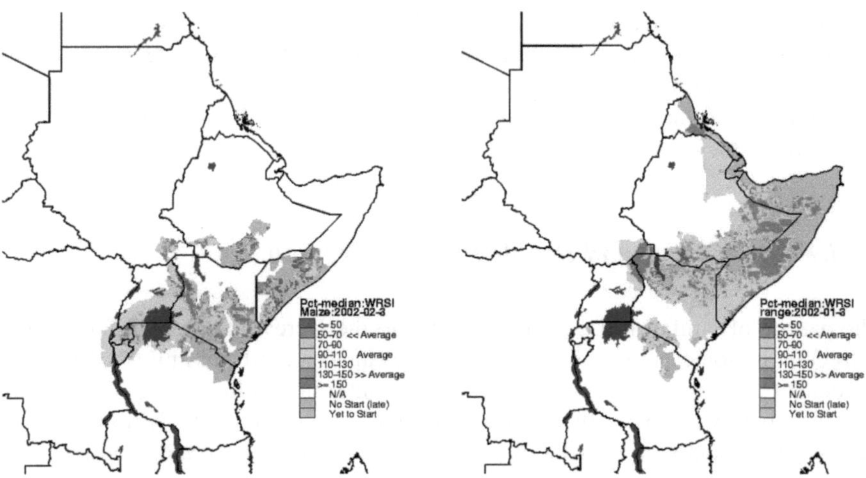

Fig. 14.6 WRSI for the Greater Horn of Africa region for fall rains, 2002. The Maize WRSI is on the *left* and rangeland WRSI is on the *right* (From USGS)

rainfall started on time in late September in southeastern pastoral lowlands of the country (South Omo, lowlands of Borena and southern parts of Somali Region). During the first dekad (10-day period) of October, most of these areas received normal to above-normal rainfall. Precipitation during the rest of the month was below to much-below normal in many of these areas. The result was below normal pasture and water availability, which became a major constraint affecting livestock performance and crop yields. In many areas the WRSI for maize was below 50%.

The seasonal rains were generally very beneficial to the growing areas in the central and southern greater Horn. However, there were isolated cases of persistent heavy rains that resulted into flooding and potential crop losses. The areas that were significantly affected by flooding were the Lake Victoria and upper Tana River (eastern Kenya) basins. The report used BERM model imagery to show water basins that were at various levels of flooding risk. Meanwhile, the cyclone that made a landfall in northeastern Mozambique, also brought heavy rainfall over southern Tanzania and there is an enhanced risk of flooding if these rains persist in the coming weeks.

As a result of the December rains, pasture conditions and water availability improved significantly in many of the pastoral areas. The forage monitoring based on NDVI anomalies provided by the Livestock Early Warning Systems (LEWS), which is an indicator of available forage, confirmed improvement in agroclimatic conditions in eastern and south eastern Ethiopia, eastern Kenya and western Tanzania. However, there was still concern over locations with poorer than normal forage conditions in pastoral areas of the Somali region of Ethiopia, parts of northern and southern Kenya and northern and central Tanzania. The latter areas were expected to improve with the on-going rains.

Field reports indicated favorable crop conditions for much of the short-rains planting regions. The Water Requirement Satisfaction Index product (WRSI) confirmed the situation. The imagery showed average to better than average crop conditions in a few of the maize growing areas. However, persistent rains might have affected the bean crop, which was in the flowering stages in parts of central and southeastern Kenya when the rains picked up again. In Rwanda, there was a significant improvement in the available moisture for crop growth in December. However, the late onset of the season and delayed planting might still have resulted in crop production losses.

14.3.1 Other Pests and Problems in the Region

The pattern of rainfall in December also led to the increased potential for malaria outbreaks in some areas. The report showed an increased possibility of malaria incidences over southern, central, northeastern and southern Kenya in 2002/3. Other affected areas were parts of northern Tanzania (around the Mt. Kilimanjaro areas), extreme southern Somalia, southern Ethiopia, eastern Rwanda and southern Uganda.

A malaria transmission risk map was shown to provide a simple indicator of changes in malaria risk in marginal transmission areas based solely on rainfall.

These malaria maps have been tested against laboratory-confirmed malaria incidence figures in Botswana where they showed a strong correlation. The impact of seasonal malaria incidence on food security in this particular crisis has not been determined, although endemic malaria has social, health and economic implications to peoples' welfare and food security.

There were isolated reports of mature locust adults, up to 16 locusts per hectare, reported in the Tokar Delta of Sudan and at two other locations along the coastal plains between Tokar Delta and Suakin during the end of November 2002. Forecasts anticipated small-scale breeding in areas of good rainfall on the Red Sea coastal plains from Karora to Port Sudan including the Tokar Delta and in Wad Oko/Diib in Sudan. However, this slight increase was not threatening and was under surveillance at the time. The Food Agriculture Organization monitors and provides updates on the desert locust locations and forecasts future movements. Figure 14.7 shows an example of the kind of FAO map that was used in 2002 to track desert locusts in Africa and neighboring regions.

Another hazard, an estimated 8.1 million Quelea Quelea birds, was identified and their reported locations in Nyanza Province and Kajiado District of Kenya identified. The government had already taken action to contain this situation through aerial spraying of invested areas. Queleas are the major animal pest of cereal crops in Africa, and international program to monitor them, coordinated by the U.N. Food and Agriculture Organization began in the 1960s.

Additional problems were anticipated from armyworms in Same district, Kilimanjaro Region in Tanzania with over 3500ha of grassland affected. Armyworms were also reported in the neighboring Taita/Taveta district in Kenya with a density of 24 larvae per square meter. Army worms are especially destructive to young crops and grassland and this required particularly careful attention and mitigation

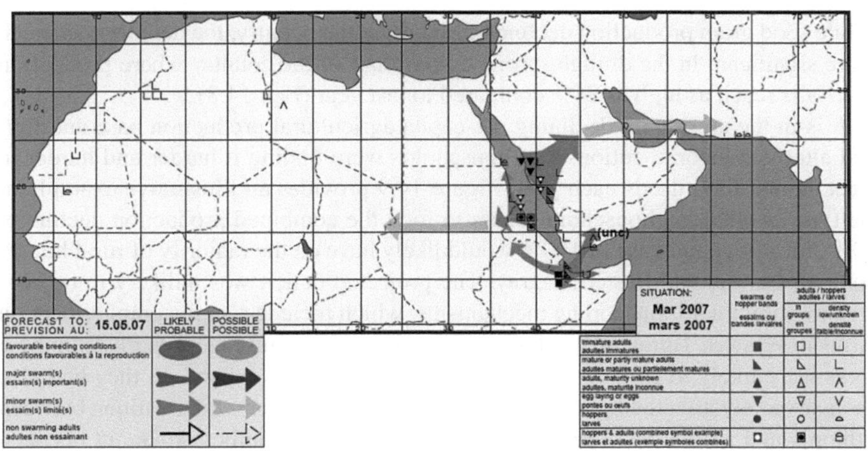

Fig. 14.7 Map of locust movements in Africa from March 2007. The *arrows* show movements of swarms and the symbols show development of locusts at various stages of their life cycle

measures. FEWS NET monitors pests for their impact on food production and ultimately food security, but does not assist with planning or implementation of mitigation steps.

14.4 The Food Security Situation

The complete failure of the rains in November 2002 and the sharp increase in food prices led to FEWS NET's December 2002 monthly report to have as its headline 'Over 1.4 Million metric tons of food aid is required in 2003 to avert famine in Ethiopia: 411,000 MT required during first three months alone'. According to the joint Appeal held at that time, the national meher (main) season food grain production declined by 21% from the previous five years' average. The main reasons listed in the report for the reported food production decline were drought:

- Early cessation of the belg season (March–May 2002) rains which affected the planting and growth of long cycle crops (maize and sorghum)
- Late onset of meher season (June–September 2002) rains by as much as 4 to 6 weeks
- Disruption of rains in July at a time when meher rains normally reach their peak and when crop water requirement is at its highest
- Earlier than normal withdrawal of meher rains in several areas and inadequate productive input use due to poor rainfall and credit constraints:
- Use of improved seed substantially declined from 11,000 MT in 2001/02 to approximately 3,000 MT in 2002/03 (70%)
- Fertilizer use in 2002 has also declined by 17% compared to the previous year
- Land preparation was inadequate in many areas because animal traction was limited during the optimal planting time when moisture was available.

While food grain production declined throughout the country, the decline was much more significant in the drought-prone eastern half of the country where production shortfalls reach as high as 81% compared to last year (Fig. 14.8).

It is in the areas with declining per capita agricultural production base and limited alternate income options that households were finding it harder and harder to meet annual food needs each year. Figure 14.9 provides an illustrative example of the type of effect on household access to food the combined production decline of 80% and a price increase of 100% would likely have on the majority of rural households in Central and Eastern Tigray. The projected deficit was unlikely to be covered through traditional coping mechanisms, which include seeking employment on farms in Western Tigray, in local towns, or in Eritrea, because labor markets in these areas were likely to contract in response to the crisis, not expand, as they had also been adversely affected by the drought. In addition, all economic relations between Ethiopia and Eritrea had been disrupted since a border conflict broke out between the two countries in May 1998.

In the mainly pastoral Afar Region and northern parts of Somali Region, including adjacent areas along eastern peripheries of Tigray, Amhara and Oromiya

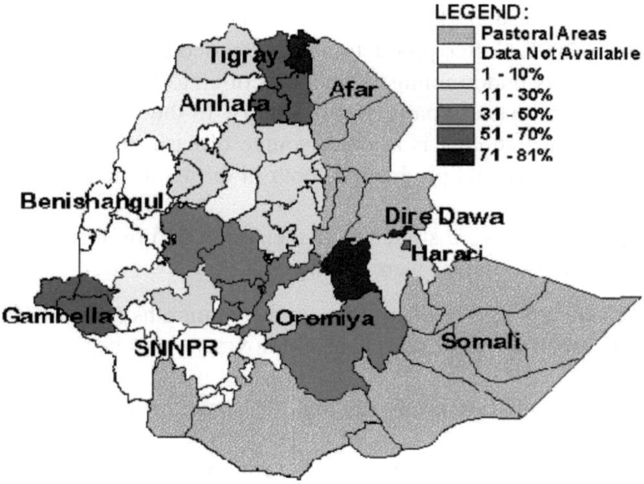

Fig. 14.8 Meher Season (Oct–Dec) Food Grains Production in 2002/03: Percent Decline from 2001/02 Source: Multi-agency Food Aid Needs Assessment Mission Reports (November 2002)

Regions, inadequate rainfall during both seasons of 2002 led to prolonged drought and exacerbated ethnic conflict over scarce grazing and water sources. As a result, hundreds of thousands of cattle died, threatening the livelihoods of pastoralists in these areas.

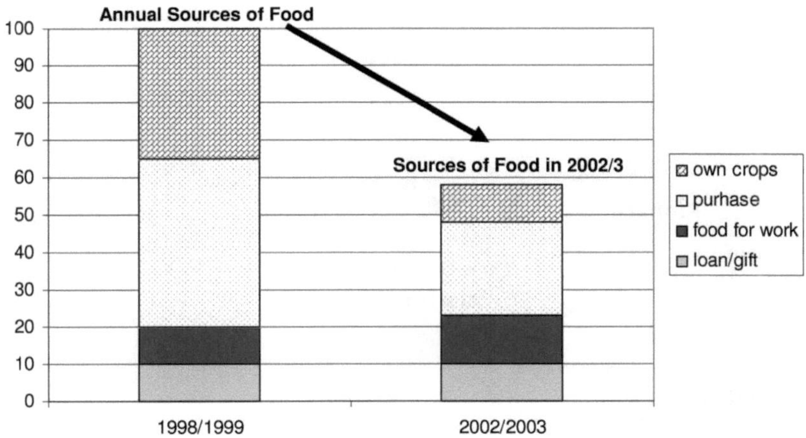

Fig. 14.9 Illustrative effect of hazards on household access to food in Central and Eastern Tigray. The 2002/2003 crop decline of 80%, combined with 100% increase in prices will lead to a critical food deficit among most households. Derived from 1998/1999 household level information obtained through field work for Oxfam/REST (1998) and Save the Children-UK (1998) in Central and Eastern Tigray combined with current year production information and price projection (from Anderson and Choularton, 2004)

Poor rainfall in 2002 also seriously reduced the production of cash crops such as coffee and chat (a mild stimulant). Coffee production was reported to have declined in 2002 by 20–30%, significantly undermining the incomes of up to 15 million people in western, southwestern and eastern parts of the country. In the midlands and lowlands of eastern Ethiopia, where chat is a major source of cash income, poor rainfall in 2002/3 adversely affected its production, both in terms of quantity and quality.

Coping strategies in many areas were stretched to their limit and by December 2002, there were signs that these strategies were failing:

- High malnutrition rates are reported in many areas in recent months (Fig. 14.10)
- Increased migration to urban areas in search of employment and begging on streets
- Increased consumption of wild 'famine' foods that are sometimes toxic if not prepared properly
- Earlier than normal withdrawal of meher rains in several areas and inadequate productive input use due to poor rainfall and credit constraints:
- Use of improved seed substantially declined from 11,000 MT in 2001/02 to approximately 3,000 MT in 2002/03 (70%)
- Fertilizer use in 2002 has also declined by 17% compared to the previous year
- Land preparation was inadequate in many areas because animal traction was limited during the optimal planting time when moisture was available.

While food grain production declined throughout the country, the decline was much more significant in the drought-prone eastern half of the country where production shortfalls reach as high as 81% compared to last year (Fig. 14.8).

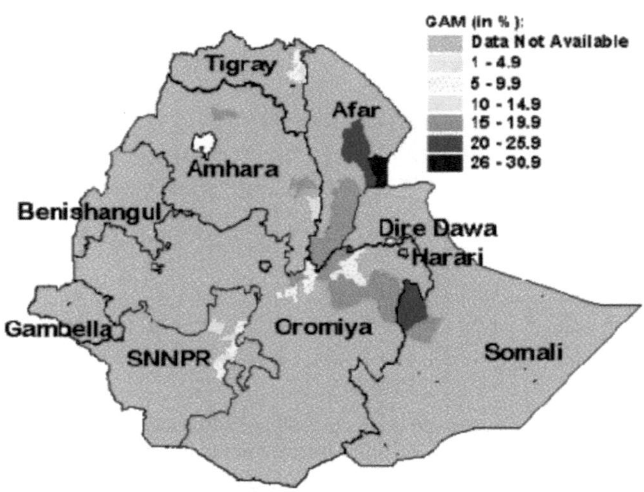

Fig. 14.10 Global Acute Malnutrition rates in Ethiopia in 2002 from a national survey conducted by the DPPC

It is in the areas with declining per capita agricultural production base and limited alternate income options that households were finding it harder and harder to meet annual food needs each year. Figure 14.9 provides an illustrative example of the type of effect on household access to food the combined production decline of 80% and a price increase of 100% would likely have on the majority of rural households in Central and Eastern Tigray. The projected deficit was unlikely to be covered through traditional coping mechanisms, which include seeking employment on farms in Western Tigray, in local towns, or in Eritrea, because labor markets in these areas were likely to contract in the current crisis, not expand, as they had also been adversely affected by the drought. In addition, all economic relations between Ethiopia and Eritrea had been disrupted since a border conflict broke out between the two countries in May 1998.

In the mainly pastoral Afar Region and northern parts of Somali Region, including adjacent areas along eastern peripheries of Tigray, Amhara and Oromiya Regions, inadequate rainfall during both seasons of 2002 led to prolonged drought and exacerbated ethnic conflict over scarce grazing and water sources. As a result, hundreds of thousands of cattle died, threatening the livelihoods of pastoralists in these areas.

According to the Emergency Nutrition Guideline of DPPC (June 2002), malnutrition rates obtained from nutrition survey (in the presence of aggrav-ateing factors such as poor household food availability, inadequate water supply, low vaccination coverage and epidemic diseases) were interpreted as follows: Global Acute Malnutrition (GAM): 15–19% = Critical, GAM: 10–14% = Serious, GAM: 5–9% = Poor. Figure 14.10 also shows the very small amount of data available, which is typical during such crises. Because the surveys required to measure GAM are expensive and very time consuming, usually only a very small fraction of the population are measured and only in the regions expected to have the worst problems. These data highlight the critical role that remote sensing data can play during such crises as they are the only spatially extensive data which can enable, in conjunction with livelihood zones, the calculation of the total number of people at risk despite the paucity of data actually measured on the ground.

14.4.1 Cereal Prices and Their Impact

In most years, the incoming main harvest should lead to a decline in cereal prices from November through February. In 2002/3, however, prices were already above both last year's prices and longer-term average levels in most markets. Given growing destitution and chronic poverty in many areas and declines in rural incomes, rising cereal prices led to deterioration of purchasing power among the rural poor.

In almost all observed major retail markets in the country, cereal prices increased significantly in November 2002 as the main season harvest failed to meet trader and consumer expectations. The reported trend of increasing prices began in April 2002 when the Belg rains started faltering (Fig. 14.11). Although farmers in traditional surplus producing areas in the western and northwestern parts of the country were

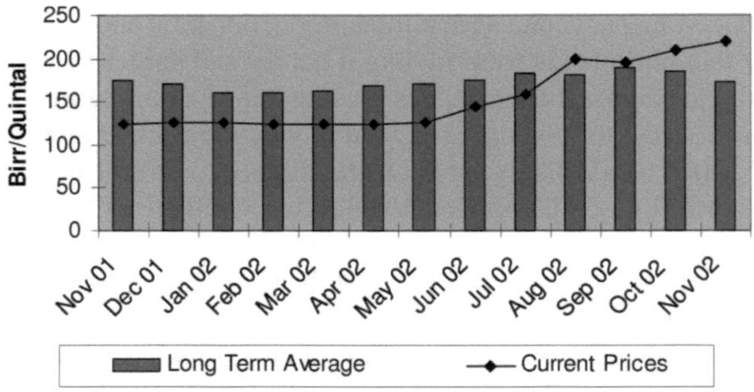

Fig. 14.11 Real Retail Prices of White Wheat in Addis Ababa: 13 Month Trend and Historical Monthly Average. Data Source: Market Information System of the Ethiopian Grain Trade Enterprise (EGTE) and European Commission Local Food Security Unit (EC-LFSU); data archives of FEWS NET/Ethiopia

expected to benefit from the unusually high prices, abnormal price increases threaten food access in the worst affected areas (especially in the eastern half of the country). Farmers in high-potential areas were adversely affected by unusually low cereal prices that persisted through most of 2000 and 2001. A further increase in prices can be expected in the next several months as cereal stocks from the new harvest start to decline, further exacerbating the current food crisis in many areas. In pastoral areas, increased sales of livestock at low prices were being widely reported in late 2002, indicating deteriorating terms of trade. Given poor pasture and water conditions that characterized many of these areas at the time, the terms of trade was expected to deteriorate even further as the dry season progressed.

14.5 Drought Tendencies in Ethiopia

Using the rainfall data developed for long-term rainfall analyses, FEWS NET colleagues were led by Chris Funk at the University of California at Santa Barbara to conduct an analysis of the drought being experience in the Horn of Africa. Although FEWS NET is an overwhelmingly forward looking organization that focuses on today and the immediate context in which today's events are occurring, it employs university researchers to provide historical analysis that can illuminate the causes and likely persistence of drought.

The analysis demonstrated that warming sea surface temperatures (SSTs), especially in the southwest Indian Ocean, may be linked to decreasing rains across East Africa (Funk et al., 2005). Rainfall from two merged satellite-gauge datasets, the Global Precipitation Climatology Project (GPCP) and the Xie-Arkin CPC Merged Analysis of Precipitation (CMAP) time series are shown in Fig. 14.12. Substantial post-1997 declines in March-September rainfall can be seen in the northeast, southeast and southwestern portions of Ethiopia. The observed rainfall reductions have been accompanied by an increase in millions needing food aid (Fig. 14.12).

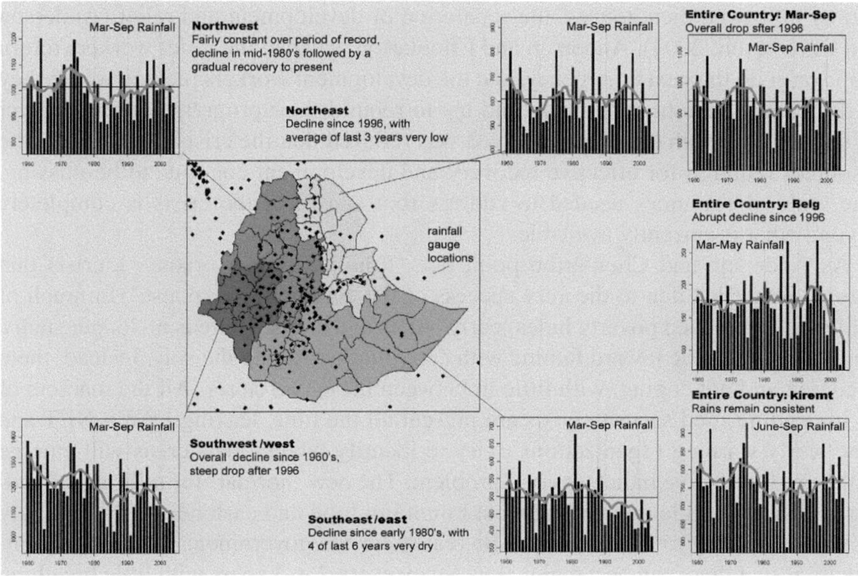

Fig. 14.12 Time series of March-September rainfall at a national scale (*right column*) and for four regions (*left column and center*). *Black bars* show seasonal rainfall. *Heavy lines through bars* show running 7-year means (from Verdin et al., (2005), figure used with permission)

The link between sea surface temperatures and Ethiopian rainfall was first identified in research performed by the Ethiopian National Meteorological Services Agency (Shanko and Camberlin, 1998). Shanko and Camberlin found that tropical storms over the southwest Indian Ocean led to lower than normal rains in Ethiopia, primarily during the Belg season, through reductions in available moisture, upper level easterly wind anomalies, and a northward shift of the subtropical westerly jet. Indian Ocean cyclonic activity played a role in the devastating drought of 1984. Recent research has identified an increase in rainfall over the southern hemisphere Indian Ocean (Hoerling et al., 2004; Hurrell et al., 2004) and there is an emerging consensus that Indian Ocean sea surface temperatures play a key role in African rainfall (Nicholson, 2003). Increasing sea surface temperatures cause a change in the off-shore flow, reducing the likelihood that rain-bearing westerly winds (Okoola, 1999) from the Congo/Zaire basin (Camberlin, 1997; Camberlin and Philippon, 2002) will come over Eastern Africa. The report concludes that global warming has caused increased convection in the southern Indian Ocean, lowered average rainfall conditions and therefore has played a role in creating current food insecurity crises in the region (Funk et al., 2005).

14.6 Responding to the New Normal: Relief and Development

In 2002/3 donors, government and NGOs worked together in an exemplary manner to resource record levels of food aid, a response most observers feel will not be replicated. After the massive emergency operation, agencies returned to an old and by now tired dichotomy – the separation of development and relief (Anderson and Choularton, 2004). Anderson and Choularton watched the relief workers return home or go to the next emergency, and the development workers pick up where they left off a few months previously and try to rebuild their programs on even fewer resources. Although everyone involved was relieved that the crisis was averted, the resources required for effective recovery and development continue to be missing. The scale of resources needed to address root causes of the crisis is completely beyond what is currently available.

As Anderson and Choularton point out, 'Ethiopia is in a crisis – a crisis that may go unheeded due to the very success of the emergency response'. In much of Ethiopia, entrenched poverty hides 'early' signs of a crisis. There is no longer such a thing as a slow slide toward famine with early markers along the way. Instead, there is coping and not coping, with little in between these two states. All the markers of an impending food security crisis are present all the time, leaving FEWS NET and other early warning organizations to try to identify when a new crisis will emerge with very insensitive indicators of a problem. The new 'normal' for millions of people is a hard struggle to meet the most minimum food and cash needs. Increasingly they have nothing to fall back on. The reality for the government and the aid community is that there can be a major emergency every 2 to 3 years with little breathing space in between for recovery (Anderson and Choularton, 2004). This is clear from the rainfall analysis presented earlier.

The root problem in Ethiopia is the increasing numbers of people depending directly upon smallholdings on the same land area. At worst, periodic shocks to production lead to outright starvation. At best, gains from marginal improvements in production (e.g. through the sustainable use of fertilizers or development of some niche products) are invested in supporting still more people on the same smallholdings or land area. Many of the development projects that are working in the region are focused on helping to maximize production on these farms. But taking a longer view, smallholding is the problem (Anderson and Choularton, 2004).

One solution must be to add value to rural labor beyond smallholding agriculture. Livelihoods analysis shows that this is what farming households themselves are trying to do. The added value comes essentially from urban/commercial demand for labor as well as products. This is the process of transition from subsistence smallholdings to an ever stronger rural-urban economic link which is not simply defined by rural-urban migration. Economic opportunity is the magnet to draw the poor away from unsustainable patterns (Holt and Rahmato, 1999).

In the vulnerable areas of the Northeast Highlands, SNNPR, Somali Region and Hararghe, food economy livelihood analysis estimates that between roughly 40–50% of the population is unable to sustain themselves with their own crop or animal production. They have already been forced to diversify into other activities. Their primary strategy is to seek wage labor opportunities (Holt and Rahmato, 1999). Many will find on-farm labor but the agricultural sector cannot absorb them all. But we must look to create labor opportunities for the majority in small towns/service centers if we are to prevent wide scale migration to Addis Ababa (especially when and if land reform takes place). For this group, access to and the development of labor markets is the key. The West African experience with urbanization holds valuable lessons for Ethiopia. In Mali for example, urbanization has been relatively moderate and controlled with small town growth the source of much non-form employment growth. The increased density of urban markets is transforming the nature of agriculture towards smaller urban and peri-urban producers who can quickly respond to and service this market. Ethiopia would do well to study the Mali experience.

There are strong and encouraging signs of a renewed and open dialogue between the Ethiopian government and the international community. Ethiopia needs radical strategies for building sustainable pro-poor growth, and mechanisms for protecting those gains should be at the center of government efforts. Most importantly Ethiopia needs a much higher level of non-emergency assistance in order to fund this transition. Finding these inputs will be very challenging in a world where aid for emergencies is much easier to obtain than aid for prevention (Anderson and Choularton, 2004).

14.7 Summary

This chapter chronicles the extraordinary response to the 2002/3 crisis in Ethiopia, which provides a good case study for how FEWS NET uses data to identify and motivate response to an impending crisis. Indicators were presented that were used in November and December 2002 to convince USAID and the donor community of the crisis that was emerging in Ethiopia. FEWS NET reporting was summarized, including agroclimatic analyses, food production projections, access to food, threats to agriculture from pests, and the context in which these events were occurring. An analysis produced by the Funk et al. (2005) was presented that seeks to understand and estimate the likelihood that the drying trends observed in Ethiopia will continue. Although the skills and datasets needed for an agroclimatic analysis conducted routinely by FEWS NET USGS field representatives and a study of long term trends are similar, the identification and understanding of the drivers of climate in the Greater Horn of Africa takes a focus on the long term that is hard to get to with FEWS NET's enormous work load. The role of collaborating institutions and organizations in FEWS NET's mission is substantial, as much synergy and learning occurs between practitioners and researchers, as this volume attests. The final section provides a retrospective analysis on how to move forward in Ethiopia after an enormous effort by the relief community to respond to the crisis, derived from Anderson and Choularton (2004).

References

Anderson, S. and Choularton, R., 2004. Retrospective Analysis 2002/3 Crisis in Ethiopia: Early Warning and Response, The USAID Regional Economic Development Services Office for East and Southern Africa (Redso), Washington DC, 64pp.

Camberlin, P., 1997. Rainfall anomalies in the source region of the Nile and their connection with the Indian Summer Monsoon. Journal of Climate, 10: 1380–1392.

Camberlin, P. and Philippon, N., 2002. The East African March–May Rainy Season: Associated Atmospheric Dynamics and Predictability over the 1968–1997 period. Journal of Climate, 15: 1002–1019.

Funk, C., Sanay, G., Asfaw, A., Korecha, D., Choularton, R., Verdin, J., Eilerts, G. and Michaelsen, J., 2005. Recent Drought Tendencies in Ethiopia and Equatorial-Subtropical Eastern Africa, Famine Early Warning System Network, USAID, Washington DC.

Hoerling, M.P., Hurrell, J.W., Xu, T., Bates, G.T. and Phillips, A.S., 2004. Twentieth century North Atlantic Climate change. Part II: Understanding the Effect of Indian Ocean Warming. Climate Dynamics, 23: 391–405.

Holt, J. and Rahmato, D., 1999. Sustainable Livelihood in North Wollo and Wag Hamra Zones, Save the Children UK, London.

Hurrell, J.W., Hoerling, M.P., Phillips, A.S. and Xu, T., 2004. Twentieth century North Atlantic Climate change. Part I: Assessing Determinism. Climate Dynamics, 23: 371–389.

Nicholson, S., 2003. The South Indian convergence zone and interannual rainfall variability over Southern Africa and the question of ENSO's influence on Southern Africa. Journal of Climate, 16: 555–562.

Okoola, R., 1999. A diagnostic study of the eastern African monsoon circulation during the northern hemisphere spring season. International Journal of Climatology, 19: 143–168.

Shanko, D. and Camberlin, P., 1998. The effects of the southwest Indian Ocean tropical cyclones on Ethiopian Drought. International Journal of Climatology, 18: 1373–1388.

Turner, B.A. and Pidgeon, N.F., 1997. Man-Made Disasters. Butterworth-Heinemann, London, 250pp.

Verdin, J., Funk, C., Senay, G. and Choularton, R., 2005. Climate science and famine early warning. Philosophical Transactions of the Royal Society B: Biological Sciences, 360: 2155–2168.

Chapter 15
Remote Sensing Data in the Mesoamerican Food Security Early Warning System (MFEWS)

The Mesoamerican Food Security Early Warning System (MFEWS) is an extension of FEWS NET and its activities into Central America. This chapter will describe the use of the remote sensing and analysis tools available in Central America in the context of the historic increase in maize prices occurring in 2007 combined with a minor and localized drought in the summer of 2006 in Honduras which increased food insecurity in the region (Fig. 15.1). Remote sensing data was used to assess variations in local production, which impact the amount of grain which must be purchased at high prices from the market. An extended drought from June to September 2006 in the southwest portion of Honduras has reduced the food security status of the country. This chapter will assess how remote sensing was used to identify the drought, evaluate its consequences and provide support to food security analyses using a variety of other datasets.

MFEWS' objective is to strengthen the ability of Central American countries and regional organizations to manage risk of food insecurity through the provision of timely and analytical early warning and vulnerability information. FEWS NET was developed in Africa, and thus its remote sensing and food security experts primarily have experience and a large network of contacts in Africa. To transfer this expertise to Central America, FEWS NET has hired new experts on Central America both locally and in Washington DC to ensure effective and nuanced analysis. African countries benefit from a network of food security officials in local, regional and national governments and non-profit organizations who are experts in understanding food security indicators and who have been working with early warning organizations for decades. In Central America, MFEWS has had to educate and build networks of officials at various levels and in multiple agencies in order for an early warning of an impending food security crisis to have an impact on the government's activities. FEWS NET has also worked with the existing teams of experts on rainfall, crop modeling and vegetation to identify new products which meet the needs of MFEWS and its very humid environment.

The system and processes that this relatively new organization has set up are focused on developing new contacts among government structures that previously had little connection. Although MFEWS works only in Guatemala, Honduras,

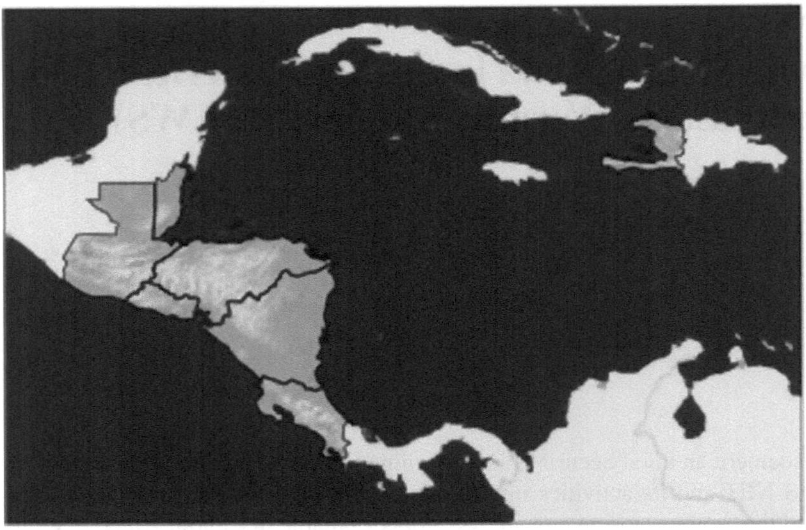

Fig. 15.1 Region covered by the weekly weather analysis conducted by NOAA

Nicaragua, and Haiti, the remote sensing products are produced for the entire region, an area with diverse ecosystems and livelihood strategies. Efforts have been made to increase the spatial resolution of the remote sensing products to better suit the geographic diversity of agriculture and small scale of each country in the region.

15.1 MFEWS and Remote Sensing

Like FEWS NET in other regions, MFEWS's primary responsibility is to create networks and to provide information – credible, understandable, relevant information – for policy makers to support better early warning and early action. It seeks to assess emerging or evolving food security problems and define their severity and spatial extent across the region and countries. The information products that MFEWS produces focus on the need for and timing of information necessary for decision-makers to intervene effectively. Remote sensing data provides the historical context, spatially extensive, and crop-specific information regarding food production and relates it to livelihood information to enable policy guidance to be provided to decision makers. Agriculture provides approximately 30% of the local gross domestic product in Central America, which is significantly less than in most African countries. Remote sensing data used in Central America has a higher resolution and is focused on the hazards that threaten food security in the region: principally droughts, floods, hurricanes, and other agroclimatic disturbances.

15.1.1 Remote Sensing for Agricultural Monitoring

Table 15.1 lists the NASA data used by MFEWS to summarize the current climatic situation and the observations that the data supports. When MFEWS was started, the RFE data did not have spatial coverage in the region. Because the RFE requires a very dense network of information from rainfall gauges to be useful, NOAA postponed the development of a new RFE product for use in the region. For the first five

Table 15.1 NASA data in monitoring products data includes precipitation gauges and gridded data from merged satellite models, vegetation data from a variety of sensors, gridded cloudiness products, global climate indicators, precipitation forecasts (24–72 hours), modeled soil moisture, gridded fire products, and seasonal forecasts

What is being monitored?	How is it being monitored?	Satellite input
Precipitation	NOAA RFE – Rainfall Estimate	AMSR-E
	TRMM – Tropical Rainfall Monitoring Mission 3b42-RT	TRMM
	GTS Station Data	
	NOAA CMORPH data	
		NOAA Outgoing Long Wave Radiation
Derived Products	SPI – Standardized Precipitation Index	TRMM
	SOS – Start of Season	AMSR-E
	WRSI – Water Requirement Satisfaction Index	
	Cyclone Monitoring	NOAA OLR
Clouds	OLR – Outgoing long wave radiation	MODIS
	IR – Infrared Temperature	
	Water Vapor	
Global Climate Indicators	MJO IR – Madden Julian Oscillation/200 h/PA velocity potential	NOAA AVHRR
	GFS Vorticity	
	ENSO Phase	
	Sea Surface Temperature Anomalies	
Precipitation Forecast	GFS Model – Global Forecast System	
	NCEP ETA model	
	NOAA WRF model	
Vegetation	GIMMS AVHRR NDVI	AVHRR
	NOAA AVHRR Vegetation health	
	SPOT Vegetation NDVI	SPOT VGT
	MODIS NDVI	MODIS
Soil Moisture	SSM/I Soil Moisture	AMSR-E
	CPC Leaky Bucket model	
	Moisture Index	TRMM
	Soil Water Index	
Fires	MODIS Fires – Rapid Response	MODIS
Seasonal Forecasts	IRI SSTA	
	COLA AGCM	

years, MFEWS used the TRMM merged satellite- gauge product 3B42-RT, produced in real time to monitor rainfall. This product is particularly well suited to the high rainfall amounts and complex tropical rainfall development that is characteristic of the region. The gridded rainfall images are obtained every ten days and are used to drive a wide variety of models, including agronomic models specifying the moisture requirements of a particular crop given an underlying soil type (Water Requirement Satisfaction Index or WRSI) and the moisture index, among others. The parameters needed to configure the WRSI have been developed for the region, including crop type, start and end of season, and cropped area.

Vegetation Index data derived from satellites remains an important source of information for the FEWS NET program because it shows results of rainfall on the vegetation. MFEWS started using MODIS data at 250 m before other regions because of the relatively compact geography and complex orography requires a higher resolution to identify and track changes in vegetation. AVHRR NDVI is also used to identify anomalies from the long term mean as in other regions. Because of the relatively high amount of cloud cover, problems in AVHRR data of sub-pixel cloud contamination reduce its utility.

15.1.2 Regional Characteristics

Like FEWS NET in other regions, MFEWS is interested in understanding how variations in food production as estimated from remote sensing data impact food security. Because Central America has significant agricultural sectors that contributes to regional food production, this section will describe the importance of food production to food security in the MFEWS countries (Brown et al., 2007). While good nutrition requires adequate amounts of protein, fats, micro-nutrients, food balance is typically analyzed in terms of caloric intake. Figure 15.2 shows average food consumption patterns of individuals in the MFEWS countries. Cereals provide approximately half of caloric requirements, showing their importance as a food source in Central America.

A closer look at the types of cereals consumed highlights the importance of maize in local diets, which provide on average between 21% (Nicaragua) and 40% (Guatemala) of total caloric intake. In Nicaragua sorghum and beans contribute more to local diets than in other countries. More diversified diets may help mitigate the possible impact of some food security shocks. For example, sorghum is typically more tolerant to below-normal rains than is maize, but as it is less preferred than maize its widespread use may be seen as an indicator of a food crisis. In Guatemala, patterns of high overall cereal consumption and low meat consumption result in high levels of iron and zinc deficiencies. Poor people tend to eat a less varied diet with more cereal and pulses (MFEWS, 2006).

Nicaragua was able to meet its cereal consumption requirements through domestic production in 2003, but El Salvador met 85% of its cereal requirements, Guatemala 70% and Honduras 67%, with the rest being imported. Any decrease in

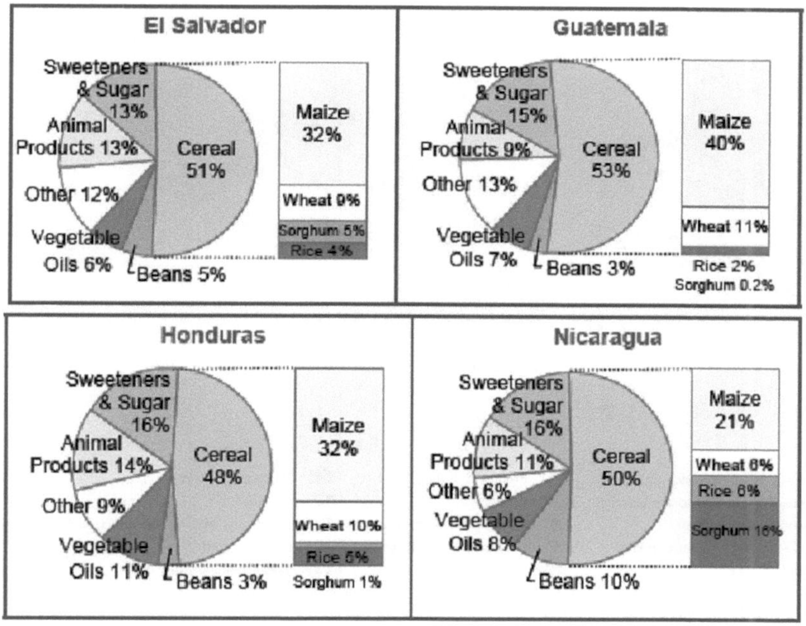

Fig. 15.2 Food consumption patterns from El Salvador, Guatemala, Honduras and Nicaragua (MFEWS, 2006)

domestic production will increase the percent of cereal that needs to be imported in order to meet consumption requirements. Ensuring food security at the national level depends not only on domestic production, but also on the capacity of countries to import food to make up any shortfalls. Food security strategies may focus on enhancing import capacity rather than trying to achieve food self-sufficiency and enhancing access to food by increasing income for the most vulnerable.

15.1.3 Livelihood Zones

Although the livelihood zones are available in Central America, their integration into the analysis of food insecurity is rather less than when analyzing food security in Africa. Their primary impact is on estimating the income for rural residents who rely on seasonal agricultural labor as their primary source of income, which usually is much reduced during times of drought (Fig. 15.3). However, the impact of variations in rainfall in the region are significant. If the May rains are delayed, there will be weak demand for agricultural laborers, a critical source of income for vulnerable families. Under normal conditions, the hunger season lasts until June, when demand for unskilled labor increases at the beginning of the agricultural cycle for staple cereal planting (Fig. 15.4).

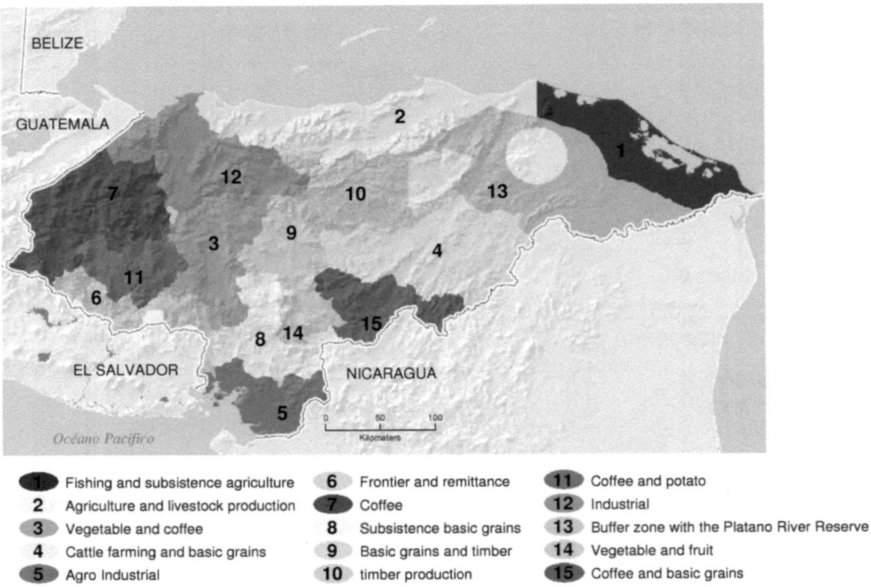

Fig. 15.3 Honduras Livelihood Zones

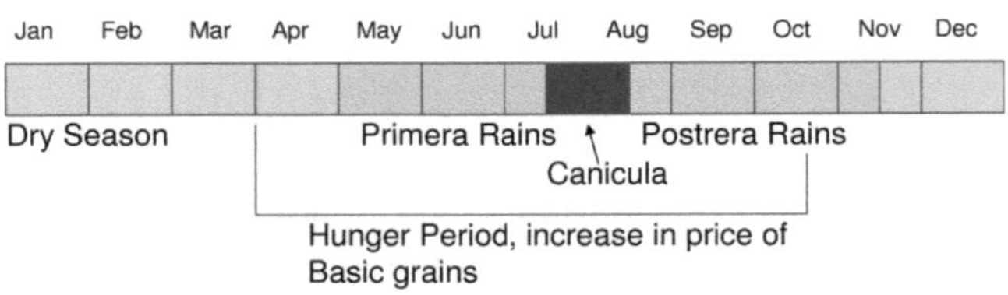

Fig. 15.4 Timing of the Primera rains, Postrera rains and the dry season in Central America. The Canicula is a period with light or no rains

Food availability through subsistence production will increase food security in August with the harvesting of the primera crop. If the beginning of the primera rainy season is delayed by the El Niño phenomenon, there could be a partial or total loss of the primera crops, and the hunger season could extend to September, when the agricultural activities for the postrera season begin. If this happens, the 6,700 families in Honduras who have been identified as chronically food insecure could require assistance until October/November. Thus MFEWS personnel were closely monitoring the start of the primera season in 2006 with remote sensing data.

15.2 Maize Production Monitoring with WRSI

In order to identify biophysical hazards, MFEWS has worked to produce crop monitoring tools that are similar to those available in Africa. MFEWS and its partners at USGS and NOAA have implemented the WRSI in Central America using the

TRMM 3B42 dataset. The WRSI requires extensive crop-specific parameters in order to be accurate, including maps of where crops are planted, start of season estimates which are generated at the start of each season, and crop parameters which are used to estimate the impact of variations in moisture availability for the plants. Thus implementing WRSI for Central America was challenging, and a great deal of work is still ongoing to ensure its reliability and to improve its accuracy through time (Fig. 15.5).

The Guatemala Ministry of Agriculture, with the encouragement of the USGS field representative based in Guatemala, has devoted significant resources to using and validating the decadal WRSI product in the region. Recently, they identified 900 farmers (3 in each municipality) who have cell phones and by calling each farmer every ten days to determine the health of their crop, they are able to increase their confidence and understanding of the WRSI product. The information from these calls is put into a database which can be accessed via the internet. By investing in the development and improvement of the WRSI measure, government representatives can clearly see the benefits satellite-based rainfall measures coupled with crop models in management of the agricultural sector. The FAO has contributed to this development by migrating this system to a web page where monitoring can go directly from the field to a central point and available immediately to the decision makers.

NASA data is also a key input to extreme weather events over Central America. NASA's TRMM data is critical for identifying the spatial and temporal extent of

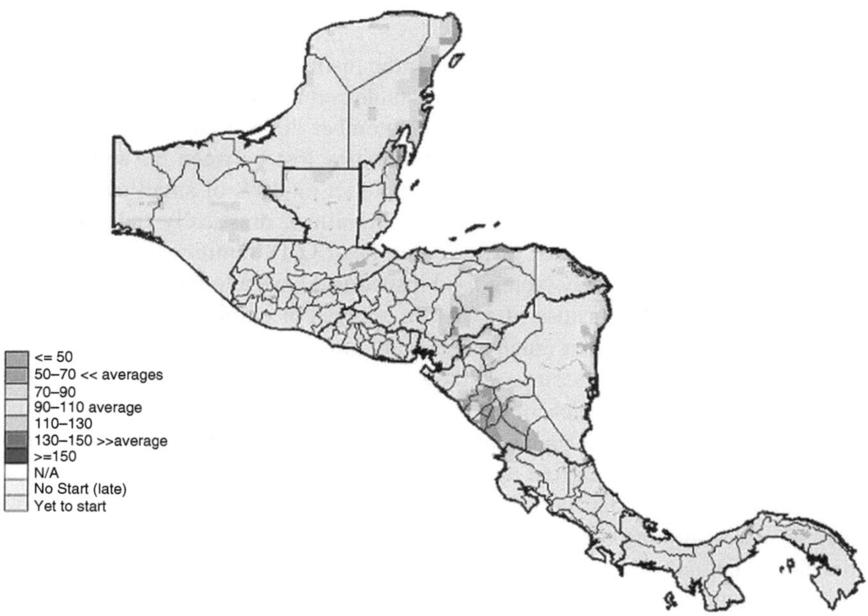

Fig. 15.5 WRSI anomaly for Central America for September 21–30, 2006

extreme weather. Although droughts are more easily assessed in their impacts on crop production, floods are far more frequent. TRMM provides critical insight into rainfall properties and their variation. These include a) frequency distributions of rainfall intensity and aerial coverage; b) the partitioning of rainfall into convective and stratiform categories; c) the vertical distribution of hydrometeors (including the structure and intensity of the stratiform region bright band); and d) variation of the timing of heaviest rainfall – particularly nocturnal intensification of large mesoscale convective systems over the oceans, and diurnal intensification of orographically and sea-breeze forced systems over land. These aspects of rainfall are critical in the Central American region. Thus identifying and providing early warning of flood events is far easier with TRMM data available from NASA.

As in Africa, FEWS NET and its partners conduct a weekly weather hazard assessment in Central America as in Africa and Afghanistan. Led by NOAA, this effort identifies at the earliest stages biophysical hazards to food production and to livelihoods. Many organizations, governmental and non-governmental, have begun to use the product as an indicator of what could occur to the food security status of particular regions over a short time frame. The governments in the region and their meteorological services have seen this product as a good initiative and every week they provide key feedback which is conveyed from the FEWS NET representatives in Central America to NOAA personnel conducting the assessment in Washington.

15.3 Impact of the Honduras Drought of 2006

Monitoring carried out in the dry corridor by the institutions of the Food and Nutrition Security Coalition showed the 2006/07 staple cereal production cycle was poor due to the extended drought from June to September 2006. Maize crops averaged a loss of 50%, and one-fourth of the communities lost between 50 and 65% of their crops. These losses have had a negative effect on 40% of small subsistence-producing households in the corridor (8,040 families), drastically reducing their food reserves. Forty-five percent of these families (3,618 families, which are part of the 6,700 families currently facing food insecurity) have a low response capacity to food deficits, mainly because of a lack of sources of employment and reduced access to productive resources (such as land and inputs).

The effects of the food insecurity in the zone are reflected in the reduced quantity and diversity of food in the diets of the most vulnerable households, as well as in the nutritional status of children under 5 years of age. The preliminary results of a nutritional monitoring study carried out by the Food and Nutrition Security Coalition, based on the nutritional data collected periodically by the Ministry of Health, show severe levels of stunting (above 5%) in some of the Choluteca, southern Francisco Morazán, El Paraíso and Valle municipalities.

Food insecurity is expected to increase in southwest Honduras beginning in April. In 2006, 8,040 families were affected by an extended drought in the zone known as the corredor seco sur, or the southern dry corridor, due to its ecological degradation and proneness to drought, which caused losses of up to 65% in their

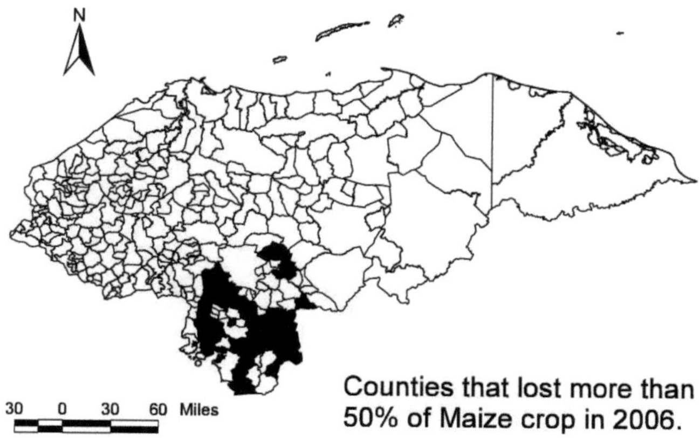

Fig. 15.6 Counties in Honduras with crop losses in 2006

postrera production. Of these households, 3,618 subsistence-farming families have a low response capacity to cope with the situation, and their food reserves have diminished drastically. When the normal hunger season begins in April, exacerbated by the end of the demand for labor in agro-industry, these families will likely face a food access crisis (Fig. 15.6). If the primera sowing starts as usual, food aid assistance will be required until June, but if El Niño delays the primera rains, assistance will be needed until October/November.

15.3.1 International Maize Prices

Although local production is important to overall food availability, the international price of maize is also a critical element. The national maize market price changes may be caused by speculation or by real supply problems. The increase of the maize price will impact poor households throughout the country, especially those that depend on purchased food. In Honduras, the poorest households purchase 70–75% of their food, making the population highly vulnerable to price increases. Thus monitoring production shortfalls due to drought, frost or floods is critical during periods with high international prices.

An additional concern is that small producers, who normally use their crops for their own consumption, will sell their crops to benefit from the current lucrative prices and will be left without reserves, not foreseeing that the price could remain high for many months. In this case, they would face a food shortage before April and May, when the usual hunger season begins. Figure 15.7 shows the dramatic increase in international maize prices since the fall of 2006. The impact of variations in maize prices will be determined by household income. If there is increased income from a bumper coffee crop, price increases will be less of a problem. On the other hand

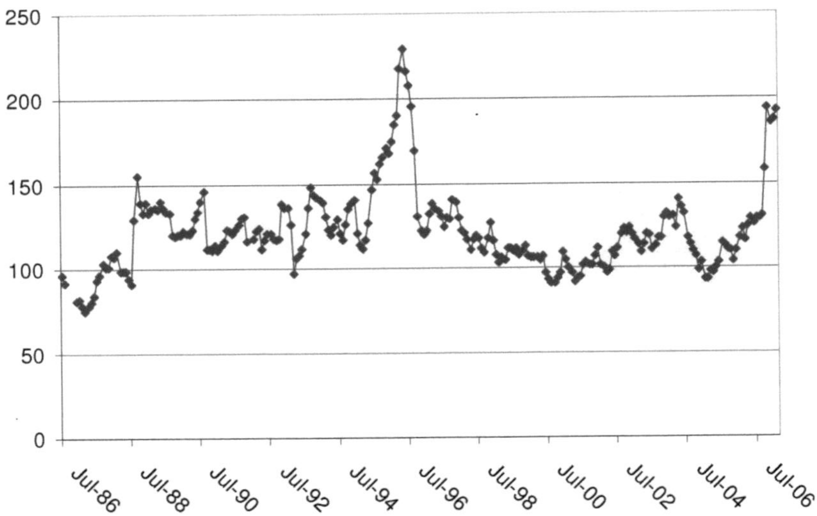

Fig. 15.7 International price for Maize from 1986 through 2006, showing the high maize prices. From the United Nations Conference on Trade and Development database

even with low prices, with a major disruption to incomes, households will not be able to access adequate food.

Despite the decreasing trend of white maize production since 2002, it has been sufficient to cover the population's consumption for the last five years, representing approximately 63% of the domestic availability. The maize price has increased by 30% since November 2006, and has limited the access of the most vulnerable populations to this product, which is the basis of nourishment for the majority of the Honduran population. This increase has not had a significant impact in the access to maize for most rural poor households, as they are mainly consuming grain that was produced on their own farms and are purchasing other items with the income from their sale of labor in agro-industry. In this same period there was a 25% deficit in the domestic availability of beans, due to exports towards markets (such as in El Salvador), which offer better prices than the Honduran market. These exports reduce this product's availability in the Honduran markets and increase the consumer price, diminishing most household's access to beans.

15.3.2 Resulting Poverty and Food Insecurity in Honduras

Most of the poor rural population in Honduras are primarily staple cereal subsistence farmers who have adequate access to the main products of their basic food basket (maize and beans), as the main source is their own postrera production that ended in December. Furthermore, the sale of their labor in coffee, sugar cane, melon, watermelon, shrimp and banana harvesting and processing provides them with income to cover any food deficit and meet other basic needs. Approximately 40,000 people (6,700 families of subsistence farmers, workers and indigenous people) from the Choluteca, Francisco Morazan, Intibuca, La Paz, Lempira, Olancho and Valle departments are facing structural food insecurity, as a result of their poverty, limited access to productive means (land, tools and inputs), dependency on markets (their main source of food), sanitary conditions, low salaries, unemployment and high cost of the basic food basket.

Child malnutrition was expected to increase above normal levels during the April–August 2007 hunger season in southern Honduras. Malnutrition rates will be higher this season than in 2006, affecting nearly 8,300 subsistence producer households who have limited staple cereal reserves due to losses of up to 50% in the 2006/07 harvest and are affected by the 31% increase in the consumer price of maize. Of these households, nearly 4,000 are currently highly food insecure, and are already reducing the quantity and quality of their food. About 1,600 children are at high risk of acute malnutrition.

The percentage of children under five years of age affected by growth faltering, a local measure of malnutrition measured in weight for age, has increased relative to 2006 in 37 municipalities in southern Honduras (Fig. 15.1), and will continue to increase during the hunger season from April to August, according to data and monitoring provided by the Monitoring Committee of the Food and Nutrition Security Coalition and the Ministry of Health. In the municipality of Duyure, primary data gathered at the end of March documented that 26% of children are affected by acute malnutrition: 19.8% experience slight malnutrition, 4.2% moderate malnutrition and 2% suffer from severe acute malnutrition. Furthermore, 67% are affected by anemia.

To respond to this crisis, FEWS NET suggested an immediate intervention of food assistance was necessary as follows:

1. Supplementary feeding for 90 days from April to June 2007 for the 1,600 children under 5 years of age currently at high risk of acute malnutrition, as well as for pregnant and lactating women;
2. Food-for-training programs (in areas such as nutrition, health and vegetable production), providing family rations for 60 days from April to May to all mothers of the 4,000 families that are currently food insecure; and
3. Food-for-work for 30 days from mid-May to mid-June for the 8,300 families affected by 2006/07 crops losses (FEWSNET, 2007).

To implement this plan, around 1,500 MT of food would be required, which would benefit nearly 68,000 people. The World Food Program (WFP) currently has 400 MT, with which it has started the intervention in the municipalities with the highest levels of food insecurity. The 1,100 MT needed to complete the intervention has been requested from international donors at the time of this writing.

The food assistance should be complemented with the distribution of high-quality short-cycle (90 days) maize seeds and inputs for the primera sowing from May to June; the installation of small water capture and irrigation systems to enable a good harvest in August; and providing training in the proper use of natural resources (water, soils and forests) and household health (hygiene, illnesses prevention and proper food and nutrition).

15.4 Summary

The Meso-american food security early warning system or MFEWS grew out of decades of work done in Africa. The adjustment to a new agroclimatological system has been accomplished with carefully tuned products and new datasets that have been validated with the help of partners in the region. MFEWS has also worked to build an understanding of the local and regional factors that affect food security. Remote sensing plays a key role in identifying regions that are influenced by weather-related variations in food production, markets, trade infrastructure and other critical economic assets. These variations may result in food insecurity in the affected zones. The food security outlook shows clearly the impact of variations in the growing season on food security for the poorest people in the region. FEWS NET thus monitors very closely the start of the season in the country, variations in labor demand, maize crop health and the progress of the rainy season every year. Remote sensing products, including the WRSI, NDVI anomaly, soil water index, and rainfall estimates from several sources are used weekly to make these estimates.

References

Brown, M.E., Choularton, R., Aguilar, G.L. and Pedreros, D.E., 2007. Biophysical Remote Sensing for Food Security Early Warning System in Central America: New Approaches, New Methodologies 33rd International Symposium on Remote Sensing of Environment Technical Program Proceedings, San Jose, Costa Rica.

FEWSNET, 2007. February to July 2007 Food Security Outlook, USAID Famine Early Warning System Network, Washington DC.

MFEWS, 2006. Mesoamerica Food Security Update, 30 June 2006, Mesoamerican Food Security Early Warning System, Guatamala.

Chapter 16
Zimbabwe's Crisis of 2006–2007

Zimbabwe has been experiencing an increasingly severe political and economic crisis during the past decade. By the summer of 2007, Zimbabwe's inflation rate was the highest in the world, reaching more than 3,000% annually. Erratic rainfall in the 2006/7 growing season, combined with the consequences of land reform and poor availability of seeds, fertilizer and other inputs resulted in a poor harvest, which caused a larger portion of the population to turn to the markets for food. This chapter will summarize the food security situation in the country and describe how FEWS NET used a remote sensing analysis to estimate production in 2007 to determine food aid needs in a time of critical shortages. The political and economic changes seen in Zimbabwe have reduced significantly the effectiveness of rainfall-based production estimates, as seen in Fig. 16.1. Having an accurate assessment of production was critical as the government was denying any shortfalls at the same time when it was becoming increasingly obvious that assistance was going to be critical for the survival of the poor. Remote sensing became the cornerstone of FEWS NET's strategy to provide accurate and effective information for decision makers in Washington DC.

The problem in Zimbabwe began with the implementation of land reform policies in the year 2000 that transferred farms from minority white commercial farmers to majority landless blacks. In practice, this has meant the complete cessation of organized farming in these formerly commercial areas that used to produce enormous amounts of surplus food every year. Now the land lies fallow, or the landless who were resettled on the farms, who are not provided with the tools, seeds, or know-how needed to tend them properly, end up barely being able to grow enough to feed themselves. Either way Zimbabwe has become a net importer of food and the consequence, both for the economy and for the overall food security of the country, has been very bad.

A meeting was held in March of 2007 in Rome, Italy by the major partners concerned with food security in Zimbabwe (USAID, World Food Program, Southern African Development Council, US Department of Agriculture, the European Commission, European Global Monitoring for Food Security and the European Union's Joint Research Council). The objective of the meeting was to come to a consensus

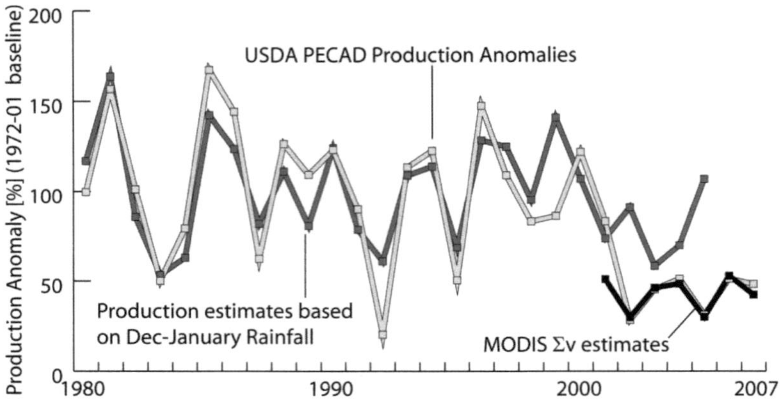

Fig. 16.1 Three estimates of Zimbabwe national maize production anomalies (Verdin et al., 2005), used with permission

about Zimbabwe's food security situation, about what could be done to assist, and to develop a general plan as to how the humanitarian community should go forward. FEWS NET brought to the meeting an estimate of area planted, yield and production for the maize sector for the 2006–2007 growing season, which ensured a consensual conclusion by everyone in the group for further action.

The situation was very complex due to a dysfunctional national government that was unwilling to admit that they had a food security problem or that its policies were the cause. The north of the country had a very good rainy season, but southern portions experienced a drought which caused crop failure in many areas. In addition, the impact of a government controlled marketing system, the lack of foreign exchange for importing grain, and severely reduced national grain production due to the elimination of functional commercial farms meant a very uncertain availability of grain for markets across the country. Having a reliable, independent estimate of national grain production estimate to bring to the Rome meeting was critical to determining the level of support that would be needed by the end of 2007.

16.1 Food Security Situation in Zimbabwe in Early 2007

Prolonged dry spells in most southern districts of Zimbabwe during the 2006/07 production season contributed to low cereal yields, particularly for maize (Fig. 16.1). Though the northern districts received comparatively good rains, yields in these areas were affected by frequent fertilizer shortages throughout the season. Total maize, sorghum and millet production for the 2006/07 agricultural season was forecast to be about 50% the previous season's production, and less than 50% of the five-year average. Estimated 2006/07 cereal production was forecast to meet between 40% and 50% of domestic consumption needs. Given the prevailing foreign currency shortages in Zimbabwe, and the many competing basic import needs (fossil fuels, electricity, medicines, agricultural inputs such as fertilizer, etc.) that require foreign currency, importing the cereals necessary to make up the consumption shortfall was a serious challenge for the Government of Zimbabwe.

Although the 2006/07 rainy season (October to April) started early in Zimbabwe, rains were inconsistent and poorly distributed for the first several weeks. It was not until mid-November that substantial rains fell throughout the country, and the majority of maize was planted in mid December. In Zimbabwe's northern and central districts, the December plantings benefited from relatively good rains at flowering and grain filling stages during January and February, though a nationwide shortage of fertilizers during these growth stages reduced potential yields in these areas. In addition, most of the maize crops in southern districts suffered irreversible damage from the prolonged El Niño-related dry spells that dominated the second half of the cropping season.

16.1.1 Economic Situation

The official annual rate of inflation in Zimbabwe, measured by the Central Statistical Office, reached 2,200% in March 2007 – a 470 point increase from the February 2007 annual rate. Such high rates of inflation wreaked havoc in the economy and severely restricted household purchasing power, while also fueling civil discontent that has led to strikes among employees and workers from several economic sectors, including Zimbabwe government departments. The overall poor 2006/07 agricultural season, coupled with the continued shortage of foreign currency were expected to fuel continued inflation for the greater part of the 2007/08 consumption year.

The ever-increasing cost of living weighed down most poor households in urban and rural areas of Zimbabwe in 2007. The cost of a household's monthly basket, monitored by the Consumer Council of Zimbabwe, rose from Z$ 686,116 in February to Z$ 1,483,324 in March 2007. Prices for all items in the basket increased last month, many by at least 70% (Fig. 16.2). Notable increases include a 158% increase in the price of bread and 142% increase in the price of fresh milk. The official annual rate of inflation was 1,098.8% in November 2006, and the economy was officially estimated to shrink in year 2007 by about 25%.

During the summer of 2007 the Zimbabwe government implemented price controls on an array of basic commodities. The result was a dramatic decline in food availability and food access, particularly in urban areas. Until recently, most basic goods, including maize meal, were available in both formal and parallel markets, albeit at rapidly rising prices. But, since the implementation of the June price controls, there has been a run on price-controlled commodities and a decline in their production due to the erosion of profit margins. In some cases the prices controls are making operations for producers non-viable altogether.

At the time of this writing, the formal market can no longer maintain a regular supply of basic goods. Sporadic deliveries of basic goods are met with long lines,

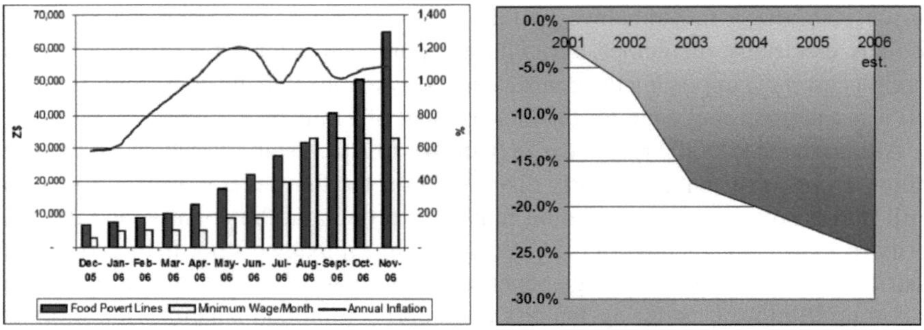

Fig. 16.2 Annual rate of inflation compared to the minimum wage rate of a low-income earner, indexed on the Food Poverty Line, and on *right* the percentage growth in Zimbabwe's major economic sectors (source CSO and ZCTU)

and not everyone makes it into the store before stocks run out. Runs on commodities are having the biggest impact on the poor, who because of their limited purchasing power are forced to make frequent purchases of smaller amounts and are not able to buy in bulk whenever commodities become available. While basic goods can still be found on the parallel markets at a substantially higher cost, these markets are constantly disrupted by more frequent police raids. Not only is the food crisis in urban areas a food access crisis, it has now become an availability crisis as well.

Minimum wages per household wage-earner can only cover about 17% of the CCZ food basket, and meager consumer wages lag behind the cost of food and non-food items. Maize prices have also increased by more than 50% from January to March 2007 (Fig. 16.2), and, due to uncertainty regarding the 2006/07 harvest, those few farmers with remaining stocks from the 2005/06 season were holding onto their grains for their own consumption, causing further shortages on local markets. Maize prices have also increased in response to inflation.

Since most districts throughout the country expected a poor cereal harvests, the decrease in maize prices normally experienced during the harvesting months of May to July, 2007 was limited to a few districts in the central and northern parts of the country. The majority of farming households will be forced to purchase available maize at high market prices as early as the beginning of the new consumption year, unless significant distributions of cheaper maize by the GoZ's Grain Marketing Board occur or substantial amounts of food aid are distributed. In 2007 about 1.7 million people were receiving food aid.

The Crop and Food Supply Assessment Mission (CFSAM) by WFP and FAO were conducted in December of 2006. Complementary vulnerability assessments, such as Zimbabwe's Vulnerability Assessment Committee (ZimVAC), as well as local NGO assessments provided updated food security information for the country and initial projections on food aid requirements for the new consumption year in 2007. Yields were very low in the southern parts of the country, but in the north the harvest was expected to be above average. Knowing how much grain was produced in Zimbabwe in the 2006–2007 growing season was critical to knowing how much food aid would be required. Although the economic situation was extremely precarious, if there was sufficient grain for the population, not much actual food aid would be required. Unfortunately, the government was not a reliable source for determining if there would be a deficit as they had been falsifying production figures to make the situation seem better than it really was. Thus FEWS NET's ability to provide accurate information on how much grain was produced from a remote analysis was central to its ability to provide reliable advice to decision makers at USAID and WFP on Zimbabwe in early 2007.

16.2 Estimating Production from MODIS NDVI

Normalized Difference Vegetation Index images are routinely used to identify areas prone to drought-related reductions in production, as well as poor pasture conditions (FEWS, 2000; Field, 1991; Hutchinson, 1998), malaria (Hay et al., 1998), epizootic diseases such as Rift Valley Fever (RVF) (Linthicum et al., 1999), and damaging pests such as locusts (Hielkema et al., 1986; Tucker et al., 1985). Traditionally, production is estimated as the product of yield and cropped area. Satellite sensors tend to convolve these two sources of 'greenness', and this can make independent assessments of yield and cropped area difficult. Moderate Resolution Imaging Spectroradiometer (MODIS) Normalized Difference Vegetation Index (NDVI) time-series were used to represent crop production anomalies (in percentages). MODIS NDVI data 'cubes' were created and a special temporal filter is used to screen for cloud contamination. Regional NDVI time-series are then composited for crop growing areas, and adjusted in time according to the timing of the onset of rains. A national index is created by taking the cropped area weighted average of the regional series. This spatio-temporal compositing allows for the identification of NDVI-green up during grain filling in crop growing areas. This metric is found to be highly correlated with US Department of Agriculture production figures, and can be used to provide an early proxy for national production.

While research applications of NDVI to crop yield and production (Bullock, 1992; Groten, 1993; Rasmussen, 1992; Rojas, 2007; Unganai and Kogan, 1998) have been made, routine quantitative analysis of NDVI is still fairly uncommon. This differs strongly with satellite precipitation (Adler et al., 1994; Arkin et al., 1994; Huffman et al., 1995; Love et al., 2004; Xie and Arkin, 1997) which routinely are used to drive numerical models of crop yield reduction based on the Water Requirement Satisfaction Index (WRSI) (Senay and Verdin, 2003; Verdin and Klaver, 2002). The WRSI model is a time, crop, and soil type sensitive ratio of actual evapotranspiration compared to the crop water requirement. FEWS NET scientist Christopher Funk and colleagues were able to show that a similar metric, the mid-season Vegetation-Sum, can provide an early and accurate quantitative proxy for crop production anomalies in chronically food-insecure Zimbabwe. The analysis summarized below was conducted by Funk and Budde in 2007 at the request of FEWS NET and USAID and presented in a technical report. The NDVI used in the analysis was Collection 4 MODIS data from 2000 to April 2007. The following is a summary of their report.

16.2.1 Vegetation-Sum Metric

The Vegetation-Sum metric was developed in analogy to the WRSI water balance model by Chris Funk at the University of California in Santa Barbara, with help from Michael Budde at USGS, in order to improve FEWS NET's ability to predict maize production in countries such as Zimbabwe that experienced significant economic changes that affect production (Funk and Budde, 2007). The seasonal WRSI is the accumulated ratio of actual evapotranspiration from the onset of rains to the end of the season. The onset of rains is typically estimated from satellite precipitation, and the end of season determined by a crop-specific length of growing period. The crop water requirement is a function of potential evapotranspiration, crop stage, and crop coefficients. Actual evapotranspiration (ET) is a function of soil water and root depth, and incorporates water holding capacity, past precipitation and past ET. In early warning applications WRSI percent anomalies are typically examined to identify areas experiencing crop water stress. Assuming that optimal seasonal water requirements vary little from year-to-year, we can express the WRSI percent anomaly as a function of actual ET.

While the WRSI estimates actual ET via extended moisture balance considerations, it has also been shown that MODIS vegetation indices can be a good proxy for actual evapotranspiration (Chong et al., 1993). This suggests that the sum of NDVI increases over the mid-to-late season growing period should be a good indicator of crop evapotranspiration.

$$\sum\nolimits_{onset}^{onset+LGP} ET_i \propto \sum\nolimits_V = \sum\nolimits_{onset+lag}^{onset+LAP+lag} (NDVI_t - NDVI_{onset}) \tag{16.1}$$

The Vegetation-Sum calculation incorporates a lag that combines delays associated with the temporal sensitivity associated with grain filling and the delayed response of vegetation to rainfall (Funk and Brown, 2006; Ji and Peters, 2003; Kerr et al., 1989; Potter et al., 1999; Richard and Poccard, 1998). A length of aggregation period was also included in the analysis. The aggregation period determines the length of the window over which the NDVI is averaged. NDVI onset is subtracted from Vegetation-Sum to remove the pre-onset influences associated with the previous dry and rainy seasons. The onset of rains dates used in this study were based on a simple rainfall accounting method defined as the 1st 10-day period in which at least 25 mm of rain fell, followed by two 10-day accumulation periods with a total of at least 20 mm of rain. These ten day onset periods were then mapped to the closest 16-day MODIS composite period (Funk and Budde, 2007).

While the equation above is physically plausible there are a number of contamination sources that can confound the potential NDVI/ET and crop productivity relationship. Temporally, cloud and moisture contamination can influence the NDVI signal. Furthermore, vegetation signals from before or after the season contain variations not related to grain filling; an onset-of-rains temporal re-alignment accounts for some of these effects. Finally, spatial filtering was used to minimize the influences of non-agricultural vegetation on the Vegetation-Sum results.

NDVI data may be affected by a number of phenomena that contaminate the signal, including clouds, atmospheric perturbations, and variable illumination and viewing geometry. Each of these tends to reduce NDVI values (Los, 1998). A time series smoothing technique developed by Swets and others (1999) was used to minimize these effects. The technique, implemented for all MODIS data used by FEWS NET at USGS EROS, uses a weighted least squares linear regression approach to "smooth" observations that are of poor quality due to clouds or other atmospheric contamination (Swets et al., 1999). A temporal window is used to calculate a regression line. The window is moved one 16-day period at a time, resulting in a family of regression line associated with each point; this family of lines is then averaged at each point and interpolated between points to provide a continuous temporal NDVI signal (Fig. 16.3) (Funk and Budde, 2007).

In order to minimize the influence of non-agricultural land cover types on Vegetation-Sum we applied a mask to the NDVI time-series using a cultivated areas map based on the Southern Africa Development Community's regional land cover database (Fig. 16.4). The classification was produced by the Forestry Commission and German Development Co-operation using 1992 Landsat imagery.

The mask was used to calculate NDVI time-series statistics for only those areas classified as cultivated lands. The end result of the temporal-spatial smoothing procedure was a set of 61 characteristic district level NDVI time-series covering the 2000–2001 season through to 49th Julian day MODIS period (March 5th) of 2007. In 2007, the 65th through 129th Julian day 16-day periods were filled by assuming climatological NDVI changes, estimated from 2000–2006 data (Funk and Budde, 2007).

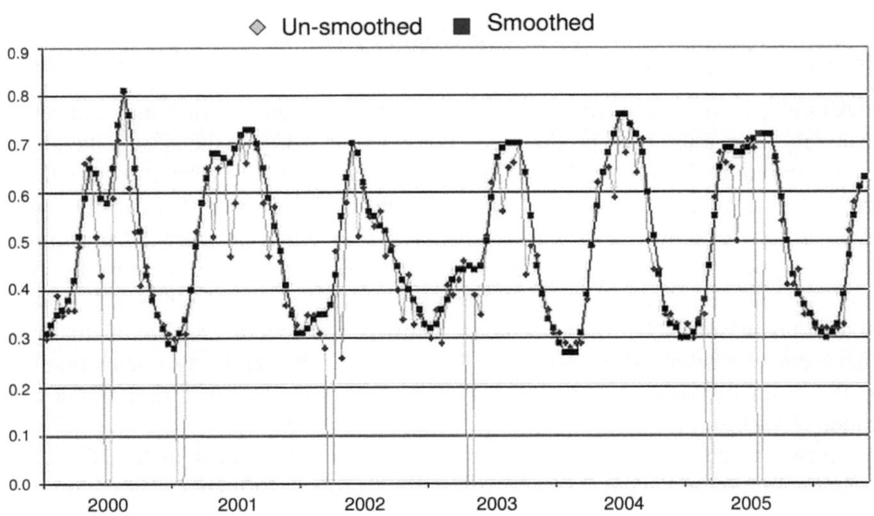

Fig. 16.3 This time series NDVI plot, for a single 500 m pixel, shows unsmoothed data in with *grey diamonds* and temporally smoothed NDVI with *black squares*. The smoothing algorithm effectively corrects these erroneous NDVI values based on characteristics of the valid NDVI curve

Fig. 16.4 Cultivated areas from the SADC landcover/landuse database (Funk and Budde, 2007)

16.2.2 Shifting the NDVI to the Onset of Rains

NDVI changes outside of the main rain vegetation response period may not relate positively to increase crop productivity. For example, early in the season the clearing of agricultural lands (reduced NDVI) may be positively related to increased production at season's end. Similarly, once the NDVI-responses associated with grain-filling are complete, increasing greenness may represent late season cyclonic activity, which can hamper seed drying and harvesting activity. Late season greenness may also represent continued green-up in non-cultivated regions. Focusing on the mid-season NDVI response helps identify crop-specific vegetation changes. This focus was achieved by estimating Vegetation-Sum over a set of dates indexed by the onset of rains, and specified by the coefficients in Eq. (16.1) (Funk and Budde, 2007).

Given that seasonal maximum and integrated or seasonal accumulated NDVI is a common metric used in the evaluation of seasonal crop performance, a brief comparison with Vegetation-Sum is presented in Table 16.1. Again, the r^2 with 2000–2006 USDA PECAD production is used as the evaluation metric. The summary statistics were calculated two different ways: i) by estimating the statistics for each district, and averaging the results, and ii) by estimating the statistics using the national-level

Table 16.1 Comparison of Vegetation-Sum, NDVI-Maximum, and NDVI integration metrics from (Funk and Budde, 2007)

	Average Full-season R^2	Average Filled-season R^2	Average NDVI Full-season R^2	Average NDVI Filled-season R^2
Vegetation-Sum	0.98	0.80	0.95	0.72
NDVI Max	0.47	0.36	0.76	0.49
Seasonal NDVI Integration	0.49	0.28	0.57	0.26

cultivated area weighted time-series. Values based on both full-season and with averaged data are also reported in the table. The full-season NDVI is the metric calculated on the entire season, from onset to the date of harvesting. Averaged NDVI is based on using the metric calculated from onset to Julian day 49, and after that date the NDVI was filled using the seven year average (2000–2006) change in NDVI for each district. This simulated the effect of basing an estimate on the data typically available in mid-March, when the seasonal assessment is usually done. While a substantial decline in performance did occur, R^2 values of about 0.8 could be obtained (Funk and Budde, 2007).

16.2.3 Recent Declines in Zimbabwe Production

To put the 2006–2007 season in context, variations in past maize production in Zimbabwe provides historical context. Figure 16.5 shows USDA PECAD production

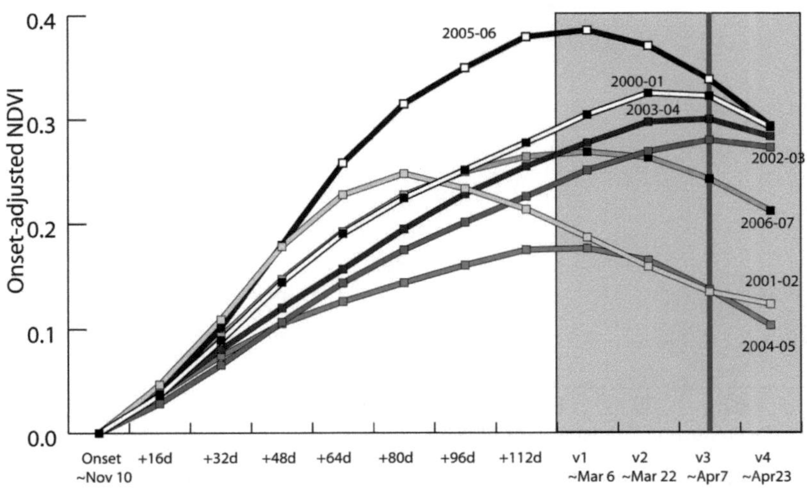

Fig. 16.5 Composited onset-adjusted MODIS NDVI time series from the onset of rains forward (Verdin et al., 2005)

anomalies, together with regression estimates of the same ($r^2 = 0.66$) based on December–January rainfall totals. The drop between 2000–2001 and 2001–2002 is precipitous, going from 83% to 28% of the historical average (also see Fig. 16.1). While below normal rainfall may have played a role, this decline appears to be mostly due to the dramatic changes in agrarian management practices that were seen in Zimbabwe. Rainfall estimates for 2001–2002 calibrated against earlier data are much too high (91% as opposed to the PECAD value of 28%). Note that the Vegetation-Sum regression estimate, calibrated over 2001–2002 to 2005–2006, underestimates 2000–2001 production (51% as opposed to 83%) (Fig. 16.1). Thus Vegetation-Sum cannot capture the systematic changes in agricultural practice. Rather, Vegetation-Sum is a satellite observed vegetation proxy that requires an external calibration in order to be related to actual maize production (Funk and Budde, 2007).

16.2.4 The 2006–07 Filled-Season Vegetation-Sum Estimates

Figure 16.6 shows a scatterplot of recent PECAD and Vegetation-Sum production anomalies, expressed as a ratio of 2005–2006 PECAD production. The six estimates

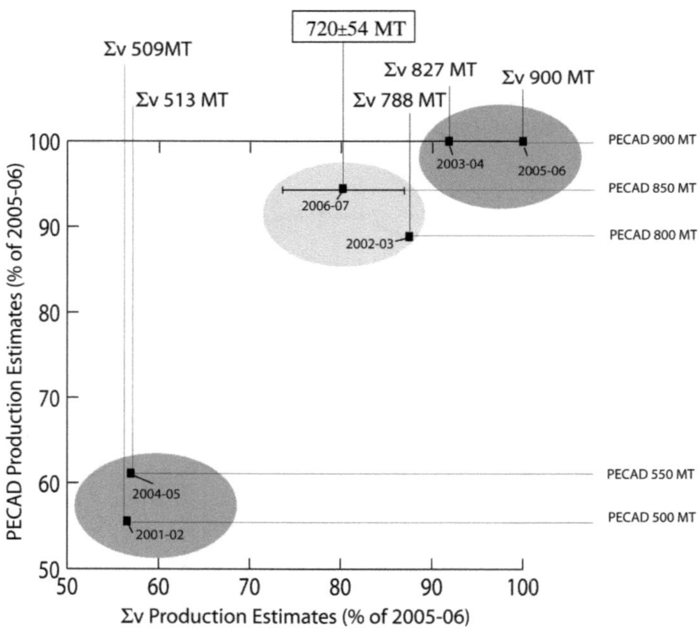

Fig. 16.6 Scatterplot of six most recent production anomalies, expressed a fraction of 2005–2006 production. The 2006–2007 estimate is also shown, together with brackets indicating ± 1 standard error (6%). In this plot, •v denotes the Vegetation-Sum based metric results (Funk and Budde, 2007)

cluster into 3 groups. This year's Vegetation-Sum estimate is about 20% below last season – and may provide an optimistic projection, given that it may not capture a likely 15% reduction in cropped area (Reynolds, 2007). Note also that the Vegetation-Sum production value presented here is based on filled NDVI data for part of February and March. The full time series provides much better results and additional estimates might be desired later in the season. These estimates may be lower than the current figure, since January-February-March precipitation was quite poor (Funk and Budde, 2007).

While this study focuses on production anomalies, and a reasonable correspondence is found between PECAD and Vegetation-Sum, it is important to note that we are mainly representing changes in yields. Over the 1999–2007 period PECAD production figures and yields are highly correlated ($R2 = 0.85$), and Vegetation-Sum tracks with both these time-series (Funk and Budde, 2007). Variations in area harvested are not captured with this method. FEWS NET is implementing a new remote sensing based operational product that will permit the estimation of area planted in countries such as Zimbabwe where grain production figures are contested.

16.3 Analysis of Zimbabwe Food Security Situation

In 2007, the southern Africa food security situation is likely to deteriorate very early in the marketing year, due to a well below average crop production after a season characterized by erratic and inconsistent rains and lengthy dry spells, accompanied by unusually hot weather. Critical food deficits are being projected for later in the year when locally produced grain runs out. The estimates derived from the study presented above suggest reduced levels of cereal production compared to last season and the past five-year average for Zimbabwe is likely to extend across most of the region. Zimbabwe will be particularly affected, since its ability to import grain with the reduced value of its currency is very low. Indications are that production decreases in the region from last year's levels range from about 20% in Botswana, to as high as 50% in Zimbabwe, while decreases from the past 5-year average range from about 20% in South Africa to just over 30% in Zimbabwe, Lesotho and Swaziland. Zimbabwe has had several consecutive years of below normal harvests and has had several critical food shortages.

The meeting held in Rome and the analysis using NDVI presented above has enabled FEWS NET to estimate the overall vulnerability of the population of Zimbabwe to hunger. Overall, most countries in southern Africa have been able to meet their cereal demands in the spring of 2007 due to cereal imports. Food aid in the region has been distributed, but overall the level of aid has been less than was planned in the spring because of resource problems at the UN World Food Program. Overall, Zimbabwe will really begin to experience significant food insecurity in the fall of 2007 as it prepares for the next cropping season. Food insecurity is expected to peak in October, as locally produced grain runs out and the country falls behind on its imports from other regions, making food extremely expensive for the average person in Zimbabwe.

Because it takes a long time – 6 to 9 months – for donors to mobilize and deliver food aid after it receives a request, having very early warning of an impending crisis is critical. Zimbabwe required assistance in October and November 2007, thus having early estimates of the harvest the previous spring is the focus of FEWS NET's research and analysis. This chapter has shown how remote sensing contributes to early estimates of production so that the formal request for food aid can be made as early as possible.

16.4 Summary

In this chapter, the food security situation of Zimbabwe in 2007 was outlined, which has resulted from a combination of political, economic and biophysical factors. The extremely poor economic condition of the country will adversely affect its ability to import necessary commodities to make up for poor production in the 2006–2007 cropping year. Land reforms that transferred highly productive commercial farms to multiple, poor farmers who have few resources and little knowledge about farming has resulted in a huge reduction in the overall ability of Zimbabwe to be self-sufficient. Thus it is forced to import grain from outside of the region at a time of record international maize prices.

An analysis conducted in April of 2007 which used MODIS NDVI data to estimate maize production was presented as a critical remote sensing input to humanitarian aid decisions. The Vegetation-Sum method was shown to be able to capture up to 80% of the production variation during the past six years that the data has been available. The NDVI has been more effective in estimating food production than traditional rainfall-based estimates because rainfall-based estimates rely on a static production system, an assumption which is not reasonable in Zimbabwe during the past few years. The estimates were a key component to determining the food aid requirement for Zimbabwe, and thus FEWS NET was able to ensure early consensus due to the analysis provided at a key moment.

References

Adler, R,F,, Huffman, G.J. and Keehn, P.R., 1994. Global rain estimates from microwave-adjusted geosynchronous IR data. Remote Sensing Reviews, 11: 125–152.

Arkin, P.A., Joyce, R. and Janowiak, J.E., 1994. IR techniques: GOES precipitation index. Remote Sensing Reviews, 11: 107–124.

Bullock, P.R., 1992. Operational estimates of Western Canadian grain production using NOAA AVHRR LAC data. Canadian Journal of Remote Sensing, 18: 23–28.

Chong, D.L.S., Mougin, E. and Gastellu-Etchegorry, J.P., 1993. Relating the global vegetation index to net primary productivity and actual evapo transpiration over Africa. International Journal of Remote Sensing, 14: 1517–1546.

FEWS, 2000. Annual performance and monitoring report for the Year 2000: FEWS NET, Chemonics International, Washington DC.

Field, J.O., 1991. Beyond relief: Toward improved management of famine. In: H.G. Bohle, T. Cannon, G.I. Hugo and F.N. Ibrahim (Editors), Famine and Food Security in Africa and Asia: Indigenous Response and External Intervention to Avoid Hunger. Verlag, Bayreuth, Germany, pp. 151–166.

Funk, C. and Budde, M., 2007. National MODIS NDVI-based production anomaly estimates for Zimbabwe. University of California, Santa Barbara.

Funk, C.C. and Brown, M.E., 2006. Intra-seasonal NDVI change projections in semi-arid Africa. Remote Sensing of Environment, 101: 249–256.

Groten, S.M.E., 1993. NDVI-crop monitoring and early yeild assessment of Burkina Faso. Remote Sensing of Environment 14: 1495–1515.

Hielkema, J.U., Prince, S.D. and Astle, W.L., 1986. Rainfall and Vegetation Monitoring in the Savanna Zone of the Sudan using the NOAA AVHRR. International Journal of Remote Sensing, 7: 1499–1513.

Huffman, G.J., Adler, R.F., Rudolf, B., Schneider, U. and Keehn, P.R., 1995. Global precipitation estimates based on a technique for combining satellite-based estimates, rain gauge analysis, and NWP model precipitation information. Journal of Climate, 8: 1284–1295.

Hutchinson, C.F., 1998. Social science and remote sensing in famine early warning. In: D. Liverman, E.F. Moran, R.R. Rindfuss and P.C. Stern (Editors), People and Pixels: Linking Remote Sensing and Social Science. National Academy Press, Washington DC, pp. 189–196.

Ji, L. and Peters, A.J., 2003. Assessing vegetation response to drought in the northern Great Plains using vegetation and drought indices. Remote Sensing of Environment, 87: 85–98.

Kerr, Y.H., Imbernon, J., Dedieu, G., Hautecoeur, O., Lagouarde, J.P. and Seguin, B., 1989. NOAA AVHRR and Its Uses for Rainfall and Evapotranspiration Monitoring. International Journal of Remote Sensing, 10: 847–854.

Linthicum, K.J., Anyamba, A., Tucker, C.J., Kelley, P.W., Myers, M.F. and Peters, C.J., 1999. Climate and satellite indicators to forecast rift Valley fever epidemics in Kenya. Science, 285: 397–400.

Los, S.O., 1998. Estimation of the ratio of sensor degradation between NOAA AVHRR channels 1 and 2 from monthly NDVI composites. IEEE Transactions on Geoscience and Remote Sensing, 36: 206–213.

Love, T.B., Kumar, V., Xie, P. and Thiaw, W.M., 2004. 20-Year Daily Africa Precipitation Climatology using Satellite and Gauge Data, American Meteorological Society, USA., pp. 5.4–5.7.

Potter, C.S., Klooster, S. and Brooks, V., 1999. Interannual variability in terrestrial net primary production: Exploration of trends and controls on regional to global scales. Ecosystems, 2: 36–48.

Rasmussen, M.S., 1992. Assessment of millet yields and production in northern Burkina Faso using integrated NDVI from the AVHRR. International Journal of Remote Sensing, 13: 3431–3442.

Reynolds, C., 2007. Results from a USDA/USAID crop tour of Zimbabwe.

Richard, Y. and Poccard, I., 1998. A Statistical Study of NDVI sensitivity to Seasonal and Interannual Rainfall Variations in Southern Africa. International Journal of Remote Sensing, 19: 2907–2920.

Rojas, O., 2007. Operational maize yield development and validation based on remote sensing and agro-meteorological data in Kenya. International Journal of Remote Sensing, 28.

Senay, G.B. and Verdin, J., 2003. Characterization of Yield Reduction in Ethiopia Using a GIS-Based Crop Water Balance Model. Canadian Journal of Remote Sensing, 29: 687–692.

Swets, D.L., Marko, S.E., Rowland, J. and Reed, B.C., 1999. Statistical Methods for NDVI Smoothing. Proceedings, American Society for Photogramity and Remote Sensing.

Tucker, C.J., Vanpraet, C.L., Sharman, M.J. and van Ittersum, G., 1985. Satellite Remote Sensing of Total Herbaceous Biomass Production in the Senegalese Sahel: 1980–1984. Remote Sensing of Environment, 17: 233–249.

Unganai, L.S. and Kogan, F.N., 1998. Drought Monitoring and Corn Yield Estimation in Southern Africa from AVHRR Data. Remote Sensing of Environment, 63: 219–232.

Verdin, J. and Klaver, R., 2002. Grid cell based crop water accounting for the Famine Early Warning System. Hydrological Processes, 16: 1617–1630.

Verdin, J.P., Funk, C.C., Senay, G. and Choularton, R., 2005. Climate science and famine early warning. Philosophical Transactions of the Royal Society B, 360: 2155–2168.
Xie, P. and Arkin, P.A., 1997. Global Precipitation: A 17-year monthly analysis based on gauge observations, satellite estimates, and numerical model outputs. Bulletin American Meteorological Society, 78: 2539–2558.

Section V
Analysis and the Future

Chapter 17
Power, Politics and Remote Sensing

In previous chapters, remote sensing has been shown to be at the center of much of the analysis that FEWS NET does to motivate a humanitarian response. Food aid is, of course, not driven entirely by the needs of the recipients but tied also to the donor's 'geopolitical, agricultural trade promotion and surplus disposal objectives' (Barrett and Maxwell, 2005). Food grain delivered during emergencies only sometimes helps people who are hungry. Roughly half of all recipient governments immediately transform the grain and other food assistance into cash which is needed to support the poor during a crisis. The fact that food is delivered at all (instead of cash directly) is much more due to the donor's commercial and political strategies then any concern with the recipient's needs. Food aid becomes then an extremely inefficient capital transfer, not a way to actually feed the hungry (Barrett and Maxwell, 2005).

FEWS NET's goal is to provide actionable policy information that will help guide decision making about when to send aid and to whom and what other interventions are required to prevent a crisis from becoming a disaster. Once a problem is identified by FEWS NET, the primary response that is available to USAID and the rest of the humanitarian community is to provide food aid. Unfortunately, expressing problems in terms of the amount of food assistance needed often obscures the complexity of the underlying situation (Trench et al., 2007). Although during humanitarian response some resources are spent on ensuring responses that addresses problems of food utilization and income, the majority is spent on the mechanics of purchase, delivery and distribution of food in problem areas. This focuses a lot of attention on food availability crises, such as those based on remote sensing, that may not be appropriate for many causes of food insecurity which do not originate with observable variations in the climate instead of the more complex picture that includes access and utilization.

This chapter will focus on the politics surrounding food, food aid and the use of remote sensing by FEWS NET in the context of providing information about food security crises. Although it strives to provide as accurate and independent information as possible, FEWS NET is influenced by politics. This chapter will explore how remote sensing is used as a political tool as well as a way to identify hazards and ameliorate their impact.

17.1 Changing Perceptions of Food Security

Early warning systems were begun with the idea of ameliorating observed food availability, with the assumption that famines were caused by a widespread lack of food (Torry, 1984) and that these problems were only going to increase as the human population increased (Malthus, 1789). It became obvious almost immediately, however, that famine is far more complicated with many more causes than a simple lack of production. Amaryta Sen's groundbreaking work brought issues of access and utilization into the forefront (Sen, 1981). These changes can be seen as a movement from a single direct cause and effect to one that recognizes far more complexity and multiple ways of coping and responding to a biophysical hazard. Livelihoods and household food economy approaches that FEWS NET uses today recognize a post-modern diversity of analysis and evidence when compared to where food security analysis began (Maxwell, 1996).

The sustainable livelihoods framework, as put forward by Turner et al. (2003), forms the basis for much of the livelihoods framework that enables FEWS NET's work. Figure 17.1 shows the sustainable livelihoods framework, with household assets, income and production at the center. Advocated by development agencies and others in the center of a great deal development research and practice during the last decade, the sustainable livelihoods framework has enabled the incredible diversity of income sources, coping strategies and vulnerabilities to hazards to be recognized and used to inform policy (Turner et al., 2003). The humanitarian community also benefited from this broadening of understanding, and FEWS NET formally incorporated the livelihoods approach into its methodologies in 2000, as has been described in previous chapters.

Recent work has shown that the sustainable livelihoods framework is not the whole picture, however. Lautze and Raven-Roberts (2006) has shown that during periods of conflict, many of the framework's assumptions do not hold. Household assets such as houses, fields and plows, for example, make their owners targets of violence and are transformed from assets to liabilities (Lautze and Raven-

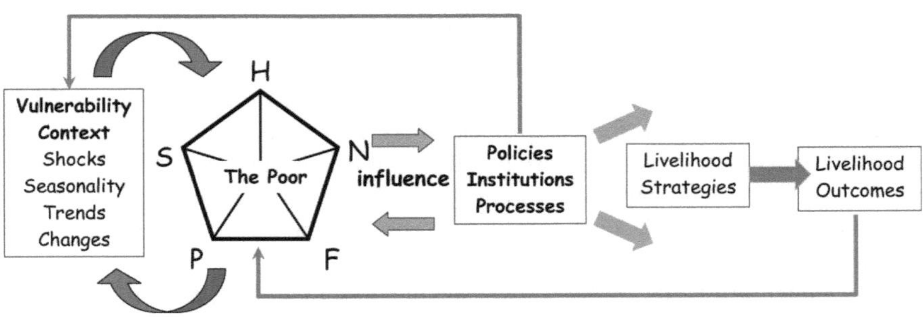

Fig. 17.1 Summary of Turner et al. (2003) sustainable livelihoods framework. The traditional livelihoods analysis is represented by the pentagon, with human (H), natural (N), financial (F), physical (P) and social (S) capital representing differing levels of household assets

Roberts, 2006). The action of the state through policies can become a major source of vulnerability instead of a source of stability, such as has been seen in Zimbabwe in 2007. Coping strategies can include acting in ways that increase short-term vulnerability to disease and malnutrition through reducing consumption now to preserve assets for later. The transformation of the normal ways people make a living in the countries where FEWS NET works by conflict disrupts the analytical framework in which it does its analysis. It is likely that FEWS NET's understanding of how people make their living and maintain food security through shocks will continue to change and evolve in the coming years.

FEWS NET is moving from being an information system whose focus is informing decisions regarding the distribution of food aid to one that can inform decisions on a suite of interventions. USAID, non-governmental organizations, the United Nations, the European Union and others have a diversity of programs that could, if allowed to move resources to respond to a crisis, respond to more of the problems that FEWS NET documents. Instead of waiting, for example, for multiple years of drought, economic marginalization and poor terms of trade to erode the resilience of pastoral livelihoods to the point when a crisis occurs, interventions to support pastoral livelihoods can be made which will prevent the crisis altogether. These include the provision of fodder and water at critical times, and subsidizing the transport of animals to markets that can provide better terms of trade. These interventions would be possible through USAID programs whose costs are far lower than those incurred through bulk cereal shipments after the crisis has occurred. Given more latitude, FEWS NET could widen its influence by ensuring that the programs which are in a position to respond to a biophysical hazard detected by remote sensing can do so in a timely fashion.

17.2 Power and Politics in the Use of Remote Sensing

Reliance on technology to estimate food production deficits has the consequence of empowering donors and national governments in their negotiations with local actors, experts and government entities, enabling a more realistic estimate of the need for intervention and the level of aid required during times of crisis. In some situations, negotiations on the level of food assistance between the local and national governments, international donors and concerned non-governmental organizations are needed because localities have been known to manipulate statistics to increase (or decrease) the perception of need and ultimately the amount of food aid they receive. Even when there is no dispute about the number of hungry people, food aid usually feeds only a portion of the needy. In some places with chronic problems, 'bad' years with significant inflow of aid see larger percentage of the needy being fed then 'good' years with aid levels are lower (Barrett and Maxwell, 2005). Only two thirds of food aid requirements are usually met on average, which tends to increase the pressure to inflate the numbers of actual needy people in hopes of improving the response.

FEWS NET's focus is to provide actionable policy information to USAID decision makers in the Food for Peace bureau. Because aid decisions involve millions of dollars and the situation on the ground is nearly always extremely complex, rapidly changing decision makers need to be able to trust the source of information. FEWS NET's primary advantage is that in most countries in which they work, the information they produce is viewed as more credible and reliable then any other source, particularly when it includes data derived from remote sensing. Even though many of its local partners have access to the base remote sensing data needed to estimate the extent and severity of a problem, an analysis conducted by these partners may not be treated as credible.

It might seem to the observer that unlike development assistance, decision making about whether or not emergency humanitarian assistance is needed should not be a political process, because such aid should be used only when absolutely essential to save lives. However, the decisions surrounding whether or not food aid is required, when to declare an emergency, the extent of the problem and its boundaries, whether or not a particular urban area is included are all fraught with politics. Although nearly all of the data that FEWS NET uses are free to the public and widely distributed, the ability to manipulate and analyze the data and produce powerful and motivating figures demonstrating the extent or absence of a biophysically-based food security problem using quantitative scientific analysis can have a profound effect on the complex power structures surrounding such decisions. Only a few actors in these situations have the ability to conduct these analyses, which restricts whether or not the conclusions derived from remote sensing can be countered with other, equally powerful information.

To demonstrate how remote sensing is used to support or weaken the position of actors at various levels, three case studies will be presented that will briefly trace how the key piece of remote sensing information was used to influence the decision making process. The examples include a discussion of the politics surrounding decision making in Zimbabwe in 2007, the use of remote sensing in estimating food needs in Ethiopia during the past decade, and the impact of remote sensing on power and local decision making in Kenya. In describing these events it will become clear that as an information source, the way remote sensing data is used highlights the unequal power relations among the various actors and the role of social networks in gaining access to and control over the development of the information.

17.2.1 Zimbabwe

The NDVI-based remote sensing analysis presented in Chap. 16 on the 2006–2007 maize production in Zimbabwe is an excellent example of the level of complexity that some of the analyses that FEWS NET needs for decision making (Funk and Budde, 2007). The analysis produced an estimate of national maize production that was based on a remote sensing analysis and historical grain production figures from USDA's Foreign Agricultural Service. FEWS NET was able to use its remote sensing expertise to strengthen the international understanding of how much assistance would be needed in Zimbabwe at a time when the government was actively obstructing clear understanding of the effect of extremely high inflation and a failing economy was having on the food security situation. 'Estimates are that in the first three months of 2008 – an election year – 4.1 million people, a third of the population, will require food aid. Most of it will be provided by the United Nations, but so far, the government is said to be in a state of "denial", refusing to make the obligatory appeal to the UN.' (Nyathi, 2007). The completely uncooperative Zimbabwe government who was, by many accounts, the source of much of the problem, has pushed remote sensing-based food production estimates to the forefront so that the UN has reliable figures with which to plan interventions.

When FEWS NET's Team Leaders, who are key representatives from each of its partner organizations (NASA, NOAA, USGS, and USAID), met in February 2007, they requested that an analysis be done on maize production figures for the 2006–2007 growing season. Much discussion revolved around how to get the necessary data, what kind of precision was required and who would conduct the analysis. Information on maize production from the Zimbabwe government itself was considered highly suspect and completely unreliable due to inconsistencies in the figures they were producing and those that the local FEWS NET representatives were collecting. It was decided that MODIS NDVI data was the best option given how poorly the WRSI had been performing during the previous few years in capturing national production (see Chap. 17). FEWS NET requested a special processing from the MODIS Land Science team that enabled this analysis, since in January 2007 they had moved to a new algorithm which rendered the data incompatible with the historical time series. This was completely unprecedented and it is only because of the importance of the task to the decision making of the US Government regarding Zimbabwe's food security that the MODIS team agreed to devote resources to the effort.

The difficulty that FEWS NET had in obtaining sufficient MODIS data for the analysis meant that any other organization without the personal and professional contacts that FEWS NET had would not have been successful. Certainly no local or regional NGO, government agency or scientist in Africa would have been able to do so, regardless of the strength of their argument. Thus FEWS NET was in a privileged position to obtain data that it later had its staff remote sensing scientists at the University of Santa Barbara transform into information needed for determining how much food aid was likely to be needed in Zimbabwe in late 2007.

A report entitled 'Review of Remote Sensing Needs and Applications in Africa' prepared for USAID by scientists affiliated with FEWS NET and other international programs at USGS EROS describes the difficulty that African remote sensing scientists have in gaining access to the appropriate data, and having the training, proper equipment and necessary software required to conduct analyses. The high turnover of properly trained remote sensing personnel makes this situation worse (Rowland et al., 2007). The consequence of this lack of expertise is that decision makers in Africa do not have a similar ability as FEWS NET to ask for and receive systematic, high quality and reliable remote sensing-based analysis upon which to make decisions. The information that African decision makers have is thus based on local knowledge of what is occurring, which does not have the weight for US and UN decision makers as a fully realized, integrated scientific analysis presented in a formal report. FEWS NET draws heavily on the local perspective for understanding what is occurring, but realizes that more is required to be persuasive than simply reports from the field. Although most of the time FEWS NET works cooperatively with its local and national partners, there are situations where the official government position (such as in Zimbabwe in 2007) contradicts the position of FEWS NET's representatives. In this situation, remote sensing is a particularly important tool.

The remote sensing analysis presented in 2007 had been conducted every year for the past three years. In previous years, it was possible to replicate the analysis because the data used was all free, available on the internet, and the methods posted on FEWS NET's web site. Because of the unique source that was used to estimate production in 2007, this replication would not have been possible. The truth is, however, that such a highly accurate and predictive remote sensing analysis was done only by FEWS NET. Not even the Europeans from FAO, JRC or GMFS were doing similar analyses thus the fact that FEWS NET has access to both the data and the expertise required gave them a leading role in determining how much and where food aid would be required in 2007.

Once the analysis predicting a maize harvest of $\sim$720 MT for 2006/7 was complete, FEWS NET called a meeting with key actors who were in a position to determine how much assistance could be set aside for Zimbabwe in the fall of 2007. The meeting was held in Rome on March 26, 2007 and included representatives from the FAO, WFP, Southern Africa Development Commission (SDAC), GMFS, Joint Research Council (JRC), USAID's Food for Peace (FFP) and key FEWS NET representatives. It was similar to those held each year for the previous two years where remote sensing-based production estimate was complemented with in-country FEWS/USDA assessments of crop conditions in the main grain producing regions. The objective was to provide a clear consensus regarding crop estimates in order to plan required food aid for needy populations, and then have the UN's FAO disseminate the conclusions. The remote sensing analysis was key to FEWS NET's ability to achieve consensus and move forward to planning interventions.

17.2.2 Politics in Kenya and Remote Sensing

Trust in the independence and accuracy of the information provided by FEWS NET is a critical element of its success (Buchanan-Smith et al., 1994). FEWS NET's reputation for solid analysis, quantitative data sources and inclusive methodologies makes its data usable and reliable for a wide variety of actors, not just US decision makers. Because FEWS NET is a US organization operating in many countries provides it credibility that local organizations may not have. This does not mean, however, that there are not a lot of politics surrounding the decision to declare a humanitarian emergency. Decision makers who must decide if a food security problem is sufficiently severe to require outside help have to make decisions in a very complex political environment.

Remote sensing data plays an important role in this situation. The USGS Field Representative in Kenya works very hard to train his colleagues in the Kenyan national government and over the several decades that FEWS NET has worked in the region, a significant understanding and trust in remote sensing datasets has grown. Familiarity with NDVI in particular is notable, with several remote sensing scientists from the region gaining international prominence. The growth and strength of the Regional Centre for Mapping of Resources for Development (RCMRD) over the past decade is testament to the importance and trust local officials have in remote sensing data in the region (Rowland et al., 2007). But not everyone has equal access to remote sensing information and it never tells the whole story.

In Kenya, food aid is one of the largest transfers of resources from outside of the country, with aid from 71,000 (1999) to 382,000 (2006) tons of cereal grain equivalents being delivered every year (INTERFAIS, 2007). Politicians at a variety of levels have connections (either direct or indirect) with the companies who own the trucks that move the grain from the port to the places where it is needed. There is often 'leakage' of grain once it is in country. As Kenyan economist James Shikwati states in a 2005 interview with der Spiegel, 'A portion of the corn often goes directly into the hands of unscrupulous politicians who then pass it on to their own tribe to boost their next election campaign. Another portion of the shipment ends up on the black market where the corn is dumped at extremely low prices.' This then damages the ability of farmers in the region to get a fair price for grain grown locally, reducing the likelihood that sufficient grain will be grown in subsequent years. These are just a few examples that show the powerful motivation politicians may have to encourage deliveries when there is a problem, even if it is not strictly necessary.

This is not to say that food aid is not needed in Kenya – there are severe and ongoing food security problems in the country. These problems are most severe in the east and northern semi-arid regions dominated by pastoral and marginal agricultural communities. Often there can be a complete failure in the long rains in the East while in the Western maize growing areas have bumper crops. The Kenyan government could purchase grain in the west and transport it to the east where it is needed, bolstering local markets and farmers while solving food security problems in the country as a whole. They have little incentive to do so, however, if the WFP and USAID are willing to ship grain from the US, store it in huge silos in local ports and transport it to the local communities that need it and pay for everything. The resources are difficult to find, but more could be done to avoid aid deliveries from international donors every time there is a problem.

The role of remote sensing and FEWS NET in this situation is to bring some independent information to the decision making process at the national level. Local officials often distrust national politicians, and the national politicians cannot tell if the information coming out of the region is accurate, inflated or minimized, depending on the situation. A decision must be made at the highest levels if an emergency should be declared and aid requested from the United Nations, without which no aid can arrive. This decision is, more often than not, based on FEWS NET reporting and on remote sensing data analysis, as this is viewed as the most independent and reliable source by everyone involved. That everyone relies on FEWS NET's information is due to its several decades- long presence in the region and to the professionalism of its personnel and analysts. Remote sensing data and analysis presented in these reports are particularly key as they provide quantitative measures of severity, extent and comparison to previous events that would otherwise be impossible. FEWS NET cannot change the underlying situation, the lack of interest by Kenyan politicians in in-country or regional food transfers to ameliorate deficits in the eastern regions. It can provide accurate information about the need for assistance in the most vulnerable and underdeveloped regions at any one time.

17.2.3 Trend Analysis in Ethiopia

Chapter 14 presents analysis that shows the tight coupling of rainfall data and food aid provisioning in Ethiopia (Fig. 14.2). The fact that there is such a tight coupling of the assessed cereal requirements and rainfall data for the past decade indicates a heavy reliance on remote sensing-based analyses to estimate these needs. The complex humanitarian situation that has existed in Ethiopia during the past decade, with millions of vulnerable and destitute people who require help every year, is difficult to understand let alone quantify. FEWS NET has recognized, however, that variation from year to year in the number of people who require immediate assistance does vary depending on the productivity of agriculture, which rests on the amount of rainfall it has received.

An Ethiopia FEWS NET monthly report issued on July 15, 2003 states that 'During the course of the crisis, non-food needs were not as fully assessed and understood as food requirements. Drought shocks cause food insecurity but they also result in a host of other economic difficulties.' (FEWSNET, 2003). Thus although there are strong arguments for relying upon rainfall as a proxy for cereal requirements, FEWS NET recognizes that there is much more that needs to be done before all the needs of the local population can be met. Unfortunately, humanitarian assistance consists primarily of grain shipments and medicine, nutritional supplements, protein sources, cooking oil, potable water and a host of other urgently needed non-grain items are often funded at a fraction of the needed levels. Thus when FEWS NET is evaluating the needs of a particular area during a crisis, its representatives focus on the amount of grain that will be needed because that is what will be available, not because that is what is truly needed by the local population to overcome their short-term crisis.

In terms of politics, by boiling down the enormously complex vulnerability issue in Ethiopia to one simply of rainfall, the humanitarian system that FEWS NET is embedded in disempowers local officials, government organizations and development programs in Ethiopia because it minimizes their understanding of the situation and reduces the ways that the system can respond. Because humanitarian aid delivered in response to an identified problem is primarily grain, other important elements of food security analysis are devalued. As the quote above shows, FEWS NET knows that understanding of the larger economic problems in Ethiopia is critical to improving the food security problem in the long run, but the underlying humanitarian aid system reduces the importance of these aspects of food security analysis, effectively removing the power of the knowledge to lead to a better response. This is very frustrating for all involved.

The close connection between rainfall and humanitarian assistance brings into focus the issue of the potential impact that the long-term trend in rainfall will have on the livelihoods of agriculturalists in the region. If some sort of transformation does not occur, Ethiopia will require more and more grain every year as less and less rain falls, if the trends documented in Chap. 14 continue. In addition, Ethiopia has been receiving less development assistance to help transform its economy and livelihood base, as most of the money it gets goes directly to programs meant to save lives in the short term, neglecting the long term in the process.

It is not only in Ethiopia that steadily declining development assistance is having a negative impact. Even in regions like West Africa, where humanitarian assistance is unusual and amounts to only a small fraction of what Ethiopia receives every year, development aid levels are extremely inadequate to address the problems that exist. Vulnerability stems from a combination of political, economic and social forces as well as the impacts of highly variable rainfall. Governments can do much more than simply responding to short term needs through humanitarian assistance programs. Longer term development is just as critical for reducing vulnerability to biophysical hazards (Trench et al., 2007). By strengthening local institutions, governments can ensure more transparent systems for gaining access to land, and for resolving disputes between different land users. They can invest in technical and financial support for small-scale irrigation activity and simple methods to trap rainwater and conserve soils which will improve the functioning of the agricultural system. Governments can also build up grain reserves for urgent food needs in case of harvest failure or invest in insurance that will help them respond appropriately to drought. Governments and donor agencies need to support local people as they try to build resilience in their families, communities and local institutions (Trench et al., 2007). By focusing too much on short term assistance, vulnerability to drought is not reduced and increases the likelihood that more aid will be needed the next time a severe drought occurs. Thus the success and centrality of remote sensing data and analysis in FEWS NET undermines some of the longer term goals that its own personnel hold dear, to provide assistance that will reduce long term vulnerability instead of increase it.

17.3 Political Challenges for FEWS NET and its Use of Remote Sensing

FEWS NET's stated goal is 'to strengthen the abilities of African countries and regional organizations to manage risk of food insecurity through the provision of timely and analytical early warning and vulnerability information'. This decision maker focus is critical to the success of FEWS NET's work. Food security analysts use remote sensing to elicit a response – maps and simple, consistent and compelling messages are at the center of their communication strategies. As FEWS NET's understanding of the complexity of livelihood response to a hazard increases, it becomes more difficult to portray these complexities clearly to decision makers. Remote sensing is the primary source of comprehensive information on biophysical hazards, but as other hazards are added, FEWS NET must work to retain its ability to create clear, compelling messages without resorting to equating hazards to vulnerability.

FEWS NET is analysis is occasionally influenced by politics in the United States. An excellent example of this occurred during the unusually heavy rains in the summer of 2007. 'Across the Sahel, torrential rain and flooding have hammered the region in the past months. The United Nations estimates the floods have affected more than 600,000 people in 13 West African countries. It's a stark increase from the 65,000 people who were affected last year at this time' (Hartill, 2007). FEWS NET reported on the impact of the floods in West Africa, which were unusually severe, and they did affect significant numbers of people.

FEWS NET came under significant political pressure to recommend intervention in West Africa, but its analysis showed that in many places increased rainfall and even increased flooding meant a much better food security situation for most in the region. As its August 2007 report states: 'Besides these waves of flooding, the agricultural situation is generally satisfactory throughout the Sahel and West Africa. Pasturelands are also rejuvenating and watering points for animals are being re-supplied' (FCPN, 2007). Through careful analysis and evaluation of the impact of the threat, it was able to establish the extent of the problem and to work with its humanitarian partners to discourage those who wanted to send food aid to the region.

Identifying hazards is a critical component to FEWS NET's work. Unfortunately, its continued focus on remote sensing analysis contributes to the perception that biophysical hazards are equated with hunger despite extensive analysis that enable the interpretation of hazards. The persistent nature of such simplistic one cause, one outcome thinking can be seen in a new livelihoods protection and crop insurance program being instituted by donors in Ethiopia. As an index-based "hunger insurance" program, it has as its focus rainfall. Payments to farmers and other vulnerable people will be linked directly to variations in rainfall. This structure overlooks other emerging hazards such as disease and health care challenges, urbanization, global cereal grain price increases, energy cost increases, trade restrictions and other threats to livelihoods. One reason for the persistent focus on biophysical hazards in Ethiopia is the perception by government and donor officials of their centrality. Rainfall is quantifiable, understandable and easily related to food insecurity in countries such as Ethiopia that have a predominantly agricultural economy. FEWS NET continues to work to include livelihoods analysis and a nuanced view of the impact of biophysical hazards on food insecurity in its work in order to provide the most accurate and effective policy guidance.

17.4 Summary

In this chapter, the politics surrounding the use of remote sensing is explored. The chapter begins with a brief summary of the politics of food aid and how ideas regarding food security have changed through time. Increased understanding of the complexity of the impact of biophysical hazards on lives and livelihoods has transformed how early warning is conducted. Three case studies of how remote sensing has been used as a political tool were presented. The first was from Zimbabwe, the second was regarding Kenya and the third Ethiopia. Finally, in the last section, a discussion of the persistence of the perception of the centrality of biophysical hazards is presented. The chapter seeks to highlight some of the more complex and difficult issues regarding the use of remote sensing in food security early warning.

References

Barrett, C.B. and Maxwell, D.G., 2005. Food Aid After Fifty Years: Recasting its Role. Routledge, New York.

Buchanan-Smith, M., Davies, S. and Petty, C., 1994. Food Security: Let Them Eat Information, Institute for Development Studies, London, England.

FCPN, 2007. Food Situation in the Sahel and West Africa: Should satisfactory agricultural and food prospects be expected? Food Crises Prevention Network (FCPN).

FEWSNET, 2003. Ethiopia Network on Food Security – Monthly Report: 15 July 2003, FEWS NET, Washington DC.

Funk, C. and Budde, M., 2007. National MODIS NDVI-based Production Anomaly Estimates for Zimbabwe, University of California, Santa Barbara.

Hartill, L., 2007. Floods Affect Thousands in West Africa, Catholic Relief Services, Baltimore.

INTERFAIS, 2007. 2006 Food Aid Flows, UN Food and Agriculture Organization, Rome, Italy.

Lautze, S. and Raven-Roberts, A., 2006. Violence and Complex Humanitarian Emergencies: Implications for livelihoods models. Disasters, 30: 383–401.

Malthus, T.R., 1789. An Essay on the Principle of Population. J. Johnson, London.

Maxwell, S., 1996. Food security: a post-modern perspective. Food Policy, 21: 155–170.

Nyathi, K., 2007. Zimbabwe: Food Crisis – A Disaster Waiting to Happen, The Zimbabwe Standard, Bulawayo.

Rowland, J., Wood, E. and Tieszen, L.L., 2007. Review of Remote Sensing Needs and Applications in Africa, USGS EROS, Sioux Falls, SD.

Sen, A.K., 1981. Poverty and Famines: An Essay on Entitlements and Deprivation. Clarendon Press, Oxford, 270pp.

Torry, W.I., 1984. Social science research on famine: a critical evaluation. Human Ecology, 12: 227–252.

Trench, P., Rowley, J., Diarra, M., Sano, F. and Keita, B., 2007. Beyond Any Drought: Root causes of chronic vulnerability in the Sahel, IIED, London, England.

Turner, B.L., Kasperson, R.E., Matson, P., McCarthy, J.J., Corell, R., Christensen, L., Eckley, N., Kasperson, J.X., Luers, A., Martello, M.L., Polsky, C., Pulsipher, A. and Schiller, A., 2003. A framework for vulnerability analysis in sustainability science. Proceedings of the National Academy of Sciences, 100: 8074–8079.

Chapter 18
The Future of FEWS NET

In 2005, the US Government commissioned a report on the effectiveness of global institutions:

> In recent years, USAID's successful Famine Early Warning System has allowed U.S. and U.N. food aid to reach potential famine zones in a timely enough fashion to avert famine.

While this was flattering, the footnote attached to FEWS was even more complementary:

> The FEWS project, designed in 1986, has been, per dollar invested, one of the most efficient and high impact efforts that Congress has ever funded, saving millions of lives by catalyzing timely aid (Gingrich and Mitchell, 2005).

FEWS NET is viewed by many as being the most effective program in existence for providing information to governments about impending food crises. It has been instrumental in ensuring that only the most in need receive assistance and that decisions are made on the basis of fact to the extent possible. Analysis has shown that in the 1990s, food aid has been allocated according to need to a much greater extent than development assistance in general (Barrett and Heisey, 2002; Neumayer, 2005).

Previous chapters have shown how satellite remote sensing data has been part of the complex decision making environment that surrounds food security decisions. FEWS NET facilitates the transformation of raw observations into useful information that can be used to ensure that appropriate decisions are made at the right time so that famines can be averted. This effectiveness is a result of FEWS NET's structure, its ability to draw on multiple disciplines of information, and its ability to influence actors at the local, regional, national and international arenas.

FEWS NET has been extremely successful during its twenty years of existence. It is one of the longest running programs at USAID, getting a sufficient funding to maintain or expand its scope year after year, even in tight budget environments. FEWS NET has high retention rate of its personnel and is able to recruit some of the best people available in the field because of its excellent reputation. Its recent move to the Food for Peace office has meant that its key personnel report directly to the people who make decisions regarding humanitarian aid. These decision makers must decide how to apportion a limited budget for assistance between multiple locations with a variety of problems. FEWS NET has worked to assist them in this process by ensuring that its key environmental monitoring products are global in extent, and that its experts can provide information about countries that are outside of its current responsibilities.

In the following sections, potential changes in FEWS NET's structure and focus will be described, with a focus on how it plans to respond to the changing circumstances and information needs of the people it serves. New remote sensing datasets, research and development for improved food security assessment will be described.

18.1 Anticipating Future Food Security Problems

In the past five years, FEWS NET has greatly increased its focus on contingency planning and preparedness for crises, particularly in regions that experience problems frequently. Clearly, anticipation of future problems makes for better integration with humanitarian aid sources that require lengthy negotiation, early purchasing, mobilization, and shipping of food aid. Observations of food production variations that are used to develop immediate food aid need requirements need to have a very high accuracy in order to be useful. The necessary precision of the information needed for planning and forecasts of future food insecurity, however, is far lower than that required to estimate current food aid needs.

FEWS NET has recently begun to assess future changes in food security status due to biophysical and socio-economic events through a twice a year projection of food security conditions. The Food Security Outlook product has pushed forecasts and statistical projections of observations into the forefront of the activities of all its representatives. The outlook will be integrated into a global process where each country's outlook can be compared in order to provide USAID with actionable advice as to where the worst problems are and which countries will need the most aid. The outlook process is based on developing food security scenarios, consisting of a current food security situation, most likely situation in the next six months and worst case scenario, which are developed with significant biophysical data input. These scenarios are then consolidated into a global outlook with estimated numbers of food insecure for each municipality and region within each country, providing a total food aid required figure for decision makers.

Integrating modeled projections of rainfall and vegetation health with remote sensing observations will greatly improve the utility and integration of forecasts into the work of food security analysts in the context of these outlooks. Some regions are leading the way with this effort – East Africa has been at the forefront of the food security outlook, having begun the effort in 2003 in order to provide guidance to USAID about the food aid needs of the region. The integration of the results of the Climate Outlook Forum into food security analysis has enabled a much improved quantification of future food availability within the context of food security analysis.

Regional Climate Outlook Forums bring together climate scientists, operational forecasters, and climate information users to formulate a consensus on model-based forecasts and to discuss the implications of probable climate outcomes for climate-sensitive sectors such as agriculture and ultimately food security. NOAA's Climate Programs Office, USAID, and the United Nation's World Meteorological Organization have supported these in Africa and other regions for nearly ten years. During this time the seasonal forecasts have significantly improved, particularly in years with a strong signal from the El Niño Southern Oscillation (Brown et al., 2007).

FEWS NET participates in and helps to fund the Forums, and through the work of its scientists transforms the probabilistic forecasts into rainfall projections (Husak, 2005), which are then used to drive models describing crop yields and pastoral conditions. This information is then combined with livelihood baselines and socio-economic monitoring data to estimate the food security situation for the next six months to support contingency and response planning efforts in a new planning product called the Food Security Outlook. Figure 2 shows the results of using the FEWS NET developed Forecast Interpretation Tool (FIT) to translate the March 2007 Climate Outlook Forum forecast into estimates of expected rainfall anomalies (Husak et al., 2007). Sector specific interpretations were analyzed to provide information on its potential impact on the pastoral, agro-pastoral and agricultural zones.

Improved food security forecasting and planning tools will need to be combined with different agricultural policies to overcome hunger in Africa and achieve the Millenium Development Goal of halving hunger by 2015. Current US aid policy focuses on physical shipments of grain, a costly and inefficient solution with little benefit to African farmers (Murphy and McAfee, 2005). Given that overall, 99% of Africa's food comes from Africa, effective and funded agricultural development policies, assisted by earth observations, could bring an end to chronic hunger.

By focusing on providing information at key decision points at USAID and the UN, FEWS NET has been able to become better able to inform decision makers as they set budget priorities throughout the year. The message that Buchanan-Smith and Davies presented in their 1995 book has had an impact, although there is little that FEWS NET can do about the underlying development-humanitarian aid structure in which it operates. New, innovative remote sensing datasets have been developed so that quantitative information will be available to guide these decisions when they occur before the end of the harvest. These include seamlessly integrated observations and modeled outputs that will permit the extension of products that can estimate crop yield, rainfall estimates and pastoral conditions up to four months into the future. These products can then be integrated into decision making by enabling a quantitative basis for budget decisions that may be made months before the actual outcome of a harvest is known.

18.2 Improving Communication Across the Center-Periphery

In order to fully take advantage of some of these developments, FEWS NET's internal structures will need to change so that a much more rapid flow of information is possible from the field to the home office. With decision making deadlines and a focus on the future, new FEWS NET food security products will need to be much more coordinated with other agencies. FEWS NET is moving towards putting all 25+ offices, its collaborators in the United States, and its Washington DC home office on an internal network that will permit the closer collaboration than is possible with email and phone alone. A global data inventory and the technological infrastructure behind it will enable the building of all reports, web site content and databases simultaneously, replacing the enormous duplication of effort that currently is made by FEWS NET personnel. Activities such as schedule management, document sharing, database development, database access, and web meetings can be coordinated if everyone is on the same local area network. This network will permit the reduction in work load for country representatives and for personnel in Washington DC because the request and delivery of data will occur at the same time as its creation.

An example of the improvement in communication that this would permit is the management of market price data. Since the end of the fourth reauthorization of FEWS, there has been no funding for the gathering, organizing and integration into a database of market prices from across all the regions that FEWS NET works in. Figure 12.2 shows the dramatic decline in the amount of price data for the Sahel available in databases starting in 2000 after the beginning of FEWS NET authorization five which was the a result of the de-funding of this activity. In many cases, there is a much more diverse and complete price analysis in the monthly reports than is available in a reproducible database online. The current mandate (the sixth FEWS NET reauthorization) renews the focus on market prices but still does not fund the personnel that would be needed to collect, collate and enter the prices into a database as was done before 2000. The solution to this problem is to collect the updated monthly prices for all markets the moment that they are entered into the computer originally by the FEWS country representative during analysis for the monthly report. By linking all reports, spreadsheets and other documents from the regional and national offices to the Washington DC office, and by having some very minimal standardization, a much more robust transfer of information can be achieved.

The network will also contribute to FEWS NET's goal of finding ways to ameliorate the center-periphery problem. As is common in global projects such as FEWS NET, it is very hard to move information from the center to the national offices and from the offices to the center. This results in much confusion and delay, particularly with remote sensing innovations and in validation of the results in the field. To reduce these communication problems, regular, programmed meetings need to be scheduled and paid for by FEWS NET to enable renewed lines of communication. Regional meetings that enable a review of what happened during the hungry season and how FEWS NET responded will enable the improvements and innovations to shine when they occur. This requires significant resources, given that FEWS NET is spread across three continents, but is essential for the continued high level of information flow. The new online networking system will also require adjustments in working, training and the development of new ways of communicating for all personnel.

18.3 Innovations in Reporting

Along with transforming how information is moved within FEWS NET, another major development is the implementation of an Integrated Phase Classification for food insecurity or IPC for short. The IPC was developed by humanitarian organizations in Somalia in their attempt to provide for vulnerable communities in that country during the past fifteen years during which there has been no functional government. The IPC is used to determine the food security status or situation in a country with an associated warning level of normal, elevated, high, which are related to very specific outcomes, such as disease mortality, malnutrition rates, and food access levels. The IPC can draw on a wide variety of methodologies and types of evidence for particular hazards. The IPC is a tool which enables the clear identification of the outcomes that may result from a particular hazard. Remote sensing data identifies a hazard, but not the outcome the hazard produces.

In collaboration with multiple partners, including the World Food Program, aid organizations from multiple nations, non-profit organizations and experts on humanitarian aid, FEWS NET is working to use the IPC to provide a standard definition for all levels of food insecurity globally. Until now, a severe humanitarian crisis in Somalia would have a completely different meaning from one in Haiti or in Afghanistan, with the same label applied to a wide variety of nutritional deficits, regional impacts, and food access problems. Standardization would require that all parties in the humanitarian aid system would need to agree on food security classifications and their details so that guidance can be produced that enable the implementation across multiple regions with few overlapping personnel. FEWS NET has been a path finder for this effort, moving forward to demonstrate the utility of such a standardized system, first within their Food Security Outlook and now other products as well. Guidance materials have been developed by FEWS NET that enables day to day support of their personnel working to integrate the new system into their work. As this system is integrated into FEWS NET, much improved decision support will enable a prioritization of crises based on analytical baselines with a common methodology by all personnel.

In parallel with the standardization of food security classifications, FEWS NET is working toward reducing the duplication of effort in its reporting. Until now, each report was written independently and produced by different subsets of people, which results in a complexity of messages, analysis and an enormous work load given the level of funding. Each report was developed to meet decision makers' needs at USAID, but until recently, no one had time to determine the overlap and coordination of the information in the reports. Similarly, the content on the web site would be in addition to the local monthly reports due to the need for a standard look and format, adding to instead of complementing other activities. By moving to a modular approach to information creation, each piece of information and analysis can be produced only once and added to reports as needed. The approach will also allow a regional/global picture of information that is available at the same level of precision for all countries that FEWS NET works in. Managers in Washington DC can identify areas in particular regions where further investment and training are needed.

Each FEWS NET representative would be responsible for a consistent set of minimum standard deliverables to be delivered each month. These include:

1. Core messages, consisting of two or three main points for each country each month
2. Summary of current situation
3. Critical events and seasonal timeline
4. Map of current food security conditions
5. Food Security Outlook map for next six months
6. Food Security Outlook scenarios for next six months
7. Market price analysis
8. Policy round-up and implications for food security
9. Food security population data for aid calculations
10. Institution context, opportunities and constraints

These analytical areas do not preclude additional analyses by the FEWS NET representatives, but they must be produced by each representative for each country. In most cases, they are analyses that are already being done, just not as systematically as will be possible with the current focus and structure. Each analysis can then be combined into the myriad of reports and web reporting and remove the need for the composition of multiple reports with slightly different content. It will also easily be integrated into a global seasonal planning calendar where all information on required decisions, field visits, deliveries and reports are integrated for a clear view of how one region or country compares with another.

New reports and products are expected to be developed in coordination with other agencies that will permit more focused decision making at USAID, particularly once all the structures and systems are set up that will enable a meaningful comparison of food security at the global level. FEWS NET hopes to improve its ability to take into account longer term problems, particularly in regions where it has a heightened presence and long history, such as in Ethiopia and Kenya. Being able to anticipate and forestall problems that will affect food security that have a relatively long lead time (years instead of months) is a long term goal of FEWS NET and other food security organizations. FEWS NET is currently working to increase analysis of policy, mostly relating to non-food security issues, in order to determine their impact on food security. A much more effective coordination and communication between the development communities and those focused on humanitarian response must occur than has previously been shown to be possible. As larger percentages of the population are food insecure due to inadequate and failing livelihood strategies, however, it is becoming more and more important to focus on the long-term transformation of vulnerable communities so that short-term assistance is less needed.

18.4 Remote Sensing Innovation

In a recent survey of FEWS NET field representatives, remote sensing scientists and affiliated personnel from multiple disciplines, rainfall data was determined to be the most important remote sensing-based observation currently provided by FEWS

Table 18.1 Results from a survey of 43 FEWS NET representatives, remote sensing scientists and affiliated field personnel regarding the usefulness of various remote sensing data products. NA means not applicable, where the respondent does not currently have access to the dataset or is not familiar with it

	Vital	Somewhat Valuable	Marginal	Not Valuable	NA
Rainfall	100.0%	0.0%	0.0%	0.0%	0.0%
Temperature	43.9%	31.7%	14.6%	7.3%	2.4%
Humidity	25.0%	47.5%	22.5%	2.5%	2.5%
Flooding	70.7%	26.8%	0.0%	2.4%	0.0%
Vegetation	82.9%	14.6%	2.4%	0.0%	0.0%
Soil Moisture	72.5%	17.5%	7.5%	2.5%	0.0%
Land Cover	41.5%	39.0%	17.1%	0.0%	2.4%
Crop Yield Estimates	85.0%	10.0%	5.0%	0.0%	0.0%

NET. Table 18.1 shows yield estimates and vegetation information are nearly as essential for identifying problems as rainfall data. The trend in FEWS NET is to provide more observation types and more derived observations such as Potential Evapotranspiration that can be used in conjunction with crop models to estimate the impact of climate on crop productivity. FEWS NET partners will also be producing standardized datasets, particularly of rainfall, humidity and precipitable water that will enable easy identification of anomalies. Previously, anomaly generation was done during the imaging step, without standard procedures that normalize the data to a particular time period or using a particular methodology. Thus although one can compare SPOT and AVHRR NDVI anomalies to the RFE anomaly, each product would have its own mean, use different methodologies and thus it was hardly surprising that each product showed a different situation. Remote sensing innovations will increase the ease of use of each of these products and provide new ways to combine them into monitoring system that is accessible by everyone.

Much of this book is focused on the remote sensing innovations and products that have come out of the twenty-year collaboration between food security analysts and remote sensing scientists. By involving the producers of biophysical data and information in the monitoring and response to food security, FEWS NET has motivated the improvement in the kind of quantitative information required to identify food security problems as early as possible. This section will focus on a few areas where FEWS NET's leadership would like to see further advancements in remote sensing to improve early warning and its effectiveness in 2007 and beyond.

Estimates of variations in food production have usually been made only with products that monitor yield. Yield is used to estimate variations from last year's production or from government estimates over the previous five or ten years. This assumes that the production figures from the government are correct and that significant variations in the amount of area cropped have not o occurred in the intervening period. Both of these assumptions have not been quantitatively verified and are often completely inaccurate. Cropped area varies substantially from year to year in semi-arid subsistence cropping systems, due to the amount of seed available from the previous year, the moisture levels and timing of planting and many other drivers. Market demand in the form of high prices also tend to increase the area planted, and particularly low prices after a bumper crop also depresses the area planted (Grabowski and Shields, 1996; Jaeger, 1992). The African farmer is very responsive to market price, but the yield-only estimation techniques FEWS NET uses tends to ignore this (Badiane and Shively, 1998; Kherallah et al., 2000).

FEWS NET recently has begun to focus on providing estimates of cropped area, as was briefly described in Chap. 6. Figure 6.11 shows the methodology used to estimate cropped area from both high resolution (∼1 m data) and from Landsat data (∼30 m) over an entire country. The method requires hand interpretation of the tens of thousands of samples across a large area, which is incredibly labor intensive. Research into developing an automated methodology which can use the spectral signatures of each dot to identify it as cropped or non-cropped is underway. If these methods can be sufficiently automated and enough high resolution data can be acquired then FEWS NET will begin to use both area and yield to estimate overall actual production annual in countries most at risk. Coupled with further research on getting the population numbers right, FEWS NET hopes to move the aid provision process from one of negotiation using soft numbers and apportionment of available assistance to a much clearer understanding of the actual local need which can be comparable across a wide variety of cultures, environments and political situations.

Another important development in FEWS NET has been the recent focus on contingency planning and preparedness for crises, particularly in regions that experience problems frequently. By providing an outlook, it is less likely that donors will wait to act until nutritional surveys and crop assessments show that there is a problem. These are very late indicators that indicate a failure to act in time (Buchanan-Smith et al., 1994).

Clearly anticipation of future problems makes for better integration with donors who have at their disposal food sources that require lengthy negotiation, early purchasing, mobilization, and shipping. The necessary precision of the information needed for planning and forecasts of possible future food insecurity is far lower than that required to estimate current food aid needs. FEWS NET has recently begun to assess future changes in food security status due to biophysical and socio-economic events through a twice-a-year projection of food security conditions. The Food Security Outlook product has pushed forecasts and statistical projections of observations into the forefront of the early warning activity. Integrating projections with remote sensing observations will greatly improve the utility and integration of climate forecasts into operational networks and FEWS NET's processes, ensuring that decision makers have the information they need when they need it.

18.5 Challenges and Opportunities

In the next few years, FEWS NET will again be up for reauthorization. The previous reauthorization instituted many new methodologies, including livelihood zoning that enables analysis on the impact of a potential hazard, an increased focus on providing information to all its partners from an online portal, and the expansion of local networks to include decision makers from a wide variety of fields. To be effective for policy development, information needs to become more relevant, timely, and accessible. It must also be demand and decision driven, drawing implications and recommending actions for preventing food crises or, should they occur, mitigating their impact. This section will outline a few areas where it is facing challenges that may transform it once again, and improve its functioning in the future.

18.5.1 P.L. 480 and FEWS NET

The ability of the US Government to provide food aid to foreign entities was initially authorized through Public Law 480. The full name for Public Law 480 is the Agricultural Trade Development Assistance Act, signed into law on July 10, 1954 by President Dwight D. Eisenhower. In signing, Eisenhower stated that the legislation's purpose was to "lay the basis for a permanent expansion of our exports of agricultural products with lasting benefits to ourselves and peoples and peoples of other lands" (USAID, 2007). Title II, the largest part of PL 480, allows donation of US agricultural products to meet humanitarian food needs, emergency and non-emergency (development) food aid activities support broader USAID objectives and can be use for both direct feeding and monetization programs.

FEWS NET's activities, however, are not funded under PL 480, which means that it must get its annual funds through the contentious, difficult and highly politicized budget process at USAID. In 2007, after nearly being defunded and having its USAID representatives spend six months trying to restore funding levels to that of 2006, FEWS NET's supporters and leadership have sought to include funding for its work under the PL 480 reauthorization effort that is currently in congress. If this is successful, FEWS NET will become institutionalized similar to USDA's Foreign Agricultural Service, one that has regular, reliable funding and that can focus on improving its services without having to focus on maintaining its funding levels in an ever-varying political environment. Since the need for information seems only to increase through time, there is significant support for this effort.

Although food aid authorized through PL 480 has improved or saved the lives of many hundreds of millions of people worldwide since its passage in 1954, there has been much discussion during the reauthorization process that describes how the law could be changed to make food aid a more effective and farther reaching tool. Barrett and Maxwell (2005) describe policy changes that would greatly improve its ability to reduce poverty and hunger while simultaneously reducing costs. There are a few of these that FEWS NET could directly address, with its use of remote sensing being at the forefront. The relevant recommendations include:

- Focus on emergencies and make the response quicker and more flexible;
- Within current budgets, adapt the resource to fit the application;
- Improve the targeting of food aid; and
- Use food aid only where it is appropriate (Barrett and Maxwell, 2005).

These are all areas where FEWS NET has worked to provide information and policy guidance for the past twenty years, and yet Barrett and Maxwell point out that they can be done better. Because the provisioning of food aid is part of a political process, information and recommendations provided to the US and other donor countries may not be followed. The information itself could be better, and the preceding sections and chapters point to ways this can be done. Overall, FEWS NET can only be as effective as the structure in which it operates.

18.5.2 Coping with Climate Change

The challenge of climate change for FEWS NET will be substantial, particularly in semi-arid regions that already experience extremely high interannual variability, such as in the pastoral areas of East Africa. Significant changes in annual rainfall, temperatures and crop yields have already been seen, which have been attributed to rising sea surface temperatures. FEWS NET employs climate scientists who have worked to identify these trends and link them to changes in the global climate system and to sea surface temperatures in specific regions (Funk et al., 2005; Thiaw and Barnston, 1996). Regions experiencing climate change seem to be less, not more, able to cope with variability due to declining investment and attention to rural development (Rosegrant and Cline, 2003).

FEWS NET has found it very difficult to find ways to ameliorate climate-related problems that undermine livelihoods systems in which they work. The structure of short term (emergency) assistance is to address only the immediate problem. Although food aid is usually forthcoming during emergencies, it is difficult to get such assistance year after year in regions with high percentage of chronically food insecure. Investment in the development of a region must occur in order to transform livelihoods from those that are marginal and vulnerable to rainfall deficiencies to those that are sustainable is the only solution over the long term. Unfortunately, during the past ten years, only humanitarian assistance has been increasing and not longer-term aid, leaving much chronic food insecurity unaddressed.

Improved food security forecasting and planning tools will need to be combined with improved agricultural policies to overcome hunger in developing regions. Growth in food production is the most important way that agricultural economies will grow and develop in order to reduce the number of people who are chronically hungry (FAO, 2006). Climate models show that future rainfall amounts in marginal areas like semi-arid East Africa are likely to continue to decline (IPCC, 2007), which means that demands for food aid will continue to increase or investment must be made in transforming the livelihoods of the people who live in the area into those that are less reliant on rain-produced resources.

FEWS NET is involved at a variety of levels in larger efforts to train, support and provide satellite remote sensing data for improved natural resource management and decision making in the regions where it works. Using the networks it has established, FEWS NET works to provide training and works to encourage and support regional organizations and national governments in their development and use of remote sensing applications. Although regional organizations such as New Partnership for Africa's Development (NEPAD), the African Association for Remote Sensing of the Environment (AARSE), and the Water Center for the Humid Tropics of Latin America and the Caribbean (CATHALAC), most African and many Central American governments have not generally supported remote sensing applications for development. This is likely to be due to the likelihood of outside donors who will support them (Rowland et al., 2007), albeit if only sporadically. FEWS NET continues to seek ways to improve the use of its data and information products for longer-term agricultural development by local, national and international actors.

Coping with long-term development issues that affect food security will be a continuing challenge for FEWS NET for the next decade. Regions that are experiencing chronic food insecurity continue to challenge FEWS NET's ability to provide accurate and reliable information that motivates an appropriate response year after year.

18.5.3 Integrating Remote Sensing and Socio-Economic Variables

Much new research is focused on developing methodologies that can be used to combine socio-economic information that will illuminate food access through integrating food prices and food production information (Brown et al., 2008) or food utilization, integrating nutrition indicators with measures of food insecurity through livelihood analysis (Mathys, 2005). Integrating the price of cereals, information on cross-border trade and transportation costs into a monitoring and predictive model that can be run operationally to provide early warning of significant changes in market prices can contribute to food security monitoring. Other researchers are working to integrate qualitative assessments of food security with quantitative measures in order to better estimate overall food security of household units (Swindale and Bilinsky, 2006).

Weakness in the information systems required to gather necessary data to support these developments is a huge area of concern for FEWS NET. With limited budgets and tight deadlines, FEWS NET has to be careful not to overwhelm its personnel and to gather data and construct locally relevant indicators that are still comparable across cultures, countries and continents (Seaman, 2000). FEWS NET needs to balance its standardized approach with methods that seek to incorporate locally appropriate indicators of a crisis. The complexity of data gathering and historical database construction in so many different areas will be an ongoing challenge, but one that is necessary to support new methodologies that seek to improve the ways that FEWS NET identifies the food insecure.

18.6 Conclusion

Remote sensing data is the cornerstone of FEWS NET and its work to provide actionable information for decision makers. This book has describe the various ways that FEWS NET has brought together the science of remote sensing with information about the livelihoods and with planning information in order to better inform decisions. By working with the international humanitarian community to develop an standardized classification of food security severity into which its remote sensing data and livelihoods analysis can be fed, FEWS NET can provide real guidance for decision makers on current and future food aid needs. Remote sensing information is the cornerstone of this work by providing quantitative information about food production.

References

Badiane, O. and Shively, G.E., 1998. Spatial integration, transport costs, and the response of local prices to policy changes in Ghana. Journal of Development Economics, 56: 411–431.

Barrett, C.B. and Heisey, K.C., 2002. How effectively does Multilateral Food Aid Respond to Fluctuating Needs? Food Policy, 27(5–6): 477–491.

Barrett, C.B. and Maxwell, D.G., 2005. Food Aid after Fifty Years: Recasting its Role. Routledge, New York.

Brown, M.E., Funk, C., Galu, G. and Choularton, R., 2007. Integrating earth observations and model results provides earlier Famine Early Warning. EOS Transactions of the American Geophysical Union, 88(39): 381–382.

Brown, M.E., Pinzon, J.E. and Prince, S.D., 2008. Using Satellite Remote Sensing Data in a Spatially Explicit Price Model. Land Economics, 84(2).

Buchanan-Smith, M., 1994. Knowledge is Power? The Use and Abuse of Information in Development. IDS Bulletin, 25(2).

Buchanan-Smith, M., Davies, S. and Petty, C., 1994. Food Security: Let Them Eat Information, Institute for Development Studies, London, England.

FAO, 2006. The State of Food Insecurity in the World, United Nations Food and Agriculture Organization, Rome, Italy.

FEWSNET, 2003. Ethiopia Network on Food Security – Monthly Report: 15 July 2003, FEWS NET, Washington DC.

Funk, C. and Budde, M., 2007. National MODIS NDVI-based production anomaly estimates for Zimbabwe, University of California, Santa Barbara.

Funk, C., Sanay, G., Asfaw, A., Korecha, D., Choularton, R., Verdin, J., Eilerts, G. and Michaelsen, J., 2005. Recent Drought Tendencies in Ethiopia and Equatorial-Subtropical Eastern Africa, Famine Early Warning System Network, USAID, Washington DC.

Gingrich, N. and Mitchell, G., 2005. American Interests and UN Reform, United States Institute of Peace, Washington DC.

Grabowski, R. and Shields, M.P., 1996. Development Economics. Blackwell Publishers, Cambridge, Massachusetts, 299pp.

Husak, G., 2005. Methods for the Statistical Evaluation of African Precipitation. University of California Santa Barbara, 221pp.

Husak, G., Michaelsen, J. and Funk, C.C., 2007. Use of the Gamma distribution to represent monthly rainfall in Africa for drought monitoring applications International Journal of Climate, in press.

INTERFAIS, 2007. 2006 Food Aid Flows, UN Food and Agriculture Organization.

IPCC, 2007. The Effects of Climate Change on Agriculture, Land Resources, Water Resources and Biodiversity, Intergovernmental Panel on Climate Change, Washington DC.

Jaeger, W., 1992. The causes of Africa's food crisis. World Development, 20(11): 1631–1645.

Kherallah, M., Delgado, C., Gabre-Madhin, E., Minot, N. and Johnson, M., 2000. The road half traveled: Agricultural market reform in Sub-Saharan Africa, IFPRI, Washington DC.

Mathys, E., 2005. FEWS NET's Approach to Livelihoods-Based Food Security Analysis, FEWS NET USAID, Washington DC.

Murphy, S. and McAfee, K., 2005. U.S. Food Aid: Time to Get It Right, The Institute for Agriculture and Trade Policy, Minneapolis, Minnesota.

Neumayer, E., 2005. Is the Allocation of Food Aid Free from Donor Interest Bias? The Journal of Development Studies, 41(3): 394–411.

Nyathi, K., 2007. Zimbabwe: Food Crisis – a disaster waiting to happen, The Zimbabwe Standard, Bulawayo.

Rosegrant, M.W. and Cline, S.A., 2003. Global Food Security: Challenges and Policies. Science, 302(5652): 1917–1919.

Rowland, J., Wood, E. and Tieszen, L.L., 2007. Review of Remote Sensing Needs and Applications in Africa, USGS EROS, Sioux Falls, SD.

Seaman, J., 2000. Making exchange entitlements operational: The food economy approach to famine prediction and the RiskMap computer program. Disasters, 24(2): 133–152.

Swindale, A. and Bilinsky, P., 2006. Development of a universally applicable household food insecurity measurement tool: Process, current xtatus and outstanding issues. Journal of Nutrition, 136: 1449S–1452S.

Thiaw, W.M. and Barnston, A.G., 1996. CCA Forecast for Eastern African Rainfall in Oct-Nov-Dec, NOAA Climate Prediction Center, Camp Springs, MD.

Trench, P., Rowley, J., Diarra, M., Sano, F. and Keita, B., 2007. Beyond Any Drought: Root causes of chronic vulnerability in the Sahel, IIED, London, England.

USAID, 2007. The History of America's Food Aid. US State Department.

Index

MIX
Papier aus verantwortungsvollen Quellen
Paper from responsible sources
FSC® C105338

If you have any concerns about our products,
you can contact us on
ProductSafety@springernature.com

In case Publisher is established outside the EU,
the EU authorized representative is:
Springer Nature Customer Service Center GmbH
Europaplatz 3, 69115 Heidelberg, Germany

Printed by Libri Plureos GmbH
in Hamburg, Germany